USING METEOROLOGICAL INFORMATION AND PRODUCTS

ELLIS HORWOOD SERIES IN ENVIRONMENTAL SCIENCE

Series Editor: R. S. SCORER, Emeritus Professor and Senior Research Fellow in Mathematics and Environmental Technology, Imperial College of Science and Technology, University of London

A series concerned with nature's mechanisms — how earth and the species which inhabit it fit together into a dynamic whole, and the means by which evolution has taught them to survive.

We are *not* primarily concerned to exploit the environment to human advantage, although that may happen as a result of understanding it.

We are interested in the basic nature of the physical world, the special forms that it takes on earth, the style of life of species which exploit special aspects of nature as well as the details of the environment itself.

ATMOSPHERIC DIFFUSION, 3rd Edition
F. PASQUILL and F. B. SMITH, Meteorological Office, Bracknell, Berks
AIRBORNE PESTS AND DISEASES
D.E. PEDGLEY, Centre for Overseas Pest Research, London
LEAD IN MAN AND THE ENVIRONMENT
J.M. RATCLIFFE, Visiting Scientist, National Institute for Occupational Safety and Health, Cincinnati, Ohio
THE PHYSICAL ENVIRONMENT
B. K. RIDLEY, Department of Physics, University of Essex
CLOUD INVESTIGATION BY SATELLITE
R. S. SCORER, Imperial College of Science and Technology, University of London
SATELLITE AS MICROSCOPE
R. S. SCORER, Imperial College of Science and Technology, University of London
GRAVITY CURRENTS: In the Environment and the Laboratory
JOHN E. SIMPSON, Department of Applied Mathematics and Theoretical Physics, University of Cambridge

On related subjects

SATELLITE MICROWAVE REMOTE SENSING
T.D. ALLAN (Ed.), Institute of Oceanographic Sciences, Wormley
ENVIRONMENTAL AERODYNAMICS
R. S. SCORER, Imperial College of Science and Technology, University of London

USING METEOROLOGICAL INFORMATION AND PRODUCTS

Edited by

AVRIL PRICE-BUDGEN M.A., Cert.Ed.

on behalf of

WORLD METEOROLOGICAL ORGANIZATION
Geneva, Switzerland

ELLIS HORWOOD

NEW YORK LONDON TORONTO SYDNEY TOKYO SINGAPORE

First published in 1990 by
ELLIS HORWOOD LIMITED
Market Cross House, Cooper Street,
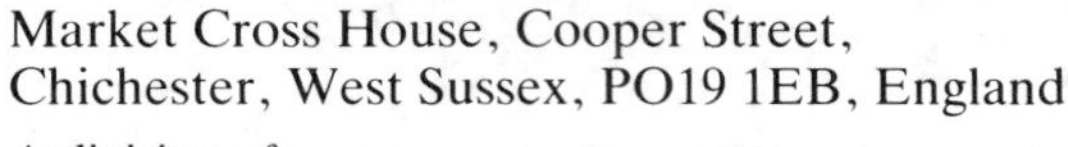
Chichester, West Sussex, PO19 1EB, England

A division of
Simon & Schuster International Group

© Ellis Horwood Limited, 1990.

All rights reserved. No part of this publication may be reproduced, stored in a retrieval system, or transmitted, in any form, or by any means, electronic, mechanical, photocopying, recording or otherwise, without the prior permission in writing, from the publisher

Printed and bound in Great Britain
by The Camelot Press, Southampton

British Library Cataloguing in Publication Data

Using meteorological information and products.
1. Meteorology
I. Price-Budgen, Avril II. World Meteorological Organization
551.5

ISBN 0–13–946914–1

Library of Congress Cataloging-in-Publication Data

Using meteorological information and products / edited by Avril Price-Budgen on behalf of World Meteorological Organization
p. cm. — (Ellis Horwood series in environmental science)
ISBN 0–13–946914–1
1. Meteorology — Information services.
2. Meteorological services. I. Price-Budgen, Avril, 1945– .
II. World Organization. III. Series.
QC866.5.C65U85 1990 67560
551.5–dc20 90–30886
CIP

Contents

CAMROSE LUTHERAN COLLEGE
LIBRARY

MEANS AND METHODS OF EDUCATION OF USERS

VARIED APPLICATIONS OF METEOROLOGY

METEOROLOGY AND WATER RESOURCES

METEOROLOGY AND THE ENVIRONMENT

METEOROLOGY AND AGRICULTURAL PRODUCTION

METEOROLOGY AND URBAN AND RURAL DEVELOPMENT

Acknowledgement

The editor wishes to acknowledge the assistance of Peter Budgen in checking the meteorological and scientific terminology and also supervising the production of the final manuscript.

Foreword

Economic, social and political pressures increasingly compel national Meteorological and Hydrological Services to become more application-oriented. At the same time, their interface with the user communities undergoes an almost continual readjustment due to rapid technological, scientific, economic and social changes. The economic and social benefits promised by a better use of meteorological and hydrological information spur efforts to achieve higher efficiency and a more effective co-operation with users. In this context a continuous dialogue between users and meteorological and Hydrological Services is of vital importance.

It is against this background that, with the support and co-sponsorship of the Food and Agriculture Organization, the International Maritime Organization and the United Nations Environment Programme, the World Meteorological Organization organized a Symposium on Education and Training with Emphasis on the Optimal Use of Meteorological Information and Products by all Potential Users.

The Symposium was held at Shinfield Park, United Kingdom, from 13 to 18 July 1987, and its purpose was to provide a forum for the interaction between meteorologists, as the producers of meteorological information and services, and non-meteorologists, as the users or potential users of such information and services. More than 150 participants from 70 countries attended the symposium at which the forty-two papers which are reproduced in this volume of the proceedings were presented. As a result of the presentations and discussions, the symposium produced a set of ten "issues" which are also included. It is hoped that this publication will serve to continue promoting the objective of the symposium.

Much credit for assisting the Secretariat with the symposium must be given to the co-sponsors, to members of the international programme committee, to the invited speakers, to all the scientists who have contributed papers, and to the session chairmen. It is with much pleasure that I place on record the thanks of the World Meteorological Organization to all these individuals and also to the Meteorological Office of the United Kingdom for its valuable support to and collaboration in organizing and hosting the symposium.

(G O P Obasi)
Secretary-General

Introduction

Economic and other benefits of meteorological and hydrological services

J. T. Houghton, Meteorological Office, Bracknell, UK

1. INTRODUCTION

The current interest in the economic benefits of meteorological and hydrological services has arisen for three main reasons: firstly the improvement in the accuracy and the range of forecasts in recent years; secondly, the explosion in information technology which is providing more effective means for collection, handling and dissemination of information; and thirdly because increased organisation and automation enables industry and commerce to make use of and respond to information in a timely fashion.

Consider the first of these, namely forecasting accuracy and range. During the past decade we have seen global forecasts out to a range of up to 10 days becoming available from a number of centres. The improvement in accuracy is such that current three day forecasts are of similar accuracy to two day forecasts of five years ago and one day forecasts of ten years ago. These improvements have come about mainly through developments in numerical modelling capability which in turn have been due to better data coverage (especially from satellites) and accuracy, advances in the mathematical and physical descriptions within the models and increases in the power and speed of computers. It remains very encouraging that significant improvement in any of these areas continues to result in a further improvement in forecast accuracy.

Developments have also taken place in the range of parameters for which forecasts are available. Predictions of height and frequency of waves or swell at sea, for instance, are important additions to the forecaster's armoury.

The second large development I mentioned is that in information technology. Much more data can now be assimilated into the models and more data can be stored, for later use in, for instance, climatological studies. Further, the variety and speed of communication channels, whether for broadcast services or for specific customised services, has expanded enormously in recent years and is still expanding rapidly. Examples are teletext and viewdata in the UK: the number of television sets with teletext facility is now about 3.5 million and is growing at 20% per annum. Private viewdata systems now number 500 - 800 and after a rather slow start there are now about 75,000 Prestel subscribers, a number which is currently growing at about 10,000 per annum.

Demand for weather information by telephone, especially in periods of bad weather, can be very high. The new Meteorological Office service 'Weathercall' enables this service to be measured. On January 12 and 13 of this year demand exceeded 250,000 calls per day. The revolution in information communication now taking place must be comparable with that which occurred when the electric telegraph was first invented nearly 150 years ago.

The data which are collected worldwide every day are the input for the forecast models. As they become historic data they go into the climatological archive. As data on the past climate become more accessible through modern means of storage and accession, analyses and interpretations of the data prepared by meteorological services are becoming valuable to a wide range of customers. If a large building, a power station, an oil rig, a road, or a new town is to be located sensibly and constructed properly and safely, information on extreme and average values of a great number of parameters, for instance winds and waves, is essential.

Shipping, aviation, agriculture and water are generally perceived to be the industries where weather sensitivity is greatest and where weather forecasts can be of most value. These remain important markets for forecasts. However, nearly all areas of industry and commerce are learning the value of climate and weather information, so that to the list can now be added the offshore industry, road and rail transport, building and substantial areas of the manufacturing and the retail industry. Market research indicates that, in most cases, carefully prepared packages of information (including past, present and future weather) are more valuable to the customer than pieces of information taken in isolation. In the following sections I shall give some examples of cost benefit in particular areas for both forecast and climatological information.

2. BENEFIT TO AVIATION

Civil aviation requires a wide range of forecasts from meteorological services, for instance forecasts of route winds, temperatures and weather for flight planning, warnings of adverse weather for inflight operations and landing forecasts of the weather expected at the destination and at suitable diversions to ensure the loading of sufficient fuel. Aircraft fuel savings can be achieved in a number of ways. The most obvious of these is to use the forecast winds and temperatures for planning the most economical route between the departure and destination airports, taking advantage of favourable winds and avoiding unfavourable ones. The freedom to choose such tracks only exists over oceanic sectors of the flights; over land the density of air traffic necessitates strict air traffic control and the confinement of air movement to predefined air lanes. Accurate forecast information can, nevertheless, allow airlines to evaluate their expected fuel burn over the land flight sectors. By taking on board an appropriate fuel load the aircraft can avoid transporting excessive quantities of fuel, or alternatively avoid making intermediate unscheduled refuelling stops because of unexpected wind conditions.

Since 1982 when the global 10-level model came into operation in the Meteorological Office there has been a marked improvement in the forecasting of upper air temperatures and winds. With the help of some of its customer airlines the Meteorological Office has estimated that the saving in fuel directly attributed to improved forecasts by the Meteorological Office amounts to at least £50M per annum (White 1987). In recognition of these savings the team from the Meteorological Office who developed the model received the Esso Energy Award from the Royal Society in 1986.

3. BENEFIT TO AGRICULTURE

Most agricultural activities are highly sensitive to weather and climate. The times of sowing, harvesting, the application of fertilisers, the applications of sprays for pest and disease control are all very dependent on the current or the recent weather. Let me give some examples taken from World Meteorological Organization (1986). In Ireland it has been estimated that £1M can be saved annually by a reduction in spraying for potato blight, using a warning system based on forecasts of weather conditions favourable to the spread of the disease. In Botswana it has been shown that crop yields can increase by about £1M per rainy season by choosing better crop cycles and planting dates according to the climatology and forecasts of the region. In New Zealand there is an economic benefit of about 1-2 per cent (NZ$ 40.6M in 1981) of the total value of agricultural production for all meteorological services to agriculture. In Canada approximately $10M (approx £5M) per year can be saved in the fruit tree sector by using short-range forecasts, the main benefits arising from reduced spraying costs and savings at harvest. In South Africa (Theron *et al*) short-range forecasts have saved R12.3M (Approx £4M) per year in preventing crop damage due to excessive high and low temperatures and untimely frosts, hail and rain.

To illustrate further the potential benefits, I give one example of the importance of accurate information in the area of pest control. The larvae of the pea moth cause substantial damage to pea crops in East Anglia. They emerge from the soil on which the eggs have been laid after 5-6 days if the weather is warm or after 15-20 days if cool. Within less than eight hours after their emergence the larvae will have found their way into a pea pod and be relatively safe from destruction. It is vital that the pesticide reaches them within that eight hour window. To predict the emergence of the larvae it is necessary to know first when the eggs were laid. This can be ascertained from observation of moth activity using scented traps. It is also necessary to predict the incubation period from a detailed temperature record. With that prediction and information from weather forecasts (especially of likely precipitation which would wash the pesticide away), a very good estimate of the best time to spray can be made. Because the cost of spraying (material and application) is typically £350 for an average size crop of 30 hectares, avoiding ineffective spraying is very important in the consideration of the farmer's "bottom-line". In East Anglia about 30,000 hectares are cultivated with peas so that, if one ineffective spray can be avoided a saving of the order of £350K is made.

4. BENEFIT TO ROAD TRANSPORT

Road transport is becoming increasingly important to many of the world's economies. In a country like the UK road transport can be beset by a number of significant weather problems, for instance, by fog, ice and snow. To salt and grit the roads in a typical county of the UK costs on average of at least £10,000 per night. Good forecasting is therefore essential. It needs to be not only accurate but also detailed so that the available resources can be concentrated into areas where they are likely to be most effective.

5. BENEFITS TO OFFSHORE INDUSTRY

Over one quarter of the world's oil and gas resources now come from wells drilled under the sea. Offshore drilling is a difficult and expensive operation, the cost of operation of a typical drilling platform being around £40,000 per day. Weather and climate information is required both for the design of the platforms and for their operation.

For oil-rig design values of maximum wind and maximum swell (height and wavelength) are essential information. One of the problems is that for some parts of the world, only scanty information regarding waves is available. The Meteorological Office in consort with other European services is trying to improve this situation by a hindcasting project. The detailed wind field over the North Atlantic from routine analyses is used as input data for a wave model so that the wave fields over a ten year period can be recovered and appropriate statistics can be generated for use in oil-rig design.

Accurate forecast information is also essential for offshore production. The Buchan Production Platform in the North Sea, for instance, is a floating rig anchored over the well which produces oil through risers connected to the seabed. If the swell exceeds 7m in height, there is severe danger of the risers breaking. Pulling the risers in advance of a storm requires several hours' notice. Each hour's oil production is worth in the region of £10,000. The penalty for pulling the risers too early on the basis of an over pessimistic forecast is therefore £10,000 an hour. The penalty for an over optimistic forecast which results in the risers breaking because precautions were not taken in time is even greater, approaching £1 million if repairs take 3 days. On 30 November 1986, when the swell was about 2m in height the Meteorological Office forecaster gave 30 hours warning of the swell reaching 7m. The people on the rig prepared to stop drilling and kept hourly contact with the forecaster. In fact the swell reached 6.8m and production was able to continue. As can be seen such forecasts can be very critical indeed.

6. OVERALL VALUE OF METEOROLOGICAL SERVICES

I have given a few examples of the value of weather and climate information. Many more could be given. Can any estimate of the overall benefit of meteorological services be made? In 1966, Sir John Mason, my predecessor as Director General of the Meteorological Office, estimated that the value of weather forecasts to the UK economy was at least ten to twenty times the cost of the service (Mason 1966). This estimate has often been quoted and has, to my knowledge, never seriously been disputed: it is generally agreed to be on the conservative side. Because of the factors I mentioned at the beginning of more and better weather information and of better means of dissemination, the benefits now should be considerably greater than in 1966. Even on the 1966 basis the benefit of weather information to the UK economy would be between £1 and 2 billion p.a. or of the order of 1/2% of the gross domestic product.

One further point needs to be made. It is strongly argued by some governments that the customers should pay for the services provided. In the case of meteorological services, it is reasonable for commercial customers to pay the full cost of specialised services organised particularly for them. For many meteorological services, however, the benefits are spread very widely. Often the general public is the customer. Further, government itself, for instance for its defence services, requires good meteorological and climatological advice and information. In these cases it is appropriate for the cost to be met from general taxation. This is, is of course, why all countries have their state meteorological service. In my view state meteorological services need not only to see their task as providing the general information required by the public and by the government; they also need to ensure that some of the returns for meteorological services which are available in the commercial world are fed back into meteorology and its infrastructure, so ensuring the continued development and improvement of those services on which so many are learning to depend.

REFERENCES

Mason, B.J., 1966. The role of meteorology in the national economy. Weather 21 pp 382-393

Theron, M.J., V.L.Matthews and P.J.Neethling, 1973. The economic importance of the Weather and Weather Services to the South African Agricultural Sector. Pretoria, Council Science Industrial Research CSIR Research Report Volume 3211 1973 Serial No fo pp 134.

White, 1987. Impact of weather forecasts on aviation fuel consumption. Met. Mag. 116 pp 29-30.

World Meteorological Organization (WMO) Commission for Agricultural Meteorology. Economic benefits of Agrometeorological Services, prepared by T. Keane, P-O Harsmar and E. Jung. Geneva 1986.

1

The effects of severe weather on aircraft take-off and landing performance

V. R. Thompson, Civil Aviation Authority, London, UK

1. The standard of meteorological services to civil aviation is very high and is steadily improving as the World Area Forecast System is progressively implemented worldwide. Further improvements seem likely within the next few years such as the numerical analysis and determination of significant weather worldwide but already pilots taking off from, say, London with a first point of landing at Tokyo or Hong Kong can expect to receive before departure remarkably accurate en-route winds and temperatures, significant weather and landing forecasts.

2. With such a comprehensive meteorological service it is easy to overlook a very important effect that weather can have on the ability of an aircraft to take-off and land and the costs of such limitations on aviation. Before an aircraft takes off on a commercial flight there is a legal requirement to ensure that it can do so safely. There are many factors which have to be taken into account in the calculations to ensure safety, such as the flying performance of the aircraft, the length of runway available and the need to clear all obstacles in the flight path of the aircraft even if an engine failure occurs. Among these safety criteria are several significant meteorological factors which may be critical during the take-off and landing phases of the flight. These include pressure altitude and temperature which affect

aircraft performance, mainly engine power output; the presence of frost which, on the aircraft, affects lift; and the presence of residual contaminants on the runway such as water, slush, snow and ice which retard acceleration and make it more difficult to bring an aircraft to a stop. Whilst they are not always critical factors they can have a marked effect, sometimes precluding a take-off or landing, with serious implications to passengers, airlines and airports. As an example, an increase of 1°C can in certain circumstances reduce the payload that can be carried by a typical jumbo type aircraft, a Lockheed 1011 Tristar, by 1.4 metric tons, equivalent to an additional 10 knot headwind for a flight between Nairobi and London. As another example, 15 mm, half an inch, of slush on a runway will preclude all take-offs.

3. This paper briefly describes the effect these weather factors can have on the take-off and landing performance of the main group of commercial aircraft - those multi-engined aircraft which have a performance capability which ensures that at whatever time an engine failure occurs after take-off, a forced landing should not be necessary. These are classified as Performance A aircraft - there are other categories covering other types of aircraft. No reference is made to the effects of wind-shear and microbursts. These have been well publicised elsewhere and do not directly affect calculations of take-off and landing performance.

AIRPORT PERFORMANCE CRITERIA

4. Before examination of the effects of weather factors on aircraft performance it is necessary to consider in very general terms, the safety factors applicable to the take-off and landing stages of flight. This is a complex subject involving aircraft structural limitations, runway length, runway slope, maximum tyre speeds, brake energy limits, obstacles on the flight path and many other factors. In order to consider how meteorological factors limit aircraft take-off performance is sufficient to know three basic performance criteria:

a. For a given runway, a decision speed, V1, is calculated. If an aircraft taking off at maximum weight on a dry runway experiences an engine failure before V1 it has insufficient residual performance to take off safely and therefore must stop: which it is able to do safely using the runway available.

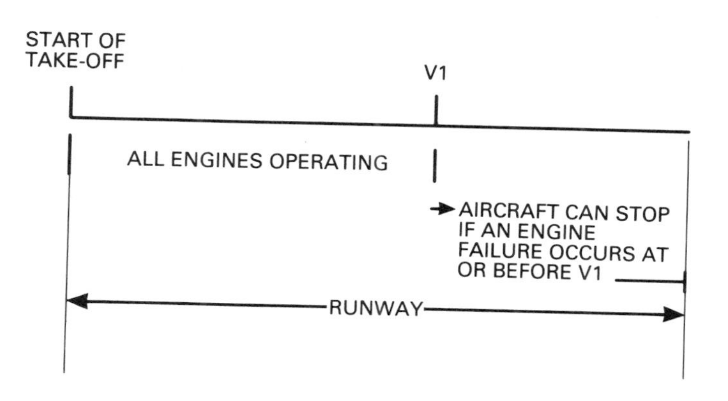

Figure 1
Decision Speed V1

b. After take-off and during the climb to 1500 ft above the runway, the aircraft must be able to maintain minimum positive rates of climb and to clear all obstacles in its flight path by a maximum of 35 ft.

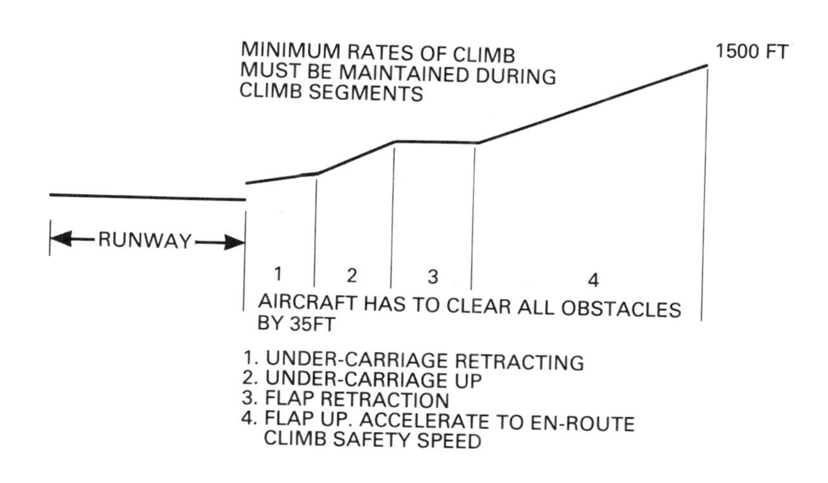

Figure 2
Take-off Criteria

c. During landing an aircraft must be able to approach from 1500 ft at an angle not exceeding 3° and stop on the destination runway within 55% of the runway available for jet engined aircraft and 60% for other aircraft. In the event of a balked approach the aircraft must be able to climb away at a minimum positive rate of climb.

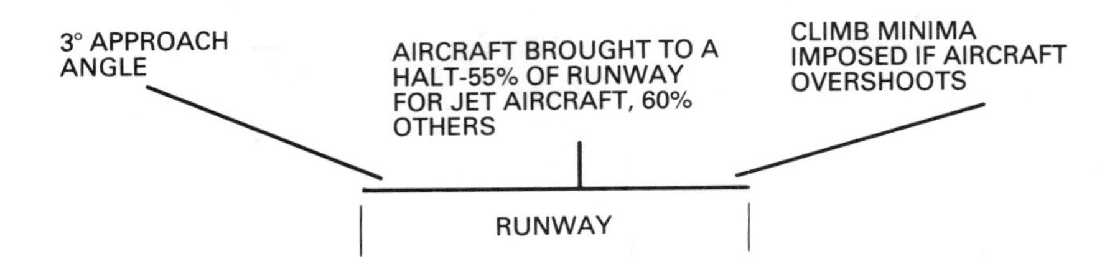

Figure 3
Landing Criteria

METEOROLOGICAL FACTORS AFFECTING AIRCRAFT TAKE-OFF PERFORMANCE

5. Before aircraft registered in the United Kingdom are given a Certificate of Airworthiness - in effect a licence to fly carrying passengers and freight, they are subject to numerous tests including the effects of take-offs and landings on runways contaminated with water, slush, snow and ice. The resultant limitations imposed by these contaminants are included in performance criteria applicable only to that aircraft, which enable the calculation of maximum permissible take-off weights when contaminants are present. However, even when they are taken into account contaminants increase risks, and can be so much of a safety hazard that in the United Kingdom it is recommended that operations from contaminated runways should be avoided whenever possible.

6. Inevitably the more contaminant present on a runway, the more adverse the effect on aircraft performance. Despite efforts made to clear runways of slush and snow there will be circumstances where a complete clearance cannot be maintained. Where this is so, at major United Kingdom airports the depth of the residual snow and slush is measured at approximately 300 metre intervals along the runway and reported in millimetres for each third of the runway.

7. The presence of water on the runway falls into two categories. Where the depth is less than 3 mm typically after heavy rain and where there is no standing water - puddles - there is no reduction made in the take-off weight of the aircraft. However, because a wet runway has a lower coefficient of friction than a dry runway it increases the distance required to stop the aircraft, so it is necessary to reduce the decision speed, V1. For a Tristar this reduction is typically 10 knots - from about 150 knots on a dry runway to 140 knots on a wet runway. This reduction means that for aircraft at maximum permissible take-off weight, on a wet runway, there will be a short period of time during the accelerating between the V1 wet decision speed and V1 dry, -about 4 seconds for a Tristar when, if an engine fails, the aircraft will be able to take off but will not achieve the full 35 ft safety margin up to 1500 ft.

8. Where there is standing water of depths greater than 3 mm due to an uneven runway or poor drainage, or wet snow of 3 mm or more, or snow more than 10 mm, there will be significant effects on the performance of the aeroplane, mainly:

a. additional drag due to the retardation effect on the wheels and spray impingement: similar to that experience when driving a car through a puddle or shallow flood.

b. the possibility of power loss due to spray ingestion

c. a reduced braking efficiency due to the lower coefficient of friction of the runway.

d. a reduced ability to maintain directional control of the aircraft.

e. the possibility of structural damage due to spray.

When a contaminant is present the inevitable result is that the maximum permissible weight for take-off is reduced. To calculate this reduced weight the exact increase in drag and the coefficient of friction are needed. Unfortunately these cannot be obtained because of the variability of the depth of contaminant both along the runway and with time, and often a variability of the contaminant if it is slush, wet snow or snow with consequent considerable variation in specific gravity. It follows that aircraft performance relative to a particular contaminated runway cannot accurately be assessed - despite much effort to do so by aviation safety authorities over the last few decades. Hence the calculations that are made must be regarded as only the best estimate that can be made of aircraft behaviour in circumstances when accurate prediction is impossible.

9. This means that when an aircraft takes off from a contaminated runway - apart from where the water is less than 3 mm there may be a time interval during the take-off when, if an engine failure occurs, the aeroplane has neither the capability to complete the take-off nor to stop within the remaining runway length - despite the use of a reduced maximum take-off weight.

10. For this reason pilots are advised in the UK not to take-off on contaminated runways. In any event there are maximum depths of contaminants for each aircraft beyond which take-off is not permitted. These vary with aircraft type but the maximum seldom exceeds 15 mm of standing water, slush or wet snow and 60 mm of very dry snow. Often the limit is much less, dependent upon the performance of different types of aircraft.

11. If the contaminant is less than the aircraft limit and a pilot decides to take-off, the reduction in maximum permissible take-off weight can be large. A few examples: a B747 has a take-off limit of 12 mm of slush. If taking off from Dusseldorf in 6 mm of slush the maximum permissible take-off weight for a flight to Anchorage would be reduced from 360 tons to 294 tons and the associated payload reduced from 60 to 15 tons. In similar slush levels the payload of a B737 on a flight from Munich to Hamburg would reduce from 11.7 to 6.6 tons. A Tristar with a normal maximum payload of 38 tons would have a reduced payload of 18 tons if it took off in 10 mm of wet snow - the limit is 15 mm.

12. There is a special hazard when a runway is covered by packed snow or ice. Packed snow and ice, unlike standing water and slush does not retard acceleration but the directional control of the aircraft is almost negligible, just like a car on an icy surface because of the low coefficient of friction. This hazard persists until the aircraft is travelling fast enough to use the aerodynamic effect of the rudder. Stopping an aircraft is also very difficult with the coefficient of friction as low as 0.05. Such conditions are extremely dangerous. To calculate their effect on maximum take-off weight, runway lengths are reduced by typically 40% in take-off calculations, which precludes most take-offs except at very low aircraft weights and on exceptionally long runways. Because of the difficulty in directional control at low speeds, the maximum permissible crosswind is reduced by approximately two thirds to, typically, 10 knots.

13. Another consideration in the calculation of take-off performance is the capability of an aircraft to maintain a maximum positive rate of climb after take-off following the failure of an engine. The rates of climb are shown in Figure 3. There are three factors: aircraft weight, atmospheric

pressure and temperature, called, in combination, the WAT limit. Aircraft (mainly engine) performance is referenced to the International Standard Atmosphere at mean sea level. Where either pressure or temperature, or both, vary markedly from that reference the performance of the aircraft is affected accordingly. The onset of the WAT limit varies with aircraft and engine type and power output characteristics. Smaller aircraft with only two engines can be limited with only small changes from ISA; however even larger three or four engined aircraft suffer severe limitation. An increase in temperature from 40 to 41°C at Athens Airport, altitude 90 ft would reduce the maximum permissible payload of a Tristar by 1.4 tons. At 3500 ft and 30°C a reduction in pressure of 13 mbs would reduce maximum payload by 2.5 tons, or an increase of 1°C by 1.4 tons.

14. The presence of frost on the surface of an aircraft, particularly wings, poses a special hazard. A hoar frost deposit equivalent in thickness to a sheet of sandpaper increases the stalling speed of a jet aircraft by typically 30 knots. Even when the aircraft has been de-iced a short delay between de-icing and take-off can be critical - the deposit must not be allowed to re-form. You may recall that an Air Florida B737 crashed on take-off at Washington in 1981. It had been delayed and the enquiry decided that after 35 minutes frost had re-formed on the leading edge flap, equivalent to a strip of sandpaper, which was sufficient to cause the crash.

LANDING

15. The WAT limit also affects landing performance as it is necessary to ensure that an aircraft has a minimum positive rate of climb if the approach is discontinued or is balked and an overshoot is made. The minimum rates of climb for an all engine overshoot is 3.2% and with one engine failed, 2.4%, both using take-off power. Since landing weight is inevitably less than take-off weight the WAT limit seldom imposes a restriction on landing but can become critical when an aircraft is carrying its maximum passenger load on a short flight to an aerodrome with a high ambient temperature. In this circumstance the aircraft is still heavy on landing and the high temperature at the destination can impose WAT landing limits.

16. Runway contaminants also tend to be less restrictive on landing than take-off. The effect of contaminant drag could even be considered to be beneficial, but is too uncertain for accurate calculation and is not taken into account in landing performance calculations. Nevertheless, as for take-off, the risks on runways heavily contaminated with water, slush and snow above 3 mm deep are high. The coefficient of friction is too low to enable the accurate determination of landing

distance and makes directional control of the aircraft very difficult below the speed for effective rudder control.

17. In addition there exists the likelihood of aquaplaning on runways where water and slush exceed 3 mm in depth. It occurs at speeds down to 0.9 x tyre pressure: about 120 knots for a Tristar - the maximum landing weight speed is 139 knots. At these speeds the inertia of the water or slush results in cushion of water or slush between the tyres and the runway surface giving a coefficient of friction similar to that for ice, about 0.05, with the attendant problems of negligible wheel braking and loss of directional control. So, in the UK, pilots are advised not to attempt to land in conditions of water, slush and ice except where there is an adequate margin of safety over the normal landing distance required and when the crosswind component is small. This means that for a Tristar at maximum landing weight the landing distance is increased from a minimum of 1760 m to 2780 m and the crosswind component is reduced to 10 knots or less.

EFFECT ON AIRPORTS

18. The clearance of slush, snow and ice from runways is a major problem and expense for airport authorities. At busy airports the penalties of runways being unusable for a few hours, because of contamination, is unacceptable for reasons of both lost traffic revenue and the implications of passenger delays, airline schedules and air traffic control congestion. The inconvenience of delays to passenger are too well known to mention. The effect of a disruption of an airline schedule may be extensive as it is often difficult if not impossible to overcome the knock-on effect of delays at one airport when the utilisation of available aircraft is already maximised to, typically, 10 to 12 hours airborne time each day in the interests of profitability and there is often little an airline can do to provide back-up reserve aircraft. The effect on air traffic control is also becoming increasingly important as in the busier air traffic areas the system is already saturated to the extent that in Europe, Air Traffic Flow Management is planned months in advance to minimise delays.

19. At airports the problems are the availability of staff to meet occasional clearance operations and the cost of providing sufficient clearance equipment to remove the contaminant from the runway area of up to 150 ft by 10,000 ft, ie 1.5 M square feet in the minimum possible time. Staff are often the critical factor. At Heathrow for example, a major snow clearance operation requires some 130 staff, drawn from existing airport staff on an emergency work basis. Meteorological advice plays a vital role in the efficiency of clearance operations. The timing and accuracy of snow

warnings given to airports is critical. If, for example, a warning is given just after daytime staff have left the airport it is extremely difficult to recall them: and in the event of heavy snow which blocks access roads, recall may not be practical. The resultant loss of revenue from landing fees alone can be very high, as much as £3,000 per landing, which quickly pays for a 24 hour snow clearance cost at Heathrow of £50K, excluding equipment costs.

20. In summary, therefore, atmospheric pressure, ambient temperature and contaminants on runways can jeopardise the safety of flight and disrupt airline schedules with a resultant inconvenience to passengers and substantial cost penalties to airlines and airport authorities. Timely warnings of the onset of heavy rain, snow and freezing conditions can alleviate these problems to the overall benefit of aviation.

2

Applications of meteorology in marine transport

J. L. Thompson and S. Felding, IMO Sub-Division for Navigation and Related Matters

1. INTRODUCTION

1.1 Apart from farmers, seafarers have probably been applying the science of meteorology in their occupation, or hobby, longer than any others. Certainly they have been taking account of meteorology on a worldwide scale longer than any others. Until the nineteenth century, except for manual propulsion, sail was the only means of propulsion for ships. Wind was the driving force - too little, no speed; too much, produced rough seas and resulting damage to sails and ship and in many cases loss of life. The seafarer therefore needed to know where in the world he could expect to find calms, trade winds and storms, to observe the barometer, sky, clouds, temperature, the sea and swell and to predict the onset of a squall, the approach of a storm or the possibility of fog.

1.2 The advent of mechanical propulsion and steel ships did not relax in any way the dependence of seafarers on meteorological information for their safety and that of the ship and its cargo. Despite the excellent weather reports, storm warnings and other information provided by world meteorologists, bad weather conditions remain both directly and indirectly the largest single cause of shipping casualties.

1.3 It was a seaman, Lieut. Maury of the United States Navy, who was responsible for organizing the first International

Conference on Maritime Meteorology in Brussels in 1853. As a result, the hosts of this Symposium, the United Kingdom, formed the British Meteorological Office in 1855 with a seaman Admiral Fitzroy as its Director, its function at that time being, to collect observations from ships, developing meteorological atlases, collecting data on ocean currents and other phenomena of benefit to shipping. International meteorology developed from this Maritime Conference with the foundation in 1873 of the International Meteorological Organization which in 1951 was superseded by the World Meteorological Organization (WMO). While maritime meteorology, through its Commission for Marine Meteorology (CMM), is only a small portion of WMO's work, WMO's contribution is of crucial importance to the safety and prosperity of all connected with ships and the sea and warrants the appreciation of seafarers worldwide.

1.4 This paper briefly outlines the meteorological and oceanographical requirements of seafarers and those connected with the sea; some of the reasons for these requirements and the training given to seafarers to enable them to use and understand the information provided by meteorological services worldwide and to prepare weather reports and transmit them to the shore for use by meteorologists. Information is also given on present and future means for transmission of meteorological information to and from ships, as well as the specialist needs of search (SAR) organizations, marine pollution operations and for dumping or incinerating waste materials at sea.

2. INTERNATIONAL REQUIREMENTS AND RECOMMENDATIONS

2.1 <u>1974 SOLAS Convention</u>

2.1.1 Chapter V of the 1974 SOLAS Convention, which applies to <u>all</u> ships includes mandatory requirements for Administrations and ships to provide meteorological and oceanographic information to assist ships and meteorological authorities ashore.

2.1.2 Regulation V/2(a) - Danger messages, requires that the master of every ship which meets with dangerous ice, a dangerous derelict, or any other direct danger to navigation, or a tropical storm, or encounters sub-freezing air temperatures associated with gale force winds causing severe ice accretion on superstructures, or winds of force 10 or above on the Beaufort scale for which no storm warning has been received, is bound to communicate the information by all the means at his disposal to ships in the vicinity, and also to the competent authorities at the first point on the coast with which he can communicate.

2.1.3 Regulations V/2(b) requires governments to ensure that the intelligence referred to in paragraph 2.1.2 is brought to the attention of ships and those concerned and to WMO Members (other interested governments).

2.1.4 Regulations V/2(c) and (d) requires ships transmitting danger messages to precede the message by the safety signal prescribed in the ITU Radio Regulations (TTT by morse or "SECURITY" by radiotelephone) and the messages concerned to be accepted by coast radio stations free of charge.

2.1.5 Regulation V/3 lays down the information that should be included in danger messages.

2.1.6 Regulation V/4 relates to meteorological services.
2.1.7 Regulation V/4(a) requires Governments to encourage the collection of meteorological information by ships at sea, their use of highly accurate instruments and to facilitate the checking of such instruments on request.

2.1.8 Regulation V/4(b) requires Governments to:

1. warn ships of gales and storms;

2. issue weather bulletins to ships containing data of existing weather, waves and ice, forecasts and information to enable weather charts to be prepared;

3. prepare meteorological publications for the use of ships;

4. arrange for selected ships to be equipped with meteorological instruments for them to take regular observations for use by meteorological services;

5. encourage ships to take observations in data sparse area and transmit the details to other ships and meteorological services particularly when in the vicinity of storms or suspected storms;

6. arrange for the reception and transmission by coast radio stations of weather messages to and from ships; and

7. conform with the technical regulations and recommendations of WMO

2.1.9 Regulation V/5 requires governments to continue the ice patrol service in the North Atlantic and regulation V/6 provides for the management and cost of the service.

2.1.10 Thus Chapter V of the 1974 SOLAS Convection lays down all the essential meteorological information that must be provided by and to ships.

2.2 1978 STCW Convection

The International Convention on Standards of Training, Certification and Watchkeeping for Seafarers, 1978, prescribes the training that should be given to masters and mates at all levels and for all sizes of ships to understand and apply the information provided by meteorological services worldwide More detailed information is given elsewhere in the paper.

2.3 1979 Convention on Maritime SAR

The specialist meteorological and oceanographic requirements of maritime search and rescue services are referred to in the International Convention on Maritime Search and Rescue, 1979. This requires that the operating plans of every SAR organization should contain details regarding action to be taken by those engaged in SAR operations including methods of obtaining essential information relevant to SAR, such as forecast of weather and sea surface condition. As will be explained later, these are important to committing SAR resources such as vessels and aircraft, determining SAR areas and cancellation of unsuccessful SAR operations.

2.4 Conventions relating to marine pollution

The 1969 International Convention relating to Intervention on the High Seas in Cases of Oil Pollution Casualties and its 1973 Protocol concerning substances other than oil and the 1972 Convention on the Prevention of Marine Pollution by Dumping of Wastes and other Matter, all contain or imply requirements for specialized meteorological information. More detailed information is given in this paper.

2.5 Resolutions and recommendations which relate to meteorology

2.5.1 The IMOSAR and MERSAR Manuals (resolutions A.439(XI) and A.387(X)), contain information, for SAR organizations and ships respectively, on the meteorological and oceanographical requirements for SAR purposes and the use of such information in SAR operations (see section 5).

2.5.2 The Standard Marine Navigational Vocabulary (resolution A.380(X), as amended) which was compiled to assist safety of navigation by standardizing the language used in communications at sea, includes in the phrase vocabulary a section on "weather".

2.5.3 Section VI of the International Code of Signals deals

with meteorology and contains code groups which may be used by ships, coast radio stations and the shore to transmit or receive meteorological information where language difficulties exist. All ships carry copies of the Code.

2.5.4 The recommendations on training and examination of fishermen (resolutions A.539(13) and A.576(14)) contain syllabi covering meteorology and oceanography similar to that for masters and mates in the 1978 STCW Convention.

2.5.5 The IMO/ILO Document for Guidance, 1985, and the FAO/ILO/IMO Document for Guidance on Fishermen's Training, 1986, contain sections which provide further guidance on the training that should be given to seafarers and fishermen in meteorology and oceanography.

3. REQUIREMENTS FOR SAFETY OF NAVIGATION

3.1 The maritime meteorological requirements for safety of navigation relate mainly to wind, fog and ice.

3.2 Wind and its effects on the sea can be recognized as the most important meteorological condition affecting ships. Those who are meteorological scientists will probably shudder at the thought that seafarers still define wind and sea state in terms used in the Beaufort Scale but this is demonstrative of the approach of all seafarers to the practical consequences of the weather on ships. The scale was originally introduced in 1908 by Admiral Beaufort who defined the numbers of the scale in terms of the effect on a sailing man-of-war of his day although equally applicable to a similar sized merchant ship. The number defines weather conditions from calm to hurricane, the amount of sail that should be carried and expected ships's speed under each condition (Beaufort's criterion) and the means to recognize from the appearance of the sea, the weather conditions prevailing. The scale has been expanded from the original to include small craft and the practical actions they, being at greater risk because of their size, should take before going to sea in and on meeting such weather conditions at sea. The more technical information - wind velocity in knots, probable wave heights and typical atmospheric pressure gradient being less important; when this information is received in meteorological bulletins and forecasts seafarers frequently mentally convert it into Beaufort criteria and thereby its practical effect on their ship.

3.3 To appreciate the power of the sea on ships in bad weather let me give you some examples. In 1957 the writer joined a 32,000 ton tanker in dry dock, at that time one of the biggest afloat, undergoing repairs for weather damage received off the Cape of Good Hope on its previous voyage from the Gulf to North West Europe, in storm/strong gale conditions, on reduced speed and taking heavy seas. It took one very heavy, possibly freak, wave which passed over its bow and over the bridge superstructure. The result: the ship's forecastle head, one of the strongest parts of the ship, was crushed from a minimum height of three metres to less than one metre, the forward flying bridge, the midships lifeboats, the standard compass on the monkey island were carried overboard and pipes, ladders and valves on the after-deck were seriously bent.

3.4 Another ship carrying steel billets, when passing through the same area, and rolling 35t degrees each way with a period of 11 seconds (port to starboard and back to port) in only near gale conditions (wind velocity 28 to 33 knots) cargo began to shift. The ship was hove to (head to wind and sea) and the cargo re-secured. Had the weather been worse, the writer of this paper might not have been here to prepare it.

3.5 By far the greatest number of search and rescue operations occur within three miles of the shore and generally involve small craft which, in the case of sailing vessels because they are unable to move upwind is frequently due to "lee drift" (the drift of the vessel downwind). The dangers of drifting onto a lee shore has always been a problem for sailing ships. Everyone has read of the difficulties for sailing ships in rounding Cape Horn; this was because the prevailing wind was westerly and to make headway into it the ship could only sail within about 67 degrees of the wind direction at the same time the effect of the wind on the hull was carrying the ship downwind If the vector of the leeward movement of the ship downwind exceeded that of the movement upwind the resultant ship motion was downwind The same principle applies on a lee shore when a sailing vessel runs aground.

3.6 Although the proper use of radar has greatly assisted, for and mist continues to be the cause of accidents at sea. The reason for this is obvious. Fog reduces visibility, renders navigation more difficult and because seafarers cannot see where they are going, collisions and groundings occur. Even with radar, ships in general have to reduce speed which reduces their efficiency and has attendant economic consequences.

3.7 Ice is the remaining main meteorological and oceanographic problem for safety of navigation whether it be icebergs, sea ice or ice accretion.

3.8 The first SOLAS Convention (1914) was the consequence of the loss of the passenger ship "TITANIC" in the North Atlantic which collided with an iceberg. The iceberg season in the

North Atlantic is from March to July, the main factor governing its severity is the frequency and direction of the wind off the Labrador coast, prolonged north-westerly winds immediately prior to and during the season, usually increasing the number. Ships are normally kept informed by surveillance by the International Ice Patrol.

3.9 Information on the extent of sea ice is of importance to seafarers operating in areas where fast and pack-ice occur both from the safety of navigation and the economic consequences of becoming icebound in a port at sea.

3.10 Ice accretion is a serious problem in particular for small ships; there have been a number of cases of vessels capsizing as a consequence of ice forming rapidly on their upper decks, superstructure, masts and rigging. Serious ice accretion normally takes place in gale conditions and can be an accumulation of frozen spray, snow and freezing rain when the sea temperature is below 0°C and the air temperature is very low. These conditions make work on deck very difficult and dangerous and manual removal of the ice impracticable. The added weight of the ice gradually reduces the vessel's positive stability until a point is reached when the dynamic effect of rolling in heavy seas and inadequate stability produces a capsizing moment and the vessel capsizes. The only action available to small vessels is to seek shelter. Warnings are broadcast to ships in areas where serious ice accretion is likely to occur.

3.11 The best action to be taken insofar as gales, storms and ice are concerned is to plan the voyage, so far as possible, to avoid them. This is dealt with under the section on voyage planning. Historical information obtained from meteorological and oceanographic records is necessary for this purpose e.g. currents, frequency of gales, fog, wind direction, ice and icebergs. 3.12 However, the best voyage planning is a compromise to obtain the best average conditions throughout the voyage. Voyage plans must always be adjusted to take account of the changing weather conditions ahead of the ship. For this purpose forecasts and weather bulletins, storm and ice warnings etc., are necessary; in the case of storm warnings, as far ahead as possible, so that action to avoid the worst of the storm can, when necessary, be taken. Thus the ship needs to know the extent, wind force, position of the centre, existing track and speed of movement of the storm in a detailed warning as well as a forecast of the storm's probable behaviour. Using this information and by applying simple practical rules the ship can avoid the worst of the storm (see Appendix).

3.13 On receipt of a storm or gale warning, action will be taken to "secure the ship" that is to ensure that all openings are closed, weather tight, deck ventilators closed, cowls removed and plugs and covers secured, derricks lashed down etc.

3.14 Forecast of the possibility of fog is essential to the efficient running of the ship. Fog generally places a burden on the ship's crew which, for economic reasons, is nowadays reduced to a minimum. Fog generally means doubling watchkeeping personnel. On the bridge, permanent lookout must be kept and constant observation of the radar is necessary. The engines must be on standby and ready to manoeuvre; this usually requires additional engineers. If all are warned in advance, this provides opportunity for all concerned to obtain proper rest, which is a watchkeeping requirement (regulation II/1.5 of 1978 STCW Convention), to ensure their efficiency is not impaired by fatigue.

4. REQUIREMENTS FOR VOYAGE PLANNING

4.1 Voyage planning has traditionally been undertaken by the deck department of the ship. However, increasingly nowadays it is being undertaken by specialists, either part of a national meteorologist department or a commercial company.

4.2 As has been said, the function of the earliest meteorological associations was to collect observations from ships on weather, ocean currents and other phenomena. This meteorological and oceanographical information has been analysed over many years and numerous publications prepared which are used by seafarers for voyage planning viz. sailing directions, meteorological atlases and climatic charts, recommended routes etc. Information used includes:

1. <u>Meteorological</u>

Average, maximum and minimum monthly

- wind; velocity and direction (number of days on each of the eight cardinal points of the compass);
- frequency of gales and tropical storms;
- frequency of fog;
- limits of ice and icebergs; type and extent;
- precipitation;
- direction, height and length of swell;
- air temperature.

2. Oceanographic

Average, maximum and minimum monthly

- current; set and drift (number of days on each of the eight cardinal points of the compass);

- tides; daily variation (speed, direction and height at all stages);

- swell (height and direction)

- sea temperature.

4.3 Using this information and his experience the navigator selects the most appropriate route to be followed, taking into account the type and size of ship (sail or power driven) and, if power driven, its engine power.

4.4 The effect of weather on ships can be appreciated from the experience of the writer on a voyage from Algiers to Port de Bouc (Marseilles) France on a 16,000 ton tanker; the total distance is 410 nautical miles, the ship's normal service speed was 16 knots, in other words, the voyage should have taken 26 hours. The actual voyage took approximately 6 days, the average speed being almost 2.5 knots. While in this case voyage planning would not have assisted, the economic consequences of the voyage can be appreciated as the time was almost 500% more than it should have been. If time can be saved this is economically desirable. The economic savings are directly proportional to the size of similar types of ship. The operational costs of some of the largest being of the order of a thousand dollars or more per hour. Therefore even a small saving of time can be of considerable advantage. For this reason shipowners and charterers are increasingly using the services of organizations providing weather routing services.

4.5 The writer's first experience of this service was on a voyage chartered to a United States company from Pakistan to Houston, Texas. After having passed through the south-west monsoon and without benefit from a weaker than normal Agulas current off the coast of Southern Africa and not having made the ship's "charter" speed, he received a telegram from the charterer pointing out this fact together with a weather forecast and the obvious information that the weather on the route ahead was favourable, which indeed it was; the ship made more than its charter speed - never achieved since its maiden voyage. Clearly the charterer had used the services of a meteorological expert in case he had to make a claim for breach of contract with regard to the ship's speed.

4.6 The advantages of using a weather routing service are obviously considerable. Whereas the ship's navigator must mentally await the meteorological information available and provided in forecasts, voyage planning services have all the necessary meteorological and oceanographic information immediately available on a computer. At the touch of a button, the properly programmed computer can provide the optimum route for a ship using its services and future changes in the route based on the up-to-date meteorological data which is being continuously fed into it.

4.7 In order to perform the weather routing task satisfactorily the expert, who is usually knowledgeable not only in meteorology but also a mariner, needs to know the performance characteristics of the ship for which a route is to be prepared under the various conditions of loading, in particular the effect of wind and waves on the ship's speed using this information and the forecasts, both historical and up-to-date, of wind, waves, ice and the possibility of fog on a voyage between the departure port and port of destination, he is able to prepare a provisional route which is provided to the ship before sailing. Once informed that the ship has sailed, he will confirm or amend the recommended route, as appropriate. Thereafter, at least every 48 hours, he will advise the ship by radio to retain the recommended route or amend it and, if requested, will provide additional meteorological information. In addition, an analysis of the weather along the ship's route is made every six hours, and if necessary routing changes made should circumstance so warrant. Ships which do not themselves normally provide weather reports are usually requested to transmit weather reports while on passage.

5. REQUIREMENTS FOR SEARCH AND RESCUE

5.1 Under the international search and rescue plan being developed by IMO in accordance with the 1979 SAR Convention the organization and co-ordination of SAR operations at sea will be the responsibility of national maritime rescue co-ordination centres (RCCs) or rescue sub-centres.

5.2 In carrying out any SAR operation the RCCs may commit any of the various SAR resources available to him, including aircraft, helicopters, rescue vessels, lifeboats, coastal rescue units and merchant ships at sea. Those used will depend on the type of operation to be undertaken and the position of search or rescue area. Coastal SAR usually involves lifeboats and helicopters; ocean SAR, long-range aircraft, rescue vessels (which may carry helicopters) and merchant ships.

5.3 Before committing these units the RCC must be sure that

they can safely proceed to the search area, carry out the search, effect rescue and safely return to base. His decision depends solely upon the weather conditions prevailing.

5.4 The meteorological and oceanographical information the RCC SAR Mission Controller (SMC) requires is considerable. He needs both historical and up-to-date information and forecasts along the route and in the search area e.g.:

1. to determine the search area, he requires the wind velocities and directions, sea current and wind current in the vicinity of the search area or along the ship's anticipated route, in the case of a missing ship, which has prevailed since the emergency occurred (historical) and a forecast covering the period for the rescue units to reach the search area;

2. to commit aircraft and helicopters - forecasts for the route and search area at the various heights the aircraft will fly. Surface forecasts i.e. visibility during search and conditions affecting helicopter rescue, medical evacuation, dropping of equipment or medical teams etc.,

3. to commit surface units - marine weather forecasts;

4. to determine when to abandon an unsuccessful SAR operation - this will depend on the possibility of survivors still being alive given the air temperatures and wind speed and the sea temperatures prevailing. A person not wearing special protective clothing can expect to survive in water at less than 2°C less than 3/4 of an hour, 15-20°C less than 12 hours. The chill factor of wind also can cause the onset of hypothermia and would be taken into account if the persons concerned were in an uncovered boat.

6. PROTECTION OF THE MARINE ENVIRONMENT

6.1 Marine pollution response operations

6.1.1 Under the 1969 International Convention Relating to Intervention on the High Seas in Cases of Oil Pollution Casualties and its 1973 Protocol Relating to Intervention on the High Seas in Cases of Marine Pollution by Substances other than Oil, an Administration may take such measure on the high seas as may be necessary to prevent mitigation or eliminate grave and imminent danger to their coastline or related interests from pollution or the threat of pollution of the sea by oil, following upon a maritime casualty or acts related to such a casualty which may reasonably be expected to result in major harmful consequences.

6.1.2 To make a decision on such intervention an Administration must have access to information on prevailing ocean currents, sea state, present and forecast wind directions and velocity, visibility and presence of ice and icebergs.

6.1.3 In the event of an actual spill the movement of spilled oil can be predicted if surface current and wind speed are known and resources mobilized in the most effective manner. Much of the available equipment has limited effectiveness in high sea states and knowledge of the sea state is essential in planning response strategies.

6.1.4 Oil spilled into the sea undergoes a number of physical and chemical changes, some of which lead to its disappearance whilst others cause it to persist. The time taken depends primarily upon the physical and chemical characteristics of the oil as well as the quantity involved, the prevailing climatic and sea conditions and whether the oil remains at sea or is washed ashore.

6.1.5 Spreading of the oil on the sea surface is initially a function of gravity and surface tension but is, after a few hours, dependent on hydrographical conditions such as currents, tidal streams and wind speeds. Evaporation is increased by high temperatures, rough seas and strong winds. Natural dispersion is largely dependent on the type of oil and the sea state proceeding most quickly in the presence of breaking waves. Many oils tend to absorb water to form water-in-oil emulsions and in moderate to rough sea conditions most oils rapidly form emulsions.

6.1.6 From the foregoing, it can be seen that knowledge of sea state, ocean currents, tidal streams and present and forecast weather is essential when planning strategies to respond to an oil spill.

6.1.7 This has been reflected in several publications such as the IMO "Provisions concerning reporting of incidents involving harmful substances under MARPOL 73/78" and the common pollution reporting system (POLREP) adopted by the contracting Parties to the Bonn, Copenhagen and Helsinki Agreements which require information in the reports on wind, current, tide, sea state and visibility.

6.2 Dumping and incineration of wastes at sea

6.2.1 Few persons are unaware of the problems of disposal of waste whether it is industrial or domestic. It has, until legislation was introduced, been the cause of pollution of the atmosphere, land, rivers and seas. The very nature of the atmosphere and seas spreads the contamination from materials dumped or burnt. The sea has historically been a convenient place for dumping. On ships when anything is thrown over the side it is known as "putting it in the big locker". Seafarers have dumped almost everything imaginable in the sea e.g. residues from cargo, timber, plastics, old nets, rope and wine etc. Ships have been specially built from disposal of waste from the shore e.g. sewage sludge tankers. The ecological consequences to human health, fauna and flora at sea or on land of past methods of disposal of rubbish have become increasingly evident in recent years and legislation enacted to reduce or prevent it, but there remains a need to dispose of waste, e.g. chemicals, radioactive materials etc.; this is an ever present problem for many industries. Such disposal needs careful control.

6.2.2 In order to promote the effective control of all sources of pollution of the marine environment especially to prevent pollution of the sea by dumping of waste and other matter that is liable to create hazards to human health, to harm living resources and marine life, to damage amenities or to interfere with other legitimate uses of the sea, the Convention on the Prevention of Marine Pollution by Dumping of Wastes and other Matter (LDC) was adopted in 1972, and the Convention entered into force on 30 August 1975. IMO is responsible for secretariat duties in relation to the Convention. The LDC Convention prohibits the dumping of certain specified wastes (annex 1) but permits dumping of others subject to issue of a special permit following careful consideration of all the factors involved; these are listed in annex III to the LDC Convention. These factors include evaluation of the characteristics of the dumping site; for this purpose meteorological and oceanographic products are required to evaluate the "dispersal characteristics e.g. effects of currents, tide and wind on horizontal transport and vertical mixing)" (annex III(B)(5)). This requirement is interpreted in the "Guidelines for Selection of Dumping Sites" as follows:

1. water depths (maximum, minimum, mean);

2. water stratification in various seasons and weather conditions (depth and seasonal variation of pycnocline);

3. tidal period, orientation of tidal ellipse, velocities of minor and major axis;

4. mean surface drift (net): direction, velocity;

5. mean bottom drift (net): direction, velocity;

6. storm (wave) induced bottom currents (velocities); and

7. concentration and composition of suspended solids.

6.2.3 Incineration of waste causes pollution initially of the atmosphere, the elements of the smoke or fumes descending to the surface either directly or, by forming or combining with the hygroscopic nuclei in the atmosphere on which water vapour condenses, eventually as contaminated rain. Consequently, pollution of the sea or land or both can be caused not only in the vicinity of the incineration site but also at considerable distances away being carried there by the effects of wind and sea currents.

6.2.4 For this reason an addendum was adopted to annex 1 of the LDC Convention, which entered into force on 11 March 1979, prescribing "regulations for the control of wastes and other matter at sea". Regulation 8 lays down the provisions to be considered in establishing criteria governing the selection of incineration sites. Those affecting meteorological and oceanographic data are, in addition to those listed in paragraph 7.2 above, as follows:

(a) "the atmospheric dispersal characteristics of the area -including wind speed and direction, atmospheric stability, frequency of inversions and fog, precipitation types and amounts, humidity - in order to determine the potential impact on the surrounding environment of pollutants released from the marine incineration facility, giving particular attention to the possibility of atmospheric transport of pollutants to coastal areas;

(b) oceanic dispersal characteristics of the area in order to evaluate the potential impact of plume interaction with the water surface".

6.2.5 In order to evaluate the effects of incineration, marine incineration facilities are required to record the meteorological condition e.g. wind speed and direction).

7. COAST GUARD

In many countries the coast guard or similar organizations provide safety services for small craft. Owners of pleasure craft are invited to contact the local coast guard station before they put to sea, giving their intended destinations and route. The coast guard station provide them with meteorological information received from the national meteorological service and, when it is unsafe due to the forecast or prevailing weather conditions, advise them not to proceed to sea.

8. PORT REQUIREMENTS

8.1 As a general rule, cargo should not be loaded and discharged during adverse weather conditions. However, some cargoes suffer more damage from, for example, rain than others, either immediately due to the rain or in transit due to condensation (see Section 9 - Cargo Care below).

8.2 In particular, the master of a ship and the berth operator within their respective areas of responsibility should not permit dangerous substances to be handled in weather conditions which may seriously increase the hazards presented by such substances. Due to the nature of explosives careful attention should be given to weather conditions, in particular thunderstorms and precipitation.

9. CARGO CARE

9.1 Goods in transit may be affected by the conditions to which they are subjected. These conditions may include changes in temperature and humidity and particularly cyclic changes that may be encountered. An understanding of condensation phenomena is desirable because condensation may lead to such damage as rust, discolouration, dislodging of labels, collapse of fireboard packages or mould formation.

9.2 Solar radiation can produce air temperatures under the inner surfaces of a cargo hold or container (cargo space) which are significantly higher than external air temperatures, while radiation at night can cause such temperatures to fall well below the external temperatures. The combination of these effects can result in a range of day and night cyclic temperature variations in the air adjacent to the inner surfaces of the cargo space which is greater than the corresponding range of temperatures just outside.

9.3 Goods closest to the steel surrounds of the cargo space will be more affected by external variations than those in the centre. If the possible extent of temperature variations of their full significance is not known, advice should be obtained from specialists.

9.4 Under the circumstances described, condensation may occur either on the surface of the cargo (cargo sweat) or on the inside steel surfaces (space sweat) both during transport or when the cargo space is opened for discharge.

9.5 The main factors leading to condensation inside cargo space are:

1. sources of moisture inside the cargo space which,

depending on ambient temperature conditions, will affect the moisture content of the atmosphere in the cargo space;

2. a difference between the temperature of the atmosphere within the cargo space and the surface temperature of either the cargo or the inner surfaces of the cargo space itself;

3. changes in the temperature of the outer surface of the cargo space which affect the two factors above; and

4. circumstances under which the cargo space is ventilated.

9.6 Warming the air in a cargo space causes it to be absorb moisture from packagings or any other source. Cooling it below its dewpoint[1] causes condensation.

9.7 If, after high humidity has been established inside a cargo space and the outside is cooled, then the temperature of the inner surface may fall below the dewpoint of the air inside it. Under these circumstances moisture will form on the inner surfaces. After forming on the underside of the deck above or the container roof, it may drop on to the cargo. Cyclical repetition of the cargo space sweat phenomena can result in a greater degree of damage.

9.8 Condensation can also occur immediately after the cargo space is opened if the air inside the space is humid and the outside air is relatively cool. Such conditions can produce a fog and even precipitation but, because this phenomenon usually occurs only once, it seldom results in serious damage.

9.9 Cargo space mechanical ventilation under carefully controlled conditions will alleviate some of the problems in holds. Cargoes likely to be damaged by condensation should not be transported in containers unless special precautions are taken such as using moisture absorbing paint or protecting the cargo from moisture dripping on it.

10. METEOROLOGICAL INFORMATION PROVIDED BY OR AVAILABLE FROM SHIPS

10.1 The requirements of regulation V/2(a) concerning the transmission by ships of danger messages containing urgent meteorological information is given in paragraph 2.1.2. In addition, a certain number of ships selected by their national

---

[1] Dewpoint is the temperature at which air saturated with moisture at the prevailing atmospheric pressure will start to shed moisture by condensation.

meteorological services (selected ships) undertake weather and sea observations on a regular basis. Whilst at sea they record the information obtained and also transmit it to the nearest coast radio station accepting meteorological information. Other ships also undertake limited observations generally at the request of the meteorological service of littoral States bordering data-sparse ocean areas which are mainly in the Southern Hemisphere.

10.2 Meteorological instruments carried on ships, other than selected ships, are in general not very sophisticated e.g. aneroid rather than mercury barometers, psychrometers rather than liaison hygrometers and few ships have a barograph. Sea temperatures are usually taken at engine room intakes rather than surface temperature taken with a proper bucket. Wind speed and direction is estimated from the appearance of the sea; particularly difficult on a dark night. Cloud observation is based on the experience of the observer or by comparison with photographs. Most ships record together with the noon position the set and draft experienced over the past day. Meteorologists should bear these facts in mind when using data observed by ships other than selected ships. Ships calling at port bordering data-sparse areas, which are requested to take observations, should have their instruments checked for accuracy.

10.3 Selected ships have over many years made an important contribution to meteorological services worldwide for many years and, from constant practice in making weather observations, provide useful meteorological and climatological data. The data they provide is as reliable as the instruments and equipment they use for observation are provided, tested and maintained by the national meteorological service.

11. COMMUNICATIONS

11.1 Present communications

11.1.1 For the purpose of promulgation of marine safety information (navigational and meteorological warnings, meteorological forecasts, distress alerts and other urgent information broadcasting to ships) and in order to co-ordinate the transmission of radionavigational warnings in geographical areas (figure 1) IMO and IHO have established the worldwide navigational warning service (WWNWS). The delimitation of such areas is not related to and shall not prejudice the delimitation of any boundaries between States. Where appropriate, the term NAVAREA, followed by an identifying roman numeral, may be used as a short title for the areas.

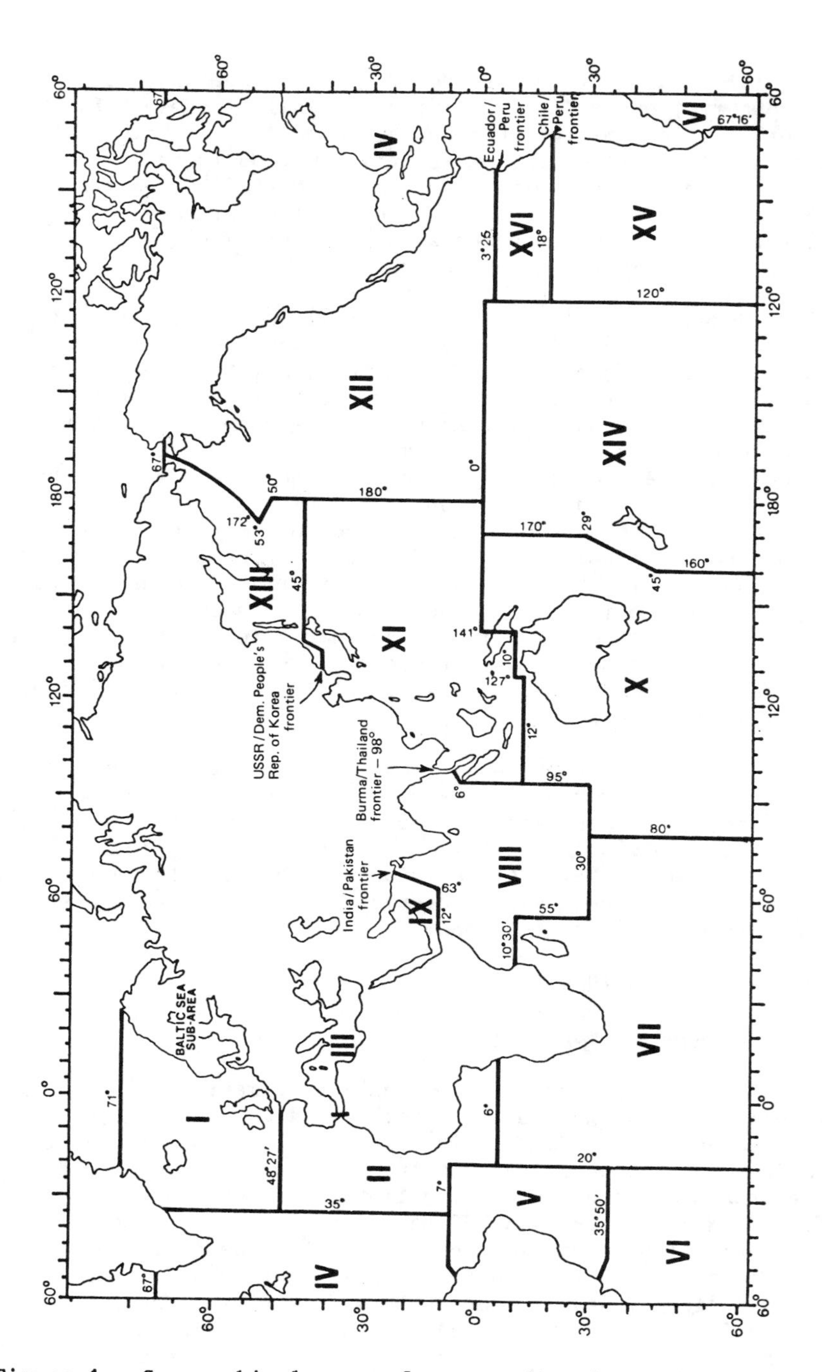

Figure 1 - Geographical areas for co-ordinating and promulgating radio-navigational warnings with area co-ordinators noted in parenthesis

11.1.2 Although guidance under the WWNWS is available to varying degrees for all three types of radionavigational warnings i.e. long-range, coastal and local), the two internationally co-ordinated services under the WWNWS concern only the long range ("NAVAREA") and coastal ("NAVTEX") warnings. Local warnings, and coastal warnings in support of purely national requirements, remain matters co-ordinated at the national level.

11.2 <u>NAVAREA Warnings service</u>

11.2.1 The NAVAREA warnings service is directed towards the timely promulgation by radio of information on hazards to navigation, in or near main shipping lanes, of concern to the international mariner. In each NAVAREA, effective and adequate facilities for scheduled broadcasts to the entire area are established, taking into account the appropriate geographic location for propagation purposes. Transmission coverage is of both the NAVAREA and as such of the adjacent areas as can be covered in 24 hours' sailing by a fast ship (about 700 miles).

11.2.2 In the existing system, promulgation of NAVAREA warnings is by HF radiotelegraphy (A1 to A1A) in all cases. In addition, other modes of emission at HF, e.g. direct-printing, are also used depending on the needs of the ships in the area and the facilities available. Broadcast times, set to coincide with at least one fixed watchkeeping period, are co-ordinated with those of adjacent or nearby NAVAREAs to ensure that a ship sailing between areas has an opportunity to copy both schedules, and that use of similar frequencies does not cause interference.

11.3 NAVTEX service

11.3.1 The WWNWS has more recently incorporated NAVTEX, an international direct-printing service for promulgation to ships of navigational and meteorological warnings and other urgent marine safety information pertaining to coastal waters up to 400 miles offshore. Unlike NAVAREA warnings, which are tailored for international sea commerce on or near main shipping lanes, NAVTEX carries information relevant to all sizes and types of vessels within a general area. It also carries routine meteorological forecasts and all storm warnings, whereas NAVAREA warnings include only advisories on major storms. A selective message rejection feature of the receiver allows the mariner to receive only that safety information pertinent to his requirements. Figure 2 illustrates the way the service is typically structured. NAVTEX service is being implemented rapidly in many parts of the world, and a fully operational service has existed throughout NAVAREA I for over two years.

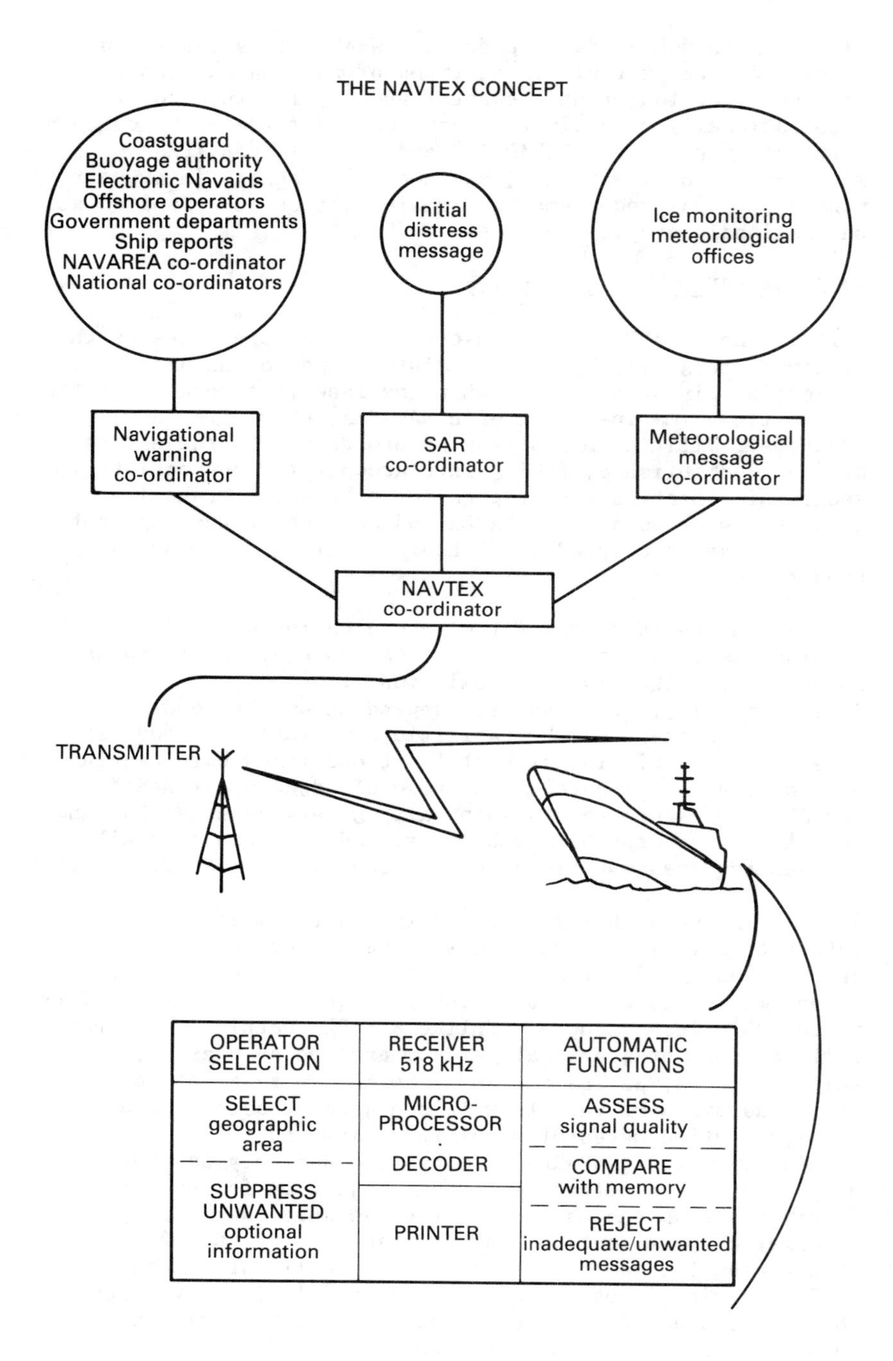

Figure 2 - Structure of the NAVTEX service

11.3.2 NAVTEX utilizes a single frequency (518 kHz) worldwide for its English language service and, where time permits, may also be used for limited non-English national language broadcasts of vital information to shipping. Mutual interference is avoided both by limiting the power of the transmitter to that necessary to reach the limits of the area assigned and by time-sharing the frequency. Careful co-ordination of broadcast schedules is imperative.

11.4 Dissemination of marine safety information in the global maritime distress and safety system (GMDSS)

11.4.1 The GMDSS communications for dissemination of marine safety information will include messages on distress and safety traffic frequencies and broadcasting messages from shore-to-ship by NAVTEX on 518 kHz, and by INMARSAT enhanced group call service or by HF (direct-printing), which services are expected to be integrated in a co-ordinated system for the promulgation of marine safety information.

11.4.2 It will also be possible to use the digital selective calling (DSC) technique on the distress and safety calling frequencies for advising shipping on the impending transmission of urgent, vital navigational and safety messages, except when the transmissions take place at routine times.

11.4.3 The enhanced group call (EGC) which is being developed by INMARSAT enables the provision of a unique global automated service capable of addressing messages to predetermined groups of ships or all vessels in both fixed and variable geographical areas.

11.4.4 The system is able to meet requirements of broadcasting anywhere in the world of global, regional or local navigational warnings, meteorological warnings and forecasts and shore-to-ship distress alerts. In addition to covering the mid-ocean areas, the EGC system could also provide an automated service in coastal waters where it may not be feasible to establish the NAVTEX service or where shipping density is too low to warrant it implementation.

11.4.5 A particularly useful feature is the ability to direct a call to a given geographical area. The area may be fixed, as in the case of NAVAREA or weather forecast area, or it may be uniquely defined. This will be useful for messages, such as a local storm warning or a shore-to-ship distress alert, for which it is inappropriate to alert all ships in an ocean.

11.4.6 EGC messages could originate from an authorized subscriber anywhere in the world and would be broadcast to the appropriate ocean region via a coast earth station. Messages will be transmitted by the coast earth station according to their priority, e.g. distress, urgency, safety, routine and commercial correspondence.

11.4.7 Aboard ship, EGC messages will be received either via a stand-alone device, via optional equipment fitted to INMARSAT Standard-A ship earth stations (SES) or as an inbuilt feature of all future SES standards.

12. CONCLUSION

12.1 As may be seen from the above, the science of meteorology still plays a very important role, even in today's world of big, fast and sophisticated ships, and foolish is the man who sets sails without carefully monitoring the weather conditions ahead which he will encounter. Millions of dollars lost may be the consequences of neglecting the weather conditions, either through damage to cargo caused by moisture or damage suffered by the ship and its crew in extreme weather conditions.

12.2 It is hoped that the continued improvement in the providing of weather reports, meteorological warnings and other information to mariners will help to reduce the losses of ships and human lives which regrettably still continues to occur. It is the hope of the authors that this presentation has shown the dependence by the maritime community on accurate, reliable and timely weather forecasts and the importance of meteorology to the seafarer.

APPENDIX

ACTION TO AVOID THE WORST OF THE STORM: NORTHERN HEMISPHERE *

RIGHT-HAND OR "DANGEROUS" SEMICIRCLE (Ship A). If under power proceed with maximum practical speed with wind ahead or on starboard bow, hauling round to starboard as the wind veers. If unable to make headway, or if the ship is under sail only, then heave-to on starboard tack.

LEFT-HAND OR "NAVIGABLE" SEMICIRCLE (Ship B). Run with the wind well on the starboard quarter (whether under power or sail) making all possible speed and haul round to port as the wind backs. If unable to make headway, heave-to on whichever tack is considered to be the safest under existing circumstances and conditions.

IN DIRECT PATH AND AHEAD OF STORM (Ship C). With the wind on the starboard quarter make all possible speed into the navigable semicircle. If unable to do this, it may be preferable to proceed into the dangerous semicircle rather than stay in the direct path, but be on the alert for possible recurvature.

VESSEL OVERTAKING THE STORM (Ship D). This may not be unusual in the fast ships of today. HEAVE-TO: the wind will then shift to the right and the barometer will rise showing that ship D is in the rear quadrant of the dangerous semicircle. She should then get the wind on the starboard bow (Ship E) and allow the storm to get clear.

* Bibliography - Maritime Meteorology - a guide for Deck Officers by C W Roberts and C E N Frankoon - copyright Thomas Reed Publications Ltd, United Kingdom, 1985

How to Avoid the worst of a Tropical Revolving Storm

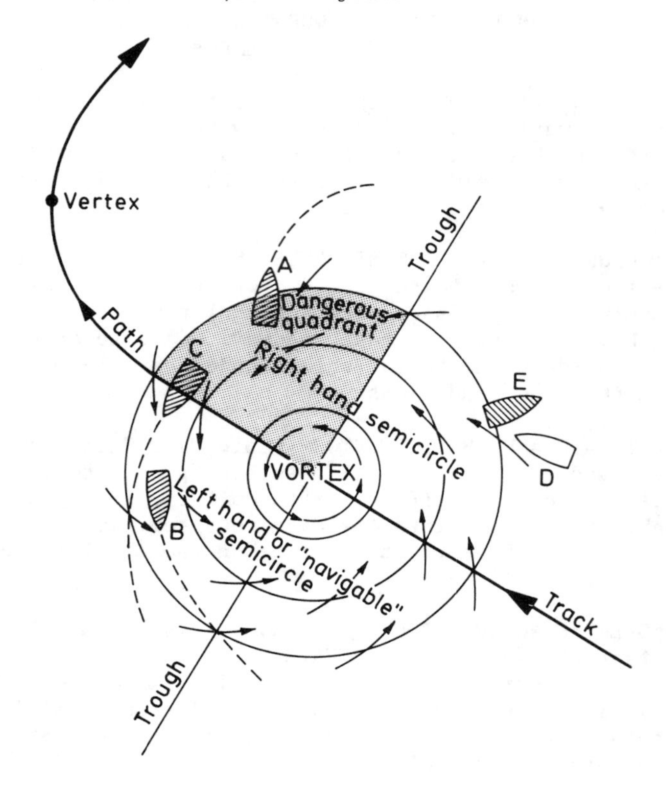

If ship D does not heave-to when the storm is first suspected and continues on course, the barometer will fall and the wind will shift to the left. This can lead to an erroneous assumption that she is in the left-hand semicircle ahead of the trough; if she then proceeds (obeying the rules) with the wind on the starboard quarter she may run into the dangerous quadrant, especially if her original course was converging with the path.

Note: In the SOUTHERN HEMISPHERE the same principles apply but, as the wind circulates clockwise, the left hand semicircle is the dangerous one.

3

Use of hydrological information and products for developing inland waterway transport on the Yangtze river of China

Ji Xuewu, Yangtze Valley Planning Office of the Ministry of Water Resources and Electric Power of the People's Republic of China, Wuhan

1. INTRODUCTION

The Yangtze is the longest river in China and in Asia. Measuring 6,300 km in length, it is the third longest in the world, after the Amazon and the Nile. From its source in the Qinghai-Tibet Plateau of West China, it flows eastwards to traverse the entire central portion of the Chinese mainland.

Figure 1 shows the geographical situation of the Yangtze River. With more than 700 sizeable tributaries, among which 437 have catchment areas of 1,000 sq km or more, the estimated drainage area of the Yangtze River is 1.8 million sq km (including 20,000 sq km of lake surface), or around 1/5 of the total area of China. Most of the Yangtze lies in the warm, humid sub-tropics. Annual average rainfall is 250 to 500 mm in the source region, 600 to 1,400 mm in the upper reaches and 800-1,800 mm in the middle to lower reaches. In most of the Yangtze basin the rainfall during the rainy season (from May to October) accounts for 70 to 90% of the total annual figure. Rain storms causing the heaviest rainfall and flooding in the middle to lower quarters of the river frequently take place in June and August. The annual average runoff of the Yangtze is approximately 1012 cu m. Runoff at the Yichang hydrometrical station is adopted as representative of the upper middle river course: the annual average is 453 x 109 cu m. Such runoffs provide great opportunities for multi purpose development of water resources. The mean annual transport suspended sediment at Yichang station, subtending a basin with an area of 1 million sq km, amounts to 521 million tons, 467 million tons of this being within the rainy season.

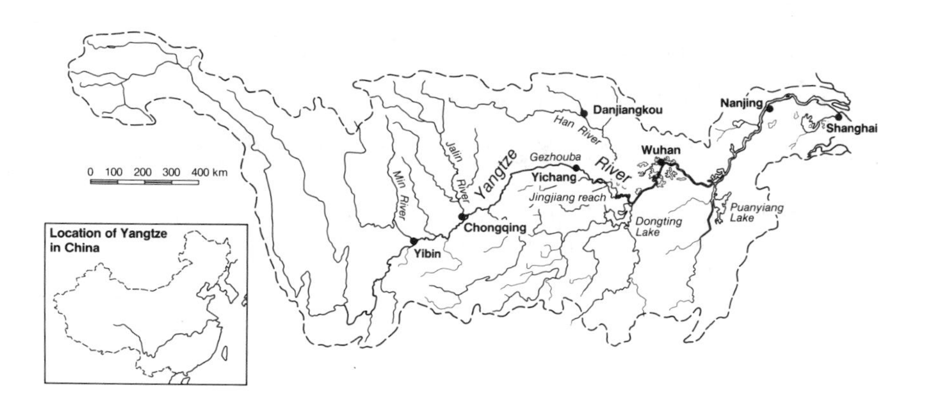

Figure 1 Geographical situation of the Yangtze River

The basin, with its 300 million inhabitants and 400 million mu (67 million acres or 27 million Hectares) of farmland, produces a large range of agricultural products. With a number of large industrial cities, such as Shanghai, Nanjing, Wuhan and Chongqing as well as many factories, mines and other enterprises, it also produces a large range of industrial products. Many famous beauty spots and historical sites are important resources for tourism. Overall, the Yangtze provides upwards of 40% of the national gross income.

There have been brilliant achievements in water conservancy in the Yangtze from ancient times up to the present. These include the Dujiangyan irrigation system, built in 250 BC on the Min River, a tributary of the Yangtze in Sichuan Province, which remains an engineering marvel to this day. The Grand Canal, built between 581 and 618 AD over 1,700 km from Hangzhou, in Zhejiang Province, to Beijing linked up China's 4 great river systems to form an inland waterway network on a national scale. Other examples are the Jingjiang flood diversion project completed in 1952 (for flood control in the middle reaches of the river in Hubei Province), cutoff projects in the lower reaches in Jingjiang, made during 1967 to 1969 for shortening (by 78 km for 3 loops) and improving the navigational channel and expanding the drainage capacity of the middle meandering river stretch in Hubei Province as shown in Figure 2.

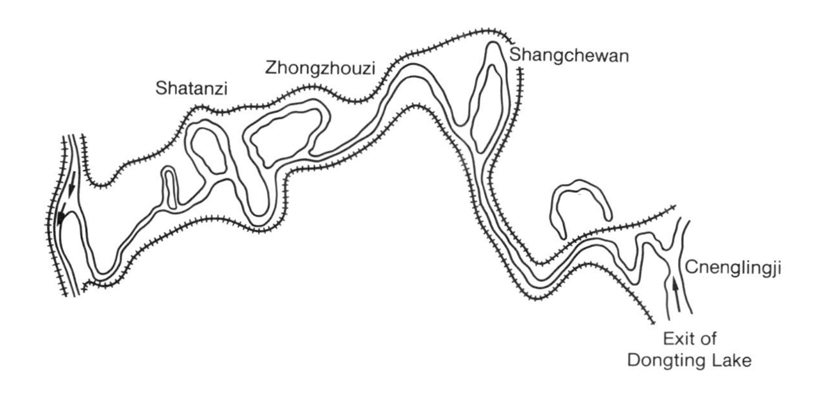

Figure 2 Cutoff projects in the lower reaches in Jingjiang

Another is the Gezhouba Project for navigation and hydroelectric power in Yichang, Hubei Province which will be discussed in more detail in section 4. During these undertakings, especially the navigation forces, hydrological advice is relied on to allow strategic decisions to be made.

2. NAVIGATION AND HYDROLOGY

2.1 Navigation Network

Obviously there is an opportunity and a need to develop inland waterway transport on the Yangtze, using it not only as an advantageous natural feature but also to help the rapidly advancing national economy. Thus the Yangtze River is an artery of China's east-west water transport system. Its navigation network consists mainly of two parts. The first, the main stream, is 2,800 km in length from Yibin to the river estuary: vessels of 200-800 tons from Yibin to Chongquing, those of 1,500 tons throughout the whole year to Yichang, 3,000 tons to Wuhan, 5,000 tons to Nanjing, and 15,000 tons to the East China Sea. The second part of the system, more than 3,800 navigable tributaries, consist of 7,000 km of navigable waterway, as well as making up a transport web with the lakes. Both parts, combine with the railway and highway network, to form the largest transportation system in China which is linked to routes over the oceans. 70% and more of the national total of freight volume is transported on inland waterways. However, navigation in the major channels is still rather poor because of numerous sharp bends, narrow channels, shoals and rapids in the Three Gorges Stretch, famous for its scenic beauty, and a series of shoals in the lower reach of Jingjiang: improvement of navigation conditions on the Yangtze would be of great significance in China's economic development.

2.2 Hydrology

The hydrological objective is to meet, in several different phases, a huge hydrological demand. This concerns ensuring that the navigation works on the Yangtze will produce suitable depth, appropriate flow velocities and reasonable radii of curvature of the waterway.

2.2a Planning and feasibility study

For the selection of the optimum scheme, the water stage, discharge, runoff, channel cross section and the sediment including its particle gradation data should be taken into account, both spatially and temporally, in specified stretches of sections. The preliminary results of the flow frequency analysis and hydrographs, with associated probabilities, need to be applied so as to optimise the alternative schemes.

2.2b Design

Usually by means of simulating stochastic processes or flow frequency the regime of the stream would be predicted for different design criteria, layout, scale and size of the works, so as to find the best possible trade-off between benefits and costs. Most navigation projects on the Yangtze, for instance channel regulation and port engineering, encounter sedimentation problems. In this case the sediment information as well as the discharge would be used in physical and mathematical models to test silting and scouring in the channel.

2.2c Operation and Management

Hydrometeorological forecasting, involving the short-term and medium to long-term, would enable water transport to be managed and damage caused by abnormally low water to be prevented. In addition, the different groups of water users could be well co-ordinated to achieve the maximum benefits from the forecasts. In this sense, the hydrological forecasting service would play an important role in monitoring the waterway.

3. HYDROLOGICAL NETWORK AND INFORMATION

3.1 Organization and network

There are 4,200 precipitation gauge stations in the basin. 4,100 of them belong to the hydrological administration and 100 to the meteorological organization. There are 900 hydrometric (stream) stations and 1,420 stage gauge stations, of which 76 stream stations and 156 stage stations of the Yangtze Valley Planning Office (YVPO), Ministry of Water Resources of Electric Power, are directly responsible for the planning, design, and management of water resources in the major reaches of the river. For data processing and the preparation of the product, YVPO has access to all hydrometeorological information through the communication systems. The Bureau of Hydrology of YVPO is responsible for a Measurement Service, Channel Service, Institute of Hydrometry, a Water Resources Laboratory and Forecasting Centre, and so on.

3.2 Hydrometric Technology

Rainfall is mainly measured manually, using a measuring jar and glass. Sometimes remote recording gauges are used. The water stage is mostly measured with a recording gauge for example the float-type or the sonar-type and is also checked manually. The stream discharge is calculated with an electric cable way controlled by microprocessor or silicon symmetrical switch and is most frequently used in the upper reaches or in mountainous regions whilst boats or the moving boat method are used in the middle to lower reaches or the tideway. Recently, a great deal of work and effort was put into improving sediment measuring devices as well as those measuring stage and stream. References 1, 2 and 3.

The data processing program covering the water stage, stream flow and sediment concentration has been written to allow computers to be used to help publish the Hydrological Year Books, year by year.

For satisfactory real time forecasting of the river flow, information is usually transmitted by telegraph from gauge stations through post offices to forecasting centres and users. Several telemetry systems at the stage and rainfall sensor stations have been established so as to extend the lead time and to improve the accuracy of forecasts. But, in the important flood control region and the key navigation channel, radio communication systems linking gauge stations directly to centres and users will have to be set up, both for convenience of use and to save time.

3.3 Review of hydrological data

In ancient times residents near the Yangtze used to make marks on stone to record when unusual flooding or drought occurred. In addition documents and writings stored in the Beijing Palace Museum have described these events in detail. Numerous flood level marks have been discovered referring to the period since 1153 AD (the Song dynasty) and low water level records for more than 100 years from 764 AD (the Tang dynasty) for the stretches from Chongquing to Yichang. Figure 3 shows the distribution of the 1788 and 1870 flood marks in the various stretches.

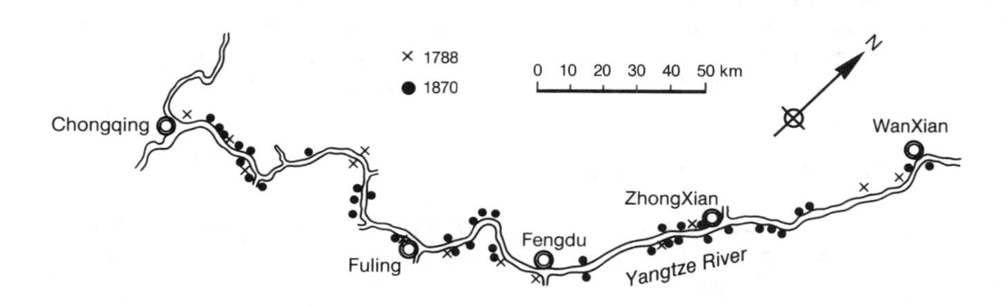

Figure 3 Location of the 1788 and 1870 carved flood marks.

According to the historic data, the estimated peak discharge of the 1870 flood was 105,000 m^3/sec at Yichang station. This was the biggest since 1153 AD or earlier. The hydrograph or the way the flood rose and fell was sometimes outlined, for example the maximum rise in 1870 in the area of the Three Gorges is illustrated in the local history of Wanshing. The records of observations at the stations have been continued for more than 30 years. Some of the stations, such as Wuhan and Yichang have been in existence for more than a 100 years. There are many data as the channel has evolved on different scales, a large volume of survey data on the waterway has been accumulated.

4. GEZHOUBA PROJECT - AN EXAMPLE OF USE

4.1 General description

The Gezhouba project, the largest in China, up to now, is located at the outlet of the Three Gorges at Yichang in Hubei Province, where there is a vital communication line between the upper and middle to lower reaches of the Yangtze. Figure 4 and table 1 illustrate its layout and characteristics respectively.

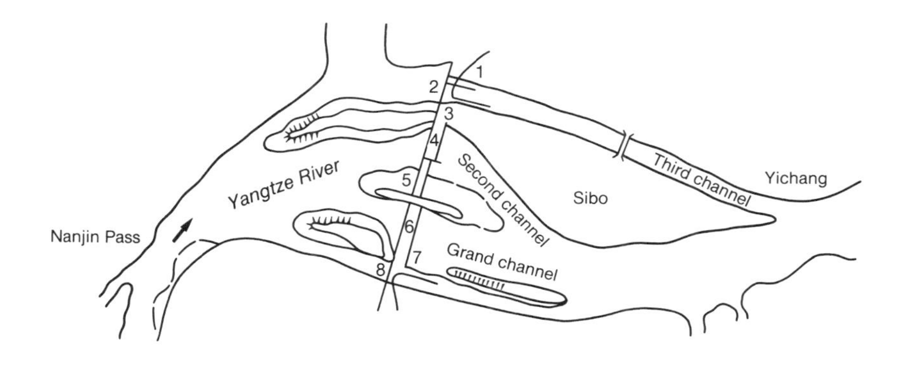

Figure 4 Schematic diagram of Gezhouba project

1 - The 3rd lock
2 - Scouring sluice
3 - The 2nd lock
4 - Power station
5 - Spillway
6 - Power station
7 - The 1st lock
8 - Scouring sluice

The 1st and 2nd locks have an effective chamber dimension of 280 x 34 x 5 metres (length x width x minimum water depth), allowing the passage of 10,000 ton vessels. The 3rd lock has dimensions 180 x 18 x 3.5 metres and can handle vessels below 3,000 tons.

Items	Units
Catchment area	1,000,000 sq km
Total length of dam axis	2,606.5 m
Reservoir storage	1.58 billion m^3
Annual mean flow	14,300 m^3/sec
Maximum peak discharge (1870)	105,000 m^3/sec
Minimum flow (1937 and 1979)	2,770 m^3/sec
Design flood discharge	86,000 m^3/sec
Check flood discharge	110,000 m^3/sec
Maximum navigable flow	60,000 m^3/sec
Minimum navigable flow	3,200 m^3/sec
Total installed generator capacity	2,715 MW

Table 1 Characteristics of Gezhouba Project

After the dam had been constructed the shoals and rapids in the 180 km stretch upstream of the dam have been left wholly or mostly within the backwater and the inundated zone, and a deeper navigable channel has been created in the reservoir.

4.2 Requirements to hydrologists for navigation

Since the 1950's the Bureau of Hydrology, has been contributing to the project, in particular, for its work on navigation. It has done this in several ways.

- Analysing the frequency of the flow (yearly, monthly, and flood peak) at the Yichang Station which is representative of the dam site, in order to determine the characteristics of the project.

- Measuring the discharge, current velocity and sediment movement, and surveying the underwater topography in the dam site region and at the end of the backwater so as to carry out the treatment, for example, engineering work or dredging, to protect the navigation channel from silting up, if necessary.

- Hydro-meteorological forecasting in the medium to long term to determine the optimum time to close a river channel for construction work to be carried out and when to put dredging the navigation channel into operation.

4.3 Approaches or procedures

4.3a Sediment and channel survey

Three stream stations are placed at the end of the backwater, 3 km upstream and 6 km downstream of the dam to keep the project informed of the current velocity and direction distribution as well as the sediment distribution and its grain gradation. Boats are moored to the overhead cable ways to survey various hydrological elements by means of current-meters, sonars, and a JX point integrating suspended load sampler, a Yangtze Type 78 bed load sampler and a Yangtze Type 80 gravel sampler which have been designed and put into operation to create better efficiency and stability, Ref. (4). There are 196 stationary cross-sections in the up-down-stream of the dam site to take the stages and the areas of movement with the measuring boats.

4.3b Flow frequency analysis

To reduce the sampling errors in estimating the daily flow frequency the observations begun in 1877 at Yichang Station are combined with supplementary data, i.e. the observations of the 1153, 1560, 1788, 1796, 1860 and 1870 flood peak discharges, as a statistical sample. After the assumption of its parent distribution function, for instance, Pearson Type-III, Log-Pearson Type-III, Extreme-Value Type-I the parameters of the functions in the population could be calculated by the moment method, maximum likelihood procedure or least square approach, curve-fitting method. Here the curve-fitting approach results in unbiassed and efficient parameters with the adopted plotting position formulas, Ref (5) (6).

4.3c Stochastic_process_modelling

To advance the existing frequency analysis approaches and to compare their results during recent years, the procedures of hydrological time series analysis have been applied to simulate the yearly and monthly data at Yichang Station, with the multivariate and periodic models, for example, the class of Auto Regressive Moving Average (ARMA) models and the linear disaggregation model, Ref (7). Among them the simple Lag-1 Auto Regressive (AR) model performed quite well for reproducing the statistics of annual flows, and the disaggregation model was shown to be superior to the Lag-1 AR Model with periodic parameters, considering the reproduction of historical statistics as well as model simplicity.

4.3d Flow_forecasting

In practice, the procedures for short-term forecasting of rainfall-runoff, the black-box model, API and the conceptual "Xinanjiang" model, might have achieved more desirable results, but the Sherman unit hydrograph was effective in accomplishing the time distribution of runoff, Ref (8). In order to deal effectively with uneven distribution of rainfall-runoff in time and in space, the catchment was divided into some smaller components and unit hydrographs for short periods are used in the medium tributaries. In addition, the relations of the stages between upstream and downstream are very useful for the forecasting in main stream considering the lateral inflow, if necessary.

The statistical procedures (such as hydro-meteorological time series and regression analysis), and atmospheric circulation analysis methods are used in long-term forecasting.

5. CONCLUSION

The water resources of the Yangtze River have developed rapidly while its navigation is more effective. It will be clear that hydrological information and products have been the firm foundation of the work. However, more progress has to be made in the improvement of hydrometric equipment and communication systems so that the optimal benefits of the multi purpose use of the resources can be achieved.

CAMROSE LUTHERAN COLLEGE
LIBRARY

REFERENCES

1. Xiang Zhian, 1986 Technique for Suspended Load Measurement on the Yangtze River. Presented to workshop between China and USA.

2. Xiang Zhian, 1986 Technique for Bed Load Measurement on the Yangtze River. Presented to workshop between China and USA.

3. Bureau of Hydrology, MWREP, China, 1986. A Report for Intercomparison Test of Suspended Sediment Samplers, presented to WMO.

4. Huang Guanghua, etc, 1983. Study on Sampling Techniques of Bed Load on the Yangtze River. Proc. of the 2nd International Symposium on River sedimentation. Nanjing, China, 1008-1015.

5. Ministry of Water Resources and Electric Power (MWREP), 1979. Flood Designing Guideline (Standard) for Hydroelectric Engineering. Water Conservancy Press, Beijing, China.

6. Ji Xuewu, Ding Jing, Shen, H W and Salas, J D, 1984. Plotting Position for Pearson Type-III Distribution. Journal of Hydrology, 74: 1-29.

7. Ding Jing, Ji Xuewu, Shen, H W and Salas, J D 1985. Stochastic Modelling of Annual and Monthly flows of the Yangtze river in China. Submitted to Journal of Hydrology.

8. Yangtze Valley Planning Office (Editor), 1979. Methodology in River Flow Forecasting, (in Chinese). Water Conservancy Press, Beijing, China.

4

Forecasting the weather for road and traffic and inland navigation

Peter Rauh, Swiss Meteorological Institute, Zürich, Switzerland

1. GENERAL CONSIDERATIONS

There are at least three reasons why we forecast the weather for road or railway traffic and for inland navigation:

There is the curiosity aspect: people are generally interested to know what the weather is going to be like when they leave their accommodation. That was true thousands of years ago when our ancestors crawled out of their caves. Not that they travelled by car, train or boat - but they wondered if they will get wet or scorched just as much as we do today.

Then there is of course the safety aspect: certain weather conditions make travelling dangerous or at least uncomfortable. If you think of icy roads, thunderstorms, sudden gales, high temperatures, snow drifts or floods you realise that all these weather elements are potentially dangerous for any kind of transport. If we can forecast the onset, the intensity and the end of some of these phenomena we are able to avoid damage or even save lives - if our forecasts or warnings reach the right people at the right time at the right place. This already shows that there is more to the problem than just a correct forecast.

The third aspect is that of the economy: you know the truth in the saying: 'What costs nothing is worth nothing'. At least in my country this is a basic and accepted way of thinking - it may be questionable in many situations but we have to consider it as a fact if we want to promote our services. On the other hand we have the potential to help many individual people and companies to work more efficiently and save money by doing so, even if our customers have to pay for the forecasts we issue.

2. SOME EXAMPLES

a. Temperature Forecasts for Transporting Bananas by Rail

Swiss people eat a lot of bananas. I would expect that in a year they consume between 50 and 70 thousand tons. One of the biggest supermarket chains is a good customer of the forecasting services of the Swiss Meteorological Institute. Every Monday in the year a cargo ship full of bananas reaches Antwerp or Hamburg or Bremerhaven. The bananas are then reloaded into insulated rail cargo carriages which are owned by the company which runs the supermarkets. It takes roughly half a day for the cargo train to reach its destination in Switzerland.

Now, you probably know that bananas are quite delicate. If they are too ripe they go brown and cannot be sold in the shops. If the bananas get cooled too much, and too much is any temperature lower than 12 degrees Celsius, they "freeze". It is true that they do not freeze in the sense that the fluid in the banana ice cream but the cells react to temperatures below 12 degrees C with irreversible chemical processes that cause the fruit to dry up. The optimum temperature for green bananas to be stored and transported at is 13.5 degrees C. If the surrounding air is 15 degrees or higher the ripening process sets in. Before the fruit is delivered to the shops it is artificially ripened at temperatures above 15 degrees. This is done in automatically controlled warming rooms.

All the way from Antwerp harbour to the destination in Switzerland the air temperature in the freight wagon has to be kept above 12 but below 15 degrees C. Most of these rail freight wagons do not have an automatic cooling or heating system which is thermostatically controlled. In the summer the correct amount of ice is put into open containers in the wagons so that the temperature of the air remains low. In the winter, gas or paraffin heaters are installed so that the temperature remains higher.

This is where we come in: in order to be able to estimate how much heating or cooling material has to be allocated, it is necessary to know the outside air temperature for the whole journey. We forecast the temperature en route for the 12 to 18 hours of the journey. This way, the company is able to transport the bananas without insuring the cargo against damage by "freezing" or uncontrolled ripening. The loss occurring to our customer by unforeseen temperature changes is a fraction of a percent of the value of the fruit.

Let us have a look at some figures to get an idea of the money involved:

One cargo boat every week means 52 boats a year. Each boat holds 20-30 rail freight wagons of bananas. Each rail wagon holds 17 metric tons of fruit in boxes. Each kilogram of banana is worth one Swiss franc. Say 25 wagon per week x 17,000 kg per wagon x 1 SFr per kg x 52 weeks per year = 22.1 million Swiss Francs a year. The transport insurance would cost 37,750 Swiss Francs a year. Regular temperature forecasts amount to 1,500 Swiss Francs a year.

There is no doubt: this company profits a lot from the forecasts issued by our office. Now you may ask several questions: How does the price of our forecasts tally with the benefit? Is it right to issue forecasts for just one company? Should private companies be served by private meteorological companies?

b) Road Conditions Forecasts for Drivers and for Road Maintenance Services

Every evening during the winter months November to March we issue a short road conditions forecast for the main lowland roads. We receive the observations of the road maintenance services, we know what the weather is going to be like during the night - or at least we think we know - and forecast the effect it will have on the road conditions.

Car and lorry drivers may get this forecast either by radio, by TV, by dialling a special telephone number. It can also be seen on Teletext, the TV one-way information services.

Of course this forecast can only be a general one and there are many local factors influencing road conditions: surface temperature, exposure, surroundings of the road, initial state of the road (e.g. snow cover, moisture, ice patches etc.) or all the meteorological factors: kind and amount of precipitation, cloud cover, air temperature, wind speed. Much depends on the density of the traffic too, and of course of the measures taken by the road maintenance services.

The driver himself has to think and watch out for dangerous road conditions. Nobody can take over this responsibility - we would be sued in court no end for damages due to unforeseen hazards - but we can help the driver to understand what effect the forthcoming weather may have on the road conditions.

I said that we issue these road conditions forecasts for the general public or at least for that part of the public which is driving on the main roads during the night. It is obvious that we have the road maintenance services in mind too, when we work out our forecasts. They get our general weather forecasts, they have all their observations of the present state of the road conditions and they get our road conditions forecasts for the night. This way they have all the information needed for planning their work during the night. If we forecast clear skies, they may allow their heavy vehicles a rest for the night and enjoy a free evening. If we forecast snow, they load up their lorries with salt and other chemicals which help to melt the snow on the road - and the metal of our cars! They are ready and may go for dinner in the canteen until the front arrives and the snow is beginning to cover the roads. It is not necessary any more to spread salt on all the roads before knocking off in the evening, just in case there might be snow or ice formation during the night. This saves money, it is better for the cars and less a threat to the precious environment.

But there is a catch in the system I have just described. There is a very slight chance that the weather does not do what we say it will do in our forecast. In this case the night shift forecaster has the option of phoning the motorway police and to issue a warning to the road maintenance services. It is also possible to amend the road conditions forecast and redistribute it to the radio stations which broadcast throughout the night.

I cannot give you any figures of how many accidents are avoided, how many lives are saved and how many people are not injured thanks to our road conditions forecasts but there must be many.

c) Precipitation Forecasts for Road Transport to Construction Sites

Imagine a main trunk road being resurfaced with tarmac. There are big machines which evenly distribute the hot and still viscous tarmac on to the road bedding. The only thing people have to do is to feed this machine at regular intervals - just like we humans have to be fed. A number of special lorries may be doing a round trip from the tarmac factory to the construction site where they unload their cargo and go back to the factory again for a new load. It is possible to work out that every so many minutes a fully loaded lorry has to arrive at the site in order to keep our surfacing machine well fed. Depending on the distance between the tarmac factory and construction site you need a certain number of vehicles to keep the chain running without interruption. Everything is well organised and running smoothly, starting on a nice summer's day at six in the morning and going on and on and on ...

But at ten to three there is the first drop of a shower or rain. It is well known that tarmac put onto a wet undercoat does not last. Therefore our dear machine has to be halted. That is not so bad. It may like to have a rest - or at least the driver may. But think of our lorries proudly keeping their timetable to the minute. At regular intervals they arrive, fully loaded. They have to wait because they cannot unload, the tarmac gets colder and colder, harder and harder. In the end the lorry is permanently loaded with a big brick of lovely solid tarmac - unless the tarmac is discharged before it is too cold to be unloaded or it is taken back to the factory for recycling. Both discharging or taking it back to the factory cost money, a lot of money in the case of discharging. 15 tons are worth 2000 Swiss Francs.

By keeping a close watch on the weather, mainly by radar, we were able to forecast the onset of showers pretty well. The chain of lorries could be broken in good time and at the factory, not at the construction site thus saving quite a number of loads of tarmac from being wasted or at least from being transported in vain.

d) Precipitation and Temperature Forecasts for Modelling Run-Off for Inland Water Ways

The runoff of rivers in my country is determined by two meteorological factors: 1. precipitation and 2. temperatures causing the snow melt. It is possible to set up runoff models which are quite accurate if the meteorological input is correct.

For about two years we have been issuing temperature and precipitation forecasts to the Federal Institute of Hydrology which runs a model for the runoff of the Rhine near Basle. This precipitation forecast has to be accurate not only in the amount of rainfall but in the amount per time unit, this is, the intensity - a task not easy to be solved.

There are two customers for runoff forecasts: hydroelectrical power stations which use the flow through their turbines to generate electricity and inland navigation companies which transport goods in cargo boats. Here we are concerned with the latter. Too little water in the river causes problems because the fully loaded barges may scrape a rock and sink, or they can only be partly loaded which increases the cost per ton of cargo transported, or they may get caught between shallow waters and have to wait for better weather: rain. Too much water is no good either: there are a number of low bridges along the Rhine through which it is impossible for a barge to pass if the water level is too high. In this case unloading does not help, in fact it would make matters worse.

Being warned of low and high water levels a couple of days ahead does help inland navigation companies to use their barges more efficiently, thus saving money and time and - since time is money - saving more money.

e) Precipitation and Temperature Forecasts for Swiss Rail

If it is cold and wet, if there is snow on the ground or if there is freezing rain, the operation of the remote controlled points may not be possible. It is not really funny if a train takes the wrong way - goes to Berne instead of Zürich - no, it is dangerous! In order to avoid any malfunction of the points the moveable metal parts are heated. That stops them from freezing. Some points of the Swiss Railway System are heated electrically. There is no problem switching on this kind of heating, as soon as the temperature is below freezing and there is water or snow on the rails.

In some main stations like Zürich for instance points are heated by gas. When this heating system was constructed some decades ago, the station and the number of points were much smaller than today and cheap labour was easily available. At the time there was no thought of the possibility of constructing an automatic ignition device for the heating of the points. Each gas fire at each point has to be lit manually - not quite with a match each, but still individually. Gas is a cheaper source of energy for heating than electricity. It takes 60 to 90 minutes to light all the heating devices of Zürich main station. There it is easy to turn the heating up or down or off altogether, but to light it yet again it takes a long time.

On request we advise the officer on duty at the station on the meteorological hazards expected to make heating necessary or when the heating can be turned off. This heating management is important: a wrong decision or action can cause havoc, cost a lot of money and be potentially dangerous.

f) Wind Forecasts for Lake Navigation

On some of the Swiss Lakes there are passenger boats transporting a lot of tourists and some commuters. There is a timetable because the boats have to arrive at the ports which have a railway station in time for a connecting train. When I initially drew up the summary for this paper I thought that in case of very strong winds these boats would not sail. I imagined that the captain of the boats would take into consideration our gale warnings before making a decision to sail or not.

However, this is not the case. The boats are so large and stable that they can cross the lakes in any kind of weather. So it is not the public transport side of lake navigation which is using our gale warnings but the pleasure boats, the wind surfers, the sailing yachts, the fishermen, but they do not belong to the domain I am talking about to you.

3. CONCLUSIONS

With the few examples I have described in more or less detail I think it is clear that we can generally say that the transport industry does benefit from specific weather forecasts issued for them. This saves goods, money, time and effort. It certainly increases safety as well.

Like most forecasting services of a national meteorological office, we issue weather forecasts for the general public. It is obvious that these forecasts are available to transport companies too, and are used for their purposes.

4. PROBLEMS

It would not be honest _not_ to talk of the problems that arise when a forecaster has to issue a bulletin or answer questions in connection with future weather. The list of problems is never-ending.

Here are just a few of them:

There are problems with forecasting methods. Often customers want to get a forecast for a meteorological element which first has to be derived from other elements - this problem can be overcome by intensive research efforts, but it takes time - or he needs forecasts in a impossible scale of time and/or space. How can you forecast the amount of rainfall in tenths or millimetres five days ahead for a specific building site? Of course you can try, but what is your rate of success? That brings us to the next item on our list.

There are problems with the reliability of the forecasts. You may argue that a forecast is always better than nothing. This is not quite true. Forecasts have to be better than persistence: anybody can say "tomorrow is the same as today" and get quite a number of hits. Forecasts have to be better than climatology, otherwise the effort is not worth while. It is therefore essential to be able to supply the customer with an estimate of the reliability of our forecasts. We must check the forecasts and keep a record.

It is all very well to produce a weather forecast which is quite reliable thanks to sophisticated forecasting techniques. There is the problem of getting the forecast to the right person at the right time in the right shape. I mean: there is a communication problem. Our customer has to be helped with understanding the content of the forecast - a teaching problem. The customer must get the forecast in good time and he must get it at the place where he is - a telecommunication problem. Often this can be solved quite easily with electronic equipment, but it takes time to find the right ways. And now we have arrived at the main problem of many forecasting services in Europe:

Manpower is limited. Weather forecasting is a job that takes time. Time to absorb information, time to think, time to communicate. In Switzerland we are legally bound to a fixed number of staff we may employ for doing all our jobs. We may have lots of very good ideas of how we could expand, we may have lots of potential customers, we may even have developed good and reliable methods out when it boils down to the daily work on the forecasting bench, then we get stuck. There is a limit to the number of forecasts one can push into an 8 hour shift. For political reasons we cannot increase the number of forecasters even if we were able to show that they would increase our income sufficiently to pay their wages.

I know there is a way out of this situation - go private - but I doubt that the community as a whole would really benefit from that possibility.

5

Practical application of meteorology – heavy-lift transportation

A. Blackham, Noble Denton Weather Services, London UK

INTRODUCTION

Since the middle 50's, the search for oil on shallow continental shelves has required platforms for exploratory drilling and, when oil is found, for production facilities. The heavy engineering required to construct these platforms is located onshore so that tow-out and emplacement is always necessary even for short distances. Of recent years, economical decisions frequently result in the finished product having to be transported for long distances by sea, and since the object being transported will be subjected to stresses, fatigue and accelerations during the voyage which it would not experience when in its final position and the expense of providing the extra strength is uneconomical it is necessary to plan such voyages carefully, particularly because damage is not acceptable. It is not acceptable because of the higher cost of replacement, the disruption of the entire schedule and the cash value of lost production (known as consequential loss).

When a company such as Noble Denton is selected as Insurance Warranty Surveyor to such a project their concern is to ensure that all reasonable steps are taken to ensure that the voyage is completed without loss or damage. A study has to be made of the project, firstly to determine its feasibility and secondly, in those cases where it is found not to be feasible, to recommend what has to be done to make it feasible. A team carries out the study. First, a meteorologist will assess the route and calculate the environmental extremes. Second, a naval architect will use the extremes to calculate accelerations and stresses. Third, a mariner will assess the route, the capabilities of the towing fleet and the actual modus operandi of the project to ensure that the normal practices of seamen are followed, and issued a Certificate of Approval before departure.

This paper will describe, in detail, two such transportations and emphasise the practical aspects of the meteorology involved, and the necessity for the mariner, the engineer and the meteorologist to function as a team. In the "offshore" world the term "meteorology" tends to cover a much wider field than one is normally used to. In the Noble Denton Group it embraces anything which is affected by the weather, as well as the weather itself and involves active participation in the projects. For this reason two widely dissimilar projects, with widely dissimilar weather problems have been selected as illustration.

1. TRANSPORTATION OF MODULES FROM OKPO (KOREA) TO CENTRAL MEDITERRANEAN

1.1 Pre-voyage investigations

1.1.1 For an oil-production platform in the Mediterranean the various component parts (known as modules) such as accommodation, processing plant, power units, drilling facilities, etc., were manufactured in Korea and were to be transported by heavy-lift vessel. There were 26 of these modules with individual values of up to 9 million dollars.

1.1.2 The first step was for the Noble Denton Weather Service (NDWS) to calculate the design conditions. Company practice for towage or transportation is to design to a "10-year storm" return period, which is both historical and arbitrary but is simple and appears to be successful. The 10-year event is defined as that value which is likely to be reached or exceeded, on average, once in 10 years during the relevant month. For towages taking less than a month it is appropriate to calculate a "reduced" extreme since reduced time implies less risk of encountering an extreme storm.

1.1.3 Long towages may be regarded as a succession of short tows through differing climate areas. It is NDWS practice to calculate a monthly extreme for the worst area and then calculate a "reduced" extreme using the exposure time in the worst area plus the exposure time in all other areas which have an extreme within 10 per cent of the worst.

1.1.4 On this particular passage it was evident from

historical data (Figure 1) that extremes were most likely in a typhoon and reference was made to the NDWS contours of extremes for that area. These contours were obtained by using the NDWS Typhoon extreme model. All cyclone tracks from 1952 to date were fed into the model and a suitable grid system was chosen. The basis of the calculation is a series of cyclone wind speed profiles given by Simpson and Riehl (1981). It is assumed that each occurrence of a storm affecting a grid point is an independent extreme event, and the average number of extreme events per year is calculated. An extreme value distribution function is fitted to the distribution of ranked extreme values and the threshold (below which values are ignored) is varied until the best least squares fit is obtained. The extreme value is obtained by extrapolation. The basis of the programme is the "peaks over threshold method" (Institute of Hydrology Flood Studies Report 1975). This study gave the following extremes:

	Aug	Oct
1-min mean wind speed (m/sec)	53	45
Significant wave height (m)	12.2	10.2
Peak energy period (sec)	13	12.5

These figures were not acceptable and could not be sustained by the unit, so it was decided to calculate non-typhoon extremes and weather-route the vessel to avoid typhoons.

The NDWS method of calculating non-typhoon wind and wave extremes again uses contour maps. These are constructed by plotting the results of analyses which involve fitting 3-parameter Weibull distribution functions to the cumulative distributions of wind speed and wave height for a series of small climatologically homogenous areas. The data source is observations from ships, which are obtained from the WMO archive of ships, which are obtained from the WMO archive of ship reports, and are then subjected to some quality control and modification.

1.1.5 The 10-year extremes, excluding Tropical Storms, were calculated as:

	Depart Aug 21	Depart Sep 4
1-min mean wind speed (m/sec)	27.5	27.5
Significant Wave Height (m)	8.8	8.2
Peak Energy Period (sec)	12.6	12.1

1.1.6 The "reduced" exposure design figures for the whole route for the season were:

	Depart Aug 21	Depart Sep 4
1-min mean wind speed (m/sec)	25	25
Significant wave height (m)	7.7	7.3
Peak Energy period (sec)	11.8	11.5

Provided that the vessel was weather-routed to avoid Tropical Storms.

1.2 Engineering calculations

The extreme figures in 1.1.6 were then used in a motion response analysis which showed that the "Superservant" vessels, operated by Wijsmuller Transport BV Ijmuiden, Holland, would roll excessively and well above the design parameters of the modules. The worst dangers were from wave slamming forces, because some of the modules extended outside of the vessel, and from immersion in the larger rolls. Further calculations showed that the greatest sea state that would generate an acceptable motion was 7.5 metres.

1.3 Weather - Routing Requirements

1.3.1 The first step was for the Noble Denton Weather Service to provide a weather-routing philosophy for the transportation at 12 knots and, in particular

· define the route (see Figure 2)

· define ports of shelter · define procedure

· define source of weather forecasts

The object of the routing was to avoid typhoons.

1.3.2 The designated route was

- east of Taiwan
- Bashi Channel
- west of Ibbayat Island
- off Cape Bojeador
- east of Macclesfield Bank

- west of Prince of Wales Bank to Singapore

with the actual courses at the discretion of the Master.

1.3.3 The routing concept was

- all Tropical Revolving Storms were to be avoided
- the vessel to be within a maximum steaming time of 48 hours from a designated safe-haven at all times
- the vessel to be equipped with an Inmarsat ground station
- the vessel must receive routing advice from an approved source, where weather forecasts were provided by meteorologists and all advice by Master Mariners. The importance of input from mariners must be stressed as some routing services have been known to route vessels over land to avoid heavy weather
- the vessel would provide:
  - time of departure from Okpo, Korea
  - daily position, course, speed and weather
  - similar reports every 6 hours if a TRS was within 400 miles

  the weather-routing service would provide:
  - continuous monitoring of the vessels passage
  - along-the-track weather forecasts each day, with expected wind and wave conditions for the next three days
  - monitoring of all Tropical Storms and give advice for avoidance
  - 6 hourly advisories if any tropical storm appeared likely to affect the vessel

1.3.4 Typhoon avoidance procedure was

- Okpo, Korea would be the safe-haven from departure until the vessel reached 29.5°N
- Chi-Lung-Chiang would be the safe-haven from 29.5°N to 21°N
- Lingayen would be the safe-haven from 21°N to 16°N
- Manila Bay would be the safe-haven south of 16°N

1.3.5 Safe-havens

The definition of a safe-haven for this project was a port or anchorage where the limiting conditions would not be experienced. In this case the limiting criterion was wave height. Normal practice for a well-found ship is to put out to sea and get "sea-room", rather than stay in a port where damage could occur from other ships breaking loose. The chosen safe-havens were only intended to give protection from high waves.

1.4 The actual voyage

1.4.1 Noble Denton Weather Services Limited were selected to provide the weather-routing service.

In the event, the sailing date was put back to mid-October and extremes had to be recalculated for that period. Excluding typhoons the reduced extremes were

1-min mean wind speed	27 m/sec
Significant wave height	8.0 metres
Peak Energy Period	12.0 seconds

The reduced figures were calculated on the basis that the length of route with extreme waves within 10% of the above was 2000 nm. Using an average speed of 12 knots and allowing for contingencies the exposure time to the extreme or near extreme wave was 10.5 days. The slightly higher wave was accepted, but put more responsibility on the shoulders of the routing service.

The actual sea temperature isotherms for October 1986 (Figure 2) showed that typhoons could originate and develop anywhere south of Taiwan. Actual tracks of Tropical Storms for October are shown in Figure 4.

1.4.2 The vessel was ready to sail on 13 October. The following table gives the vessel's daily position, the location and track of typhoons and the advised action. This information is also plotted out on the map at Figure 3.

Date	Position	"Ellen"	"Georgia"	Advice
13.10.86	OKPO	16°N 118°E		Delay sailing
14.10.86	OKPO	17°N 118°E		Delay sailing
15.10.86	OKPO	18°N 117°E		Commence at half-speed. Stay north of 30°N until 17th.
16.10.86	33°N 127°E	19°N 117°E		Proceed normal speed
17.10.86	30°N 124°E	20°N 115°E	10°N 128°E (convection)	Dead slow. Remain north of 23°N
18.10.86	26°N 122°E	CHINA	12°N 127°E	Proceed normal speed towards San Bernardino Strait.
19.10.86	23°N 123°E		12°N 126°E	Note 1. On 19.10.86 vessel experienced heavy beam Easterly swell and altered course to 120°True for 12 hours before resuming passage. Routing service advised of this action.
20.10.86	16°N 124°E		14°N 117°E	
21.10.86	16°N 124°E		15°N 112°E	
22.10.86	13°N 123°E		Dissipated over land	

1.4 3 It will be seen that the high risk of typhoon encounter, as calculated by the Weather Service, proved realistic, the routing strategy designed by the Weather Service was effective, the vessel's Master co-operated magnificently and, as was his right, took his own action at one stage and immediately communicated that action to the Routing Service. This was on 19 October when heavy swell was generated by "Georgia" nearly 600 miles to the south.

1.4.4 The vessel and cargo arrived at its destination intact.

2. DECK MATING IN HARRISON BAY, NORTH ALASKA

2.1 The Project

This project involved

a) the towage of a MAT unit from Japan to a site in Harrison Bay (Fig 5)

b) the towage of a drilling unit from McKinley Bay to site (Fig 6)

c) the mating of the two units

d) towage of the mated units to a drill site and installation

For the purposes of this paper the towage from Japan and Cape Lisburne, which took place without incident, will not be discussed. The paper will concentrate on the operations in ice and weather conditions on site.

2.1.1 The key information supplied was

a) the MAT unit would arrive Cape Lisburne 21 August

b) mating to be completed by 20 September

c) limitations for mating -

Wind	-	12 knots ahead
	-	8 knots abeam
Wave	-	1.0 metre
Current	-	0.15 m/sec

d) towage would be hampered if more than $^5/_{10}$ of ice on any part of the route

2.1.2 Noble Denton Weather Services Ltd were asked in May 1986 to investigate the environmental conditions between Cape Lisburne and McKinley Bay to determine what level of icebreaker assistance, if any, would be required.

2.1.3 The data available for this kind of investigation are very sparse indeed. Most of it is in the form of Ice Atlases, published by a number of authorities, which can be interpreted in any way the investigator feels inclined. The Alaska Maritime Atlas gave maps of ice conditions at various times of the year and from these it was deduced that between August 15 and September 1 there was a 25% probability of $^5/_{10}$ of ice close to the coast and a 50% probability at 30 miles off the coast. Several studies had been commissioned by the operating company, Canmar, and these were also studied. The most useful document was a table of severity indices for 30 years, issued by the US Navy/NOAA Joint Ice Centre. This gave, among other information, the initial date when the entire route to Prudhoe Bay had $^5/_{10}$ or less ice concentration and the dates when the combined concentration and thickness dictated the end of prudent navigation. Care had to be taken in using these data as they referred to ocean-going, ice-strengthened vessels which could manoeuvre under their own power, rather than to unwieldy barges being towed.

On 6 occasions out of 31 concentration of ice did not fall below $^6/_{10}$ until after 21 August. On 7 occasions out of 31 ice thickness and concentration dictated the end of prudent navigation before 20 September. On only 2 occasions did the late start and early finish coincide. Therefore, in 31 years, unfavourable ice conditions would have been experienced in 11 years, or 1 in 3.

2.1.4 Having determined the long-term risk it was then necessary to address the problem of the probable ice conditions in 1986. Current practice is to compare the MSLP charts for the past 30 days with the long-term average and to make a similar comparison with temperatures.

An empirical method developed by NOAA is to extract the April mean 1000 mb height at two particular points. If the sum exceeds 290 there is a likelihood of a severe summer and if the sum is less than 290 then the summer will be less severe. The system only works for those 2 points and only for April. Application to past years was said to yield a high success rate. AUS Naval Polar Oceanography Center report indicated that 1986 was an average year and that there would be a milder than usual season. The project could have been delayed by ice, but an important factor to remember was that the absence of ice could also lead to higher seastates.

2.1.5 Further delays to the project could occur through adverse weather conditions at the mating site and again it was difficult to find adequate records. The nearest was OLIKTOK and here winds exceeded 12 knots for 33% of the time in August and 42% of the time in September. Wave heights for the same period exceeded 1.0 metres for 39% of the time. Surface currents were weak and dependent on local winds.

2.1.6 However, a better idea of likely delays to an operation can be obtained if it is known how persistent an exceedance will be and how frequently it occurs. The chance of success of an operation may well be greater if the criterion is exceeded only once during a month for a long period, rather than being exceeded several times for short periods. It has been shown that the occurrence and duration of events where some criterion is exceeded can be modelled by statistical distribution functions (Houmb 1981). A consequence of this is that once the parameters of the distribution functions have been defined then a time series of events can be constructed. These parameters are site and season specific. (Kusashima and Hogben 1984). A method was developed for use in the North Sea (Graham), and although the North Sea is affected by twice as many cyclones as the Beaufort Sea coast of Canada, the mean and standard deviation of wind speed in both areas is quite similar. The Graham method was used without modification and a long series of wind speed events was constructed. Using a computer program the series was inspected to estimate the probability of different durations of waiting on weather and the following results obtained:

Waiting Time (hrs)	Probability of a Spell of Winds less than 12 knots Starting
6	0.188
12	0.236
18	0.284
24	0.332
48	0.507
72	0.629
96	0.714
120	0.788
144	0.849

2.1.7 Summing up the pros and cons

a) there would be about 30 days available for the operation

b) the Canmar Marine staff and personnel on the towing and handling fleet were all competent and professional men, with good experience of working in Arctic conditions

c) failure to complete the job on time would cancel the drilling contract, and possible jeopardise future contracts

d) the two weather factors which could delay the project were sufficient ice to make towage to the site difficult, and weather conditions being adverse for the mating operation

e) the time required to complete towage, mate and tow to the drilling site was estimated to require 10 days, with allowance for mechanical contingencies. It was estimated that up to 8 days could be spent waiting on weather at the site. This left a safety margin of 8 days in the towage through ice.

2.1.8 Noble Denton were required to assess the weather problems and make recommendations on the level of ice-breaking support required. The assessment concluded that, although here was an 8 day margin and experienced government ice experts were predicting a milder than average season, weather conditions at the time were also important. A strong northerly wind could bring down the not-so-distant ice into the path of the tows and the absence of ice could lead to a higher sea-state at the mating site. Local opinion also suggested that ice forecasts were of variable accuracy. After considering all aspects it was recommended that icebreaker assistance should be available.

The cost of hiring icebreakers is very high, currently 150,000 dollars a day with a 10-day minimum. The operating company were reluctant to add that expense to their costs and some lively and lengthy discussions were held between the operator and the Noble Denton team of a marine surveyor and a marine meteorologist. It was finally agreed that the operating company would ensure that icebreaker assistance would be available in the area and would be called in at fairly short notice should circumstances at the time justify it.

2.1.9 The tightness of the schedule was due to the presence of bow-head whales and the insistence of an Environmental lobby that the removal of the guides on the MAT unit by explosive charges should be completed before the whales' mating season.

Subsequently, after the emplacement of the drilling unit, the Environmentalists were able to prevent the drilling operation until the whales had returned to the Bering Sea in November.

2.1.10 The Actual Operation

The MAT unit departed Osaka on 10th July and arrived safely in the vicinity of the mating site on 22nd August. The SSDC unit departed McKinley Bay on 20th August at a speed of about 4.5 knots. On the afternoon of the 21st a WNW wind blew up and by evening was gusting to 40 knots in squalls. Tow speed dropped to 1.5 knots. The wind continued at gale force during the 21st and 23rd. Ice was encountered on the afternoon of the 23rd and an additional Class 2 tug was sent to assist and a helicopter provided for aerial reconnaissance. On 24 August the two stopped for further reconnaissance and for the mobilization of a Class 4 icebreaker, since there were bands of $^{8}/_{10}$ ice ahead. The tow resumed on 25th and encountered various concentrations of ice before arriving in the vicinity of the chosen mating site on 28th August.

Since the arrival of the MAT unit on 22nd August a generally unfavourable wind regime prevented any of the possible mating sites being ice-free. Adverse weather on 28th and 29th held up the operation until the 30th. Although the moderation of wind was correctly forecast, the ensuing light winds prevented dissipation of ice from around the MAT unit until late on the 30th. The MAT was ballasted down to the sea-bed early on the 31st, the SSDC unit was manoeuvred into position, the MAT deballasted and full contact was made. The towers were then detonated and the combined unit was towed to the drilling site and emplaced on 2nd September.

CONCLUSIONS

From the two foregoing examples it will be seen that, in a complicated marine project in which weather conditions are the key to its success, the meteorologist can play a vital part, both in the planning of the operation and in its execution. In each case he is a member of a large team of professionals and has to be keenly aware of what is involved in the project.

ACKNOWLEDGEMENTS

Thanks are due to Mr S K Morgan, Chief Executive, Noble Denton and Associates Limited for permission to present this paper, and to Captain M Walsh, Canmar Drilling, Alberta, and Mr Frank van Hoorn, Wijsmuller Transport BV, Holland, for permission to describe the two projects which involved their fleets. The author also wishes to acknowledge the great amount of help from within the Group and particularly Captain M Jacobs, Houston, Captain T W Hughes, London; Mr Howard Lawes, Miss Trudy Kelly and Mrs N Chakravarty from the Environmental Studies section of the Noble Denton Weather Service.

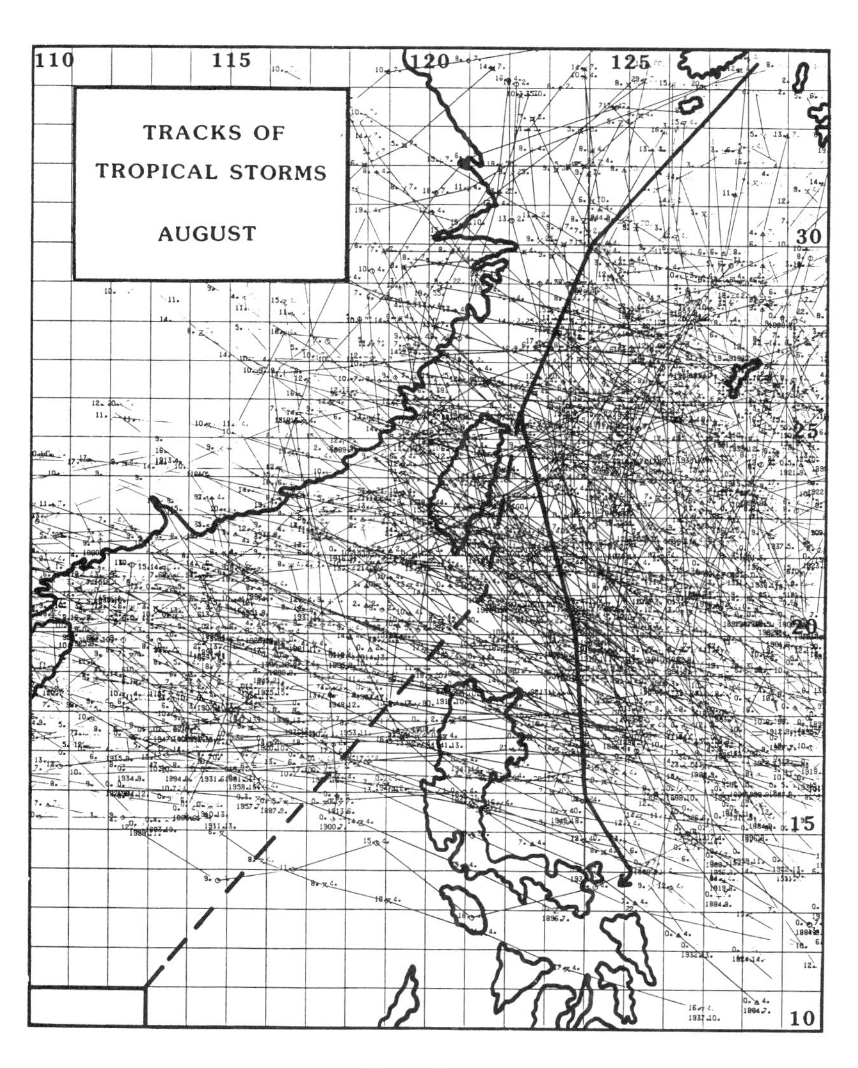
TRACKS OF
TROPICAL STORMS
AUGUST
110
115
120
125
30
25
20
15
10

Figure 1

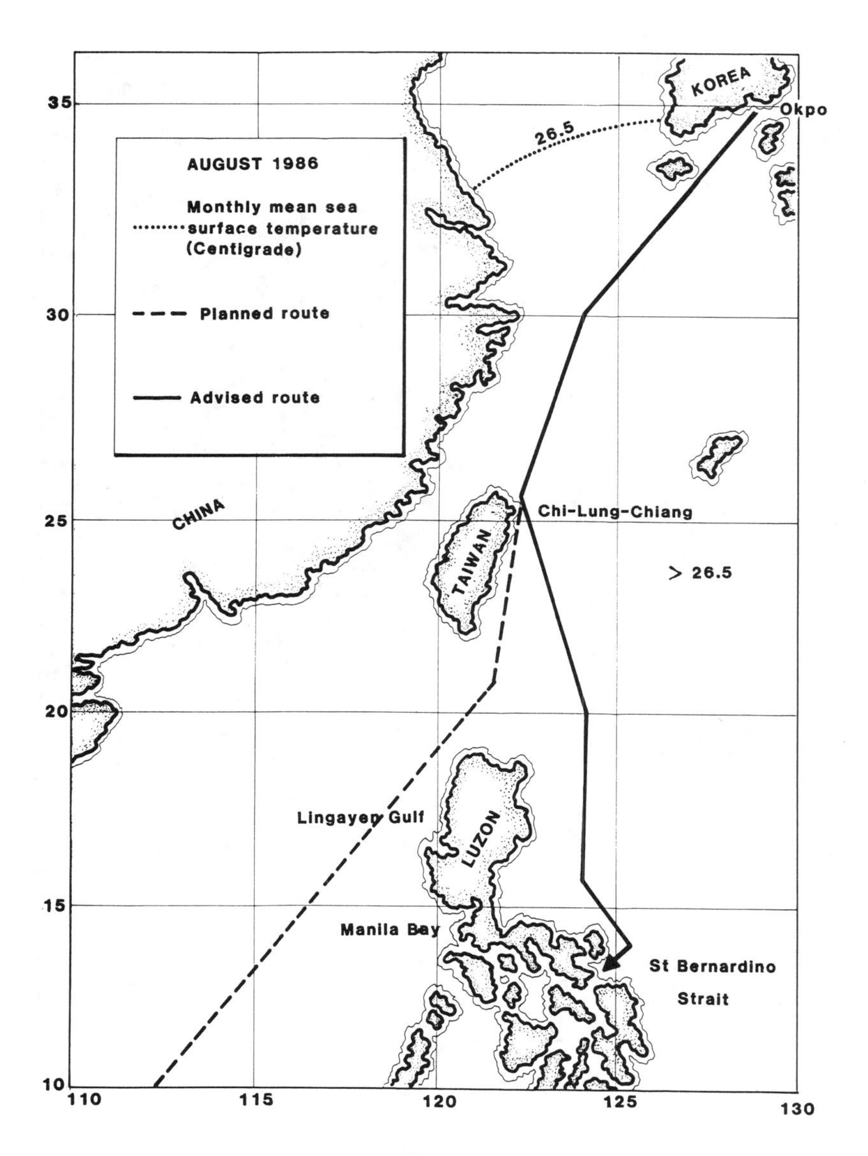
AUGUST 1986
Monthly mean sea surface temperature (Centigrade)
Planned route
Advised route
KOREA
Okpo
26.5
CHINA
TAIWAN
Chi-Lung-Chiang
> 26.5
Lingayen Gulf
LUZON
Manila Bay
St Bernardino Strait
35
30
25
20
15
10
110
115
120
125
130

Figure 2

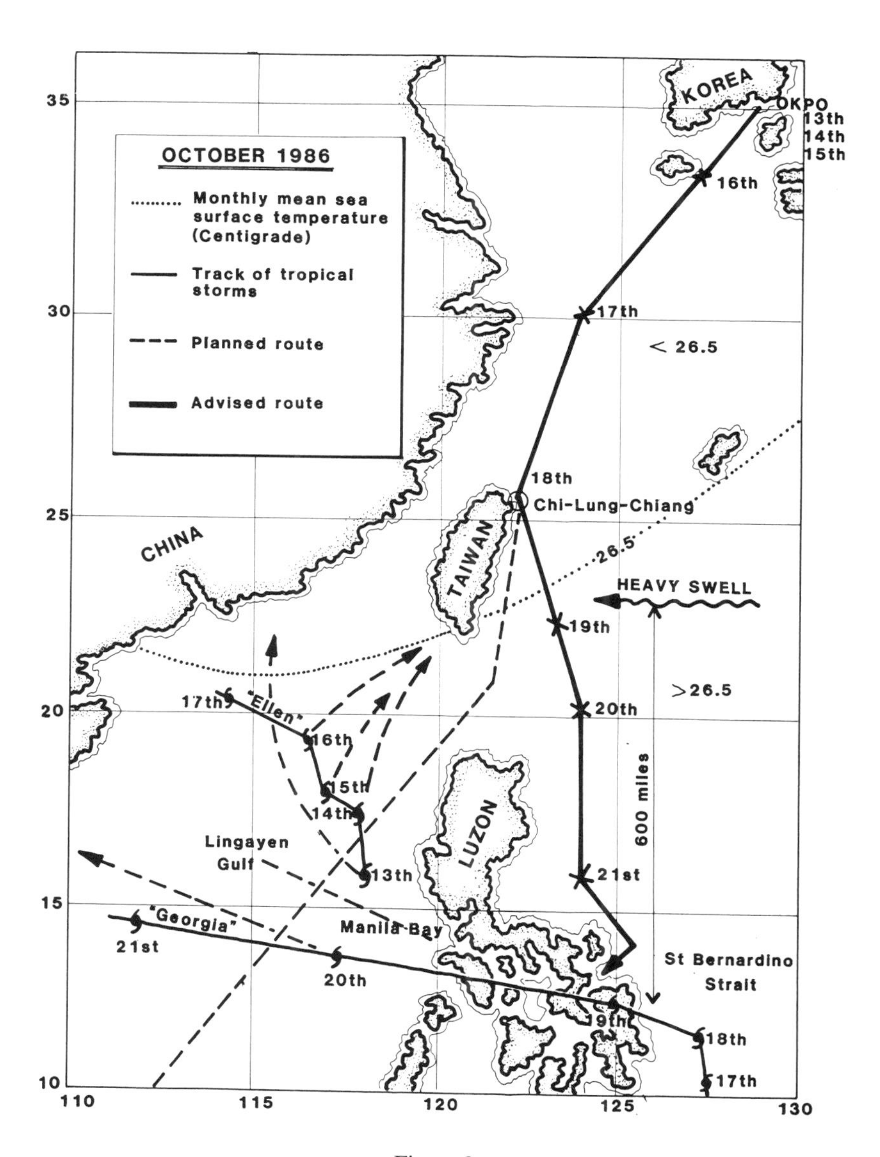
OCTOBER 1986
Monthly mean sea surface temperature (Centigrade)
Track of tropical storms
Planned route
Advised route
KOREA
OKPO
13th
14th
15th
16th
17th
< 26.5
18th
Chi-Lung-Chiang
CHINA
TAIWAN
26.5
HEAVY SWELL
19th
>26.5
20th
17th
"Ellen"
16th
15th
14th
13th
LUZON
600 miles
Lingayen Gulf
21st
"Georgia"
21st
Manila Bay
20th
St Bernardino Strait
19th
18th
17th
35
30
25
20
15
10
110
115
120
125
130

Figure 3

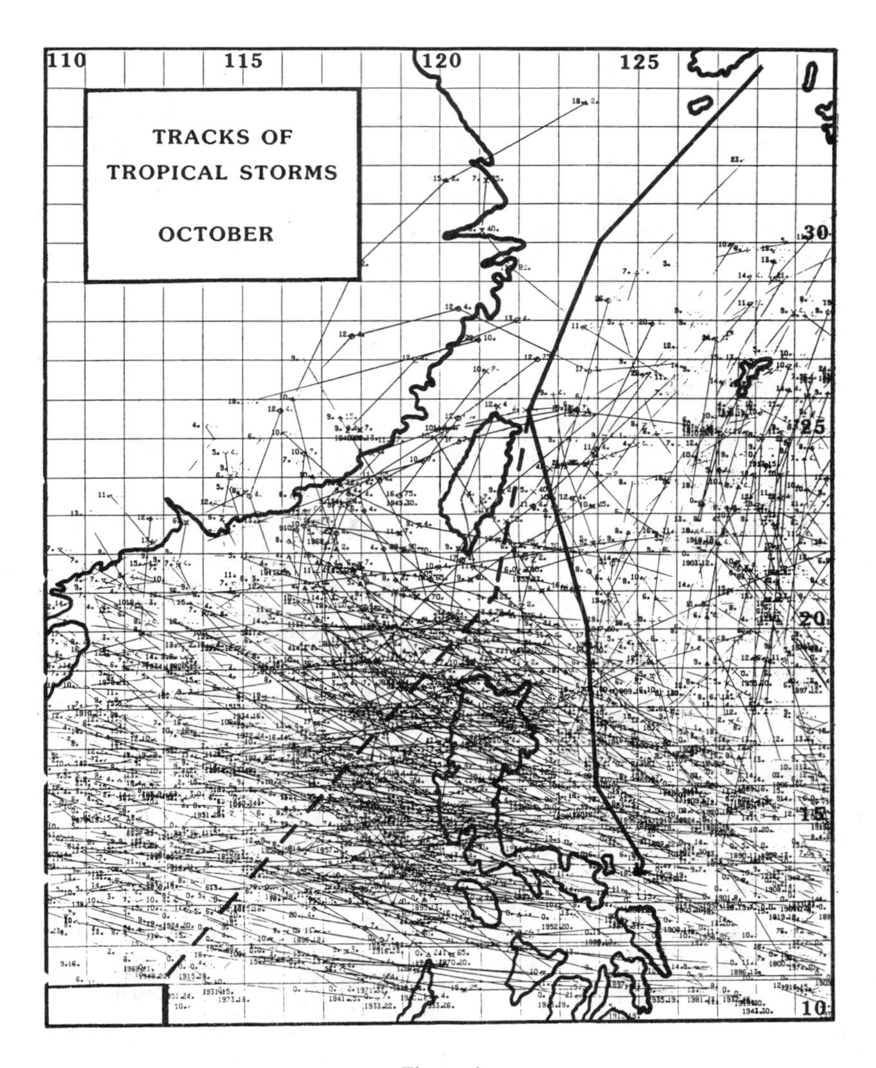
TRACKS OF
TROPICAL STORMS
OCTOBER
110
115
120
125
30
25
20
15
10

Figure 4

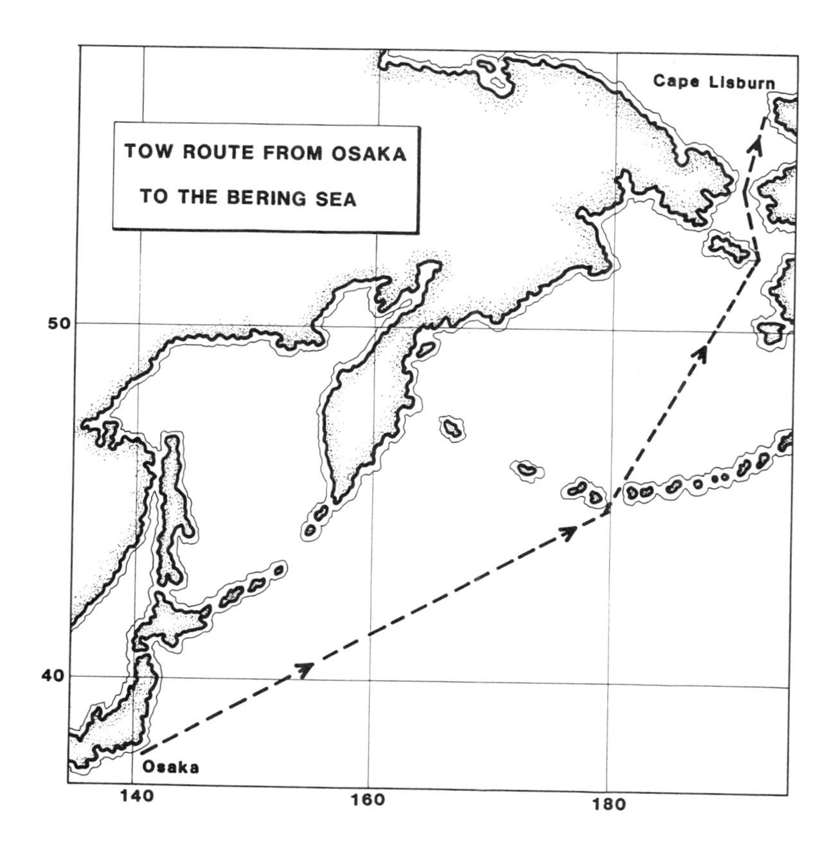
TOW ROUTE FROM OSAKA
TO THE BERING SEA
Cape Lisburn
Osaka
50
40
140
160
180

Figure 5

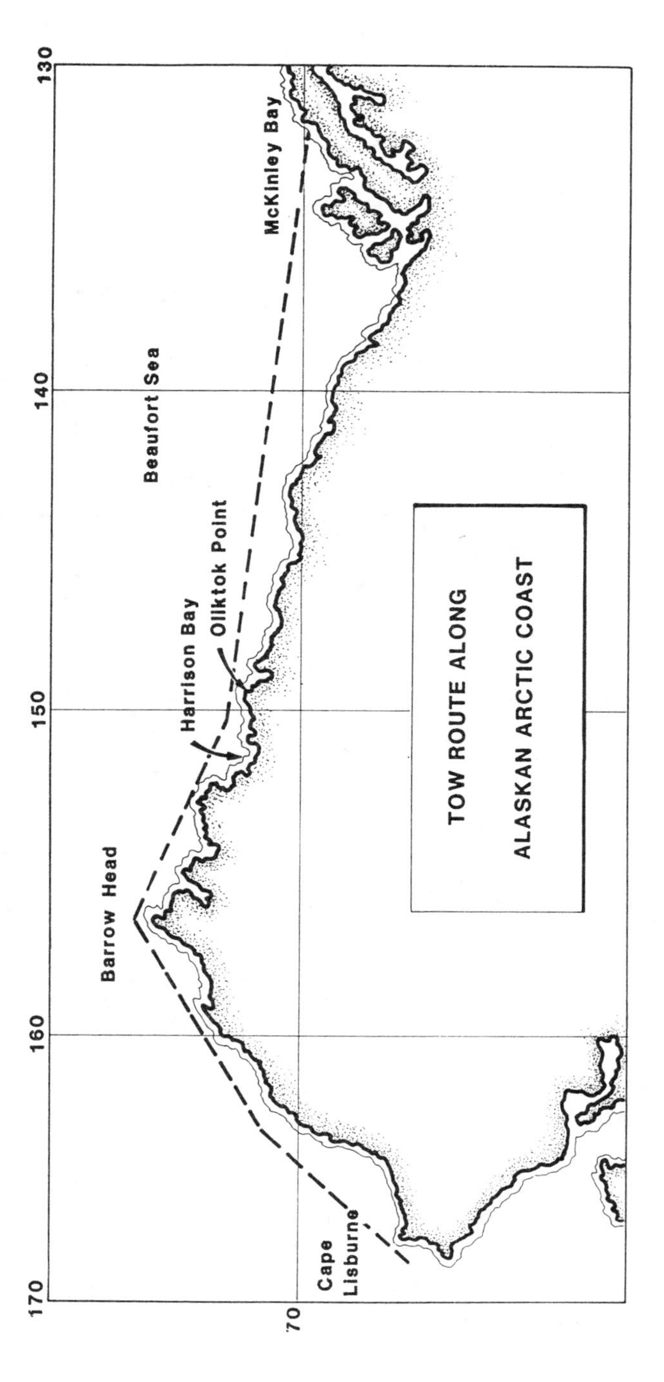
TOW ROUTE ALONG
ALASKAN ARCTIC COAST
Beaufort Sea
McKinley Bay
Oliktok Point
Harrison Bay
Barrow Head
Cape
Lisburne
130
140
150
160
170
70

Figure 6

6

United Kingdom specialised marine meteorological services

R. J. Shearman, United Kingdom Meteorological Office

1. INTRODUCTION

In the United Kingdom, the Meteorological Service owes its very existence to the needs of the shipping community, being founded in 1854 with the intention of reducing loss of life at sea due to adverse weather. In the intervening years, the needs of aviation tended to overshadow the maritime requirement, although the dissemination of gale warnings, sea area forecasts, sea-ice analyses and similar information connected with safety at sea was continued. Comparatively recently, meteorological expertise has been applied to general shipping operations in the form of ship routeing services, representing an attempt to provide a specialised meteorological service in a form similar to some of those developed for aviation purposes.

Services to general shipping placed few specialised demands upon the staff of the Meteorological Office, because the maritime community are well versed in the terminology and practice of weather forecasting. Basic meteorology forms an essential part of the training of Deck Officers and is reinforced by years of practical experience. Communication is relatively easy, and the recipient of the information is well above to identify those parts of the meteorological advice which are most relevant to his immediate needs.

However, the main impetus for specialised marine

meteorological services has come from the vigorous development of the offshore oil and gas industries. As exploration and production has moved steadily into deeper and more hostile waters, the need for accurate knowledge of environmental parameters and the penalties for failure to apply that information have grown steadily. The Meteorological Office discovered that its 'new' customers did not have the meteorological knowledge of the traditional maritime community, but had a demanding and varied set of requirements and constraints. It was necessary for meteorologists to discuss and understand the problems of engineers with a wide range of specialisms. In some cases, it must be admitted that inappropriate meteorological ideas were firmly established amongst the engineering specialists due to inadequate communications during the early stages of contact, and proved extremely difficult to dislodge during later discussions. It is fair to say that engineering techniques have also evolved to minimise the weather sensitivity and, where possible, to bring operational time scales within the validity of existing forecasting expertise.

The exploitation of any offshore oil field involves many ship-borne operations beginning with seismic survey vessels towing a string of hydrophones and a sonic transmitter. Surveying becomes impossible, even at depth, if wind speeds exceed force 6. Both the oil company and the seismic surveyor have an interest in minimising delays introduced by adverse weather, therefore the operation is usually planned for a period which will give the best chance of obtaining the required conditions, based on climatological records. Immediately before the operation, forecasts of wind speed and sea state will be required, with some estimate of duration, since there is little point in sailing if work on site is likely to prove impossible for long periods.

When a detailed survey has identified a suitable location, it is necessary to carry out exploratory drilling. This will usually be done from a self-elevating barge or 'jack-up', a semi-submerisible rig, or a drill-ship (Figure 1) depending upon the depth of water. In each case, the certificating authority must compare the expected environmental conditions in the proposed area of operation during the drilling season, with the design strength of the structure, before indicating that it may be safely used. In some cases the structure may be towed considerable distances, perhaps from one ocean to another, to be located on a new site.

In such cases, a detailed weather forecast may be required for departure from sheltered waters guaranteeing sufficient time to cross the inshore area and gain sea-room. During a long tow, the advice of experienced ship routeing staff is invaluable, but those officers must have detailed planning discussions with the customer, to establish the tow's limitations and sensible criteria for departure and seeking shelter.

In addition to the structural integrity of the installations, it is often necessary to estimate the likely amount of time when drilling will be impossible due to adverse weather, so that operations can be planned as cost-effectively as possible. When drilling has started, detailed forecasts of weather and sea state are required because there is a minimum time from the decision to cease drilling until the structure can be put into a safe state; this may be as much as 8 hours.

When the decision has been made to exploit a large offshore oil field identified by exploratory drilling, it is necessary to design and build production platforms from which a number of wells will be drilled and oil recovered. In the North Sea these platforms have tended to be steel or concrete structures resting on the sea bed, held in place either by steel piles or by the weight of the flooded lower sections as shown in Figure 1. The platforms must be designed to last for the foreseeable life of the oil field, and considerable care is taken to estimate the once in 50 or once in 100 year extreme value for relevant environmental parameters. Under-design would result in loss of life and of a considerable investment but over-design must also be avoided because of escalating costs. This is best illustrated by the figure which is often quoted of £1 million per extra metre of platform leg length. Marginal under-design can also cause operating problems because it may be necessary to evacuate the work force as a precautionary measure, resulting in lost production.

The platforms are built in large docks or on land, and either floated out upon internal buoyancy or loaded onto large barges. Although climatological advice is used to plan the operation for a favourable time of year, the actual date of 'tow-out' is determined by many other factors. The towing and deployment of the structure is extremely weather sensitive with restrictions upon the wind speed and wave height which should be encountered. An expensive task force of specialists and equipment must be assembled and held in readiness for the operation. A large tug may cost £30,000 per day and this must be balanced against the vulnerability of the capital investment at this stage of the project. Accurate and specialised weather forecasting is essential both at this stage and during the completion of the platform when a number of large production and accommodation modules must be lifted into position to form the working deck.

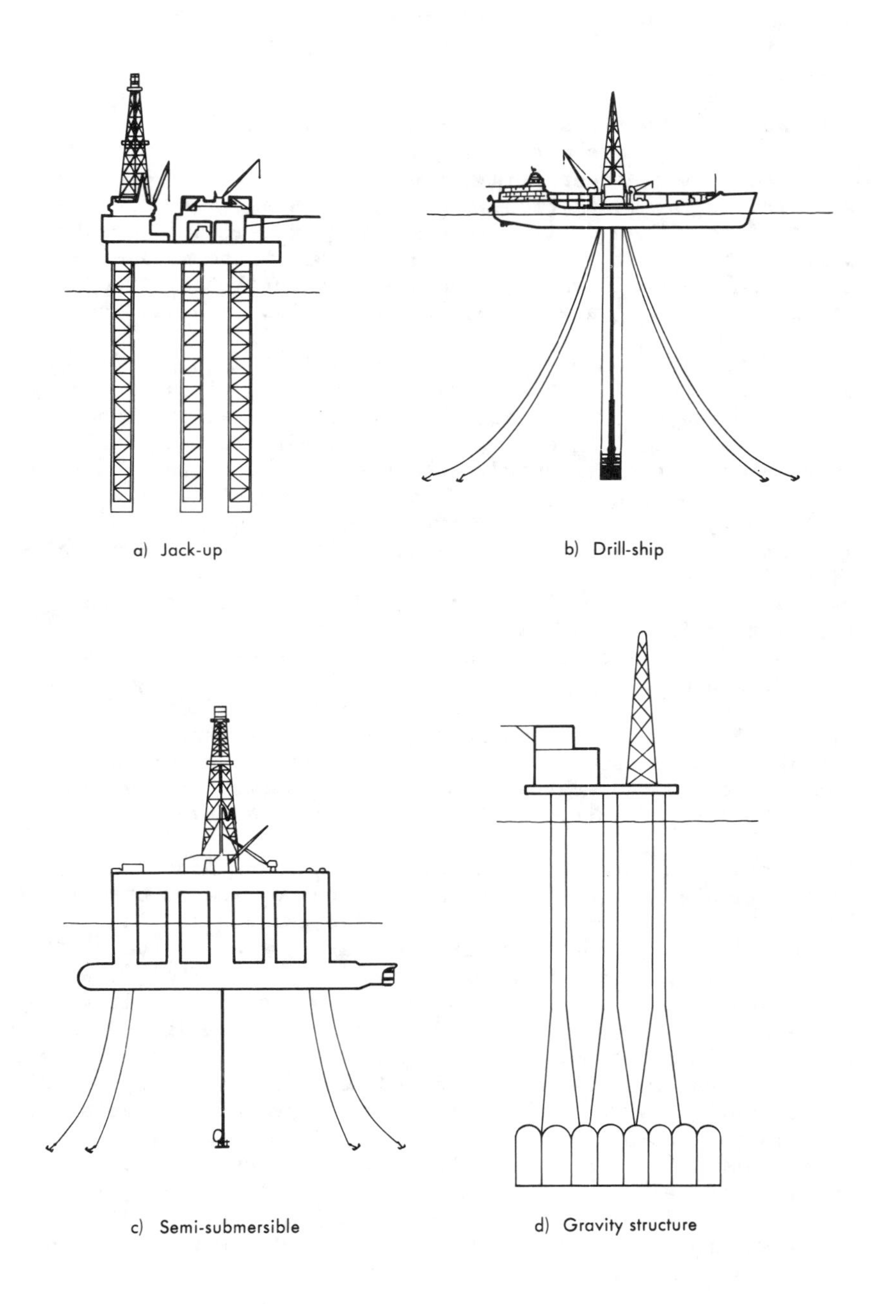

Figure 1. Types of drilling platforms.

During normal operation, the platform must be supplied by sea, unloading taking place with little or no shelter. Many weather sensitive maintenance tasks must also be performed. Good planning and accurate forecasting are essential if cost-effective and safe operating conditions are to be achieved. Special forecasts may also be required for tanker operations in confined waters e.g. berthing at Sullum Voe in the Shetlands, once again a detailed knowledge of the operation and its constraints is the only way to ensure that the forecast is appropriate.

Despite every effort to achieve high safety levels in the large and complex offshore industry, the hostile environment is such that accidents do occur. Forecasting support is necessary both for air/sea rescue work and prediction of the movement of oil slicks. Accidents and contractual delay can also lead to litigation in which past weather conditions may be considered relevant.

Having looked briefly at the offshore oil industry's requirement for meteorological support, as a typical example of a major customer purchasing meteorological services, it is appropriate to describe those services and how they are provided. However, it is necessary to remember that other customers such as coastal engineers, dredging companies and pipelayers all have specific problems which must be taken into account when producing meteorological information for their use.

2. MARINE METEOROLOGICAL SERVICES

2.1 Marine Climatology

2.1.1 The Data Source

The majority of requests for climatological advice are connected with design studies or planning operations. In virtually every case a complete and reliable climatological data base is required to provide an answer. Fortunately, meteorologists have recognised the importance of data from the oceans for many years, albeit for scientific purposes rather than answering commercial enquiries.

The Fourth Congress of the World Meteorological Organization in 1964 adopted Resolution 35 which asked all members operating merchant fleets to send observations made by those fleets to the responsible member for the area where the observation was made. This data exchange has been very successful, resulting in a good data base for most areas from 1961 to date.

The United Kingdom Meteorological Office decided towards the end of the 1960s decade to merge the available Resolution 35 data with historic data from the USA, similar data from United Kingdom ships and several smaller collections from other countries. The data set created by this action was designed mainly to be used by synoptic climatologists working upon the problems of long range forecasting. With a growing awareness of the commercial value of the data, the entire data set was re-ordered to produce a format more appropriate for commercial enquiries (Shearman 1983). This data set was adequate for enquiries in waters around the United Kingdom and also in the North Atlantic, but the density of observations elsewhere was variable and sometimes poor. An increasing number of urgent enquiries were received for advice on conditions in other oceans of the world and the decision made to incorporate data from all areas of responsibility for the period 1961 to date. This was done by a series of bilateral exchanges, and the data bank now contains observations from all areas except that of the USSR. Data from the Indian Ocean are also incomplete but will be added as soon as they become available. At present there are approximately 60 million observations held in the data bank, and this is increasing at the rate of one to one-and-a half million per year.

The data from merchant ships are semi-randomly distributed in time, so that it is impossible to construct a continuous time series at a given location for a chosen parameter. With this requirement in mind, data from Ocean Weather ships, light vessels and fixed platforms/buoys have been archived for the North Atlantic and Continental Shelf areas adjacent to the United Kingdom. Data from these sources have been stored in essentially the same format as that from the voluntary observing fleet to facilitate quality control and the development of universal enquiry programmes.

Management of the data bank is a substantial task particularly when collections of exchange data are added. A major problem concerns the avoidance of duplication because observations can appear in more than one exchange data set. Software has been developed to identify and reject both exact and near duplicates. The data could not be used with confidence unless information was available concerning their quality. Considerable effort is invested in quality control, with such obvious errors as impossible locations or meteorological parameters being outside credible ranges being identified first. Each observation is then checked for internal consistency and subjected to a more selective series of range checks. During this second stage of quality control, both the reason for a change and the rejected data are stored in a separate but cross-referenced data set, so that the original data may be restored if necessary.

Areal quality control is not attempted because the observing network is in a state of flux as ships move between ports. Thus the selection of known reliable 'neighbouring' observations is almost impossible, though it would be possible to use analysis techniques from the numerical modelling schemes to perform some areal checking. A pilot study involving approximately 50 specially selected ships is being planned under the auspices of the Ocean Observing Systems Development Project (OOSDP) and CMM. The study has the objective of producing an archive of more accurate meteorological data from a subset of ships by considering details of instrument exposure, air flow over the ship, and using a far more stringent quality control system than those currently applied, which will probably be based on a numerical weather prediction model analysis.

It is unlikely that these methods could ever be applied to all ships because they are labour intensive, and their use without adequate manual intervention could result in the loss of genuine climatic extremes.

2.1.2 The Meteorological Office Enquiry Service

The Main Marine Data Bank is the foundation upon which a successful and thriving enquiry service has been built. This service is aimed at the design and planning phases of marine projects. Experienced members of staff discuss the customer's problem before selecting the best combination of analysis technique and data to answer the question. The choice is rarely straightforward, and despite the quality control routinely applied, the analyst must be aware of factors such as anomalous data, non-standard observing practices and data distributions biased by stationary or slow moving vessels. A comprehensive and flexible suite of computer programs has been developed to analyse the data and present them, making full use of automated graph plotting and line drawing techniques. However, these programs are merely specialised tools to be used by the analyst, and every effort is made to avoid giving a stereotyped service.

As already mentioned above, engineers engaged upon the design of large fixed structures need to know the possible severity of environmental conditions which their design may have to withstand. The most important parameters in this respect are wind speed and wave height, although factors such as structural icing do demand a knowledge of other parameters, for example temperature.

Extreme environmental conditions are usually estimated by fitting appropriate mathematical functions such as those derived by Weibull (Weibull 1951) or Fisher-Tippett 1 (Gamble 1958) to a long record and extrapolating the function to provide a value with an average return period of once in 50 or once in 100 years. The choice of function is determined by the source of data, better results being given by functions fitted to annual maxima. However, this technique requires a continuous record such as that provided by ocean weather ships, light vessels and coastal stations. The Weibull function is applied to the whole distribution of data, and used where sampling is random in space and time, making acquisition of annual maxima impossible. It is generally used with merchant ship data but is known to underestimate extreme values so that it is desirable to relate the result to a set of reference values such as those provided by the Department of Energy Guidance Notes (1977) or analysis of the record from a neighbouring coastal station. There are many pitfalls in estimating extremes, and considerable subjective judgement, based on experience, is necessary to achieve credible results.

An important function of the enquiry service is to provide definitive advice to the Regulatory Government Department for the Design of Offshore Structures. A working group under the auspices of the Department of Energy, and composed of design engineers, underwriters, meteorologists and oceanographers has just completed a revision of the Department of Energy Guidance Notes using all available data and state of the art techniques. The final document is a statement of what is thought to be the acceptable extreme environmental conditions to be generally applied. Individual relaxations of these values must be justified for chosen sites, and the enquiry service can and does become involved in this work.

At the planning phase of the project it is necessary to estimate the likely effect of adverse weather upon the operation. This is done, first, to establish whether it is reasonable to carry out the operation at all, at the planned time. Secondly, an estimate is often made of likely 'down-time' both to add a realistic amount to the cost of the operation and also to establish sensible completion dates and weather 'clauses' for any contractual agreement. Where a continuous long record is available, it is possible to analyse that record and produce the frequency of occurrence of spells of unfavourable conditions having varying durations. However, most offshore enquiries including those relating to long tows can only be answered by analysing data from merchant ships and producing the frequency of occurrence of observations which signify unfavourable conditions. This analysis normally gives some idea of the occurrence of 'down-time' but does not include the information on duration which would indicate how the 'down-time' is distributed.

In addition to the main marine data bank, it may be necessary to use data from coastal stations for projects planned close inshore or in estuaries where observations from merchant ships are not representative. Such coastal data may need some adjustment to allow for local exposure effects, or alternatively a remote but long period record may be calibrated using a shorter but more local series of measurements.

In many cases there is substantial interpolation and interpretation of data, and it is important that a concise report is provided which states clearly the assumptions made, level of confidence, and any reservations concerning the final answer. Usually, the use of technical and meteorological terms is kept to the essential minimum, but where considered necessary will normally be included in an appendix. The enquiry service is consultative, and customers are encouraged to discuss any difficulty they may have with individual reports, so providing vital feedback which allows improvements to be made.

Inevitably there are occasions when adequate observations of wind and wave conditions are not readily available. In these cases, the analyst may use hindcast values from a numerical weather prediction model analysis routine, coupled with a numerical wave model. Provided that the wind fields can be produced with acceptable accuracy, improvements in wave models indicate that a long time series of wave data can be produced which would allow the prediction of extreme values with some confidence. A major study has just commenced in Continental Shelf waters around the United Kingdom, which will hindcast 25 years of data. The co-operative venture is funded by a consortium of oil companies and serviced by a consortium of meteorological services. The latter are each bringing different expertise to the project. These techniques are applicable to other areas of the world provided that funds can be made available to support an expensive operation of this type.

2.2 Forecasting Services

2.2.1 Numerical Weather prediction

All of the specialised forecasts issued by the Meteorological Office for marine purposes have their origin in atmospheric forecast products derived from integrations of two versions of the Meteorological Office's 15-level primitive equation prediction model. The global model, with a horizontal resolution of about 150 km is used to compute forecasts for up to 6 days ahead, twice daily, within about 5 hours of datum time (i.e. 00 or 12 GMT). A more detailed version with a horizontal resolution of about 75 km is used to produce forecasts for an area between latitudes 30 degrees north and 80 degrees north and longitudes 80 degrees west and 40 degrees east for 36 hours ahead. This is done twice daily within about 3 hours of datum time.

Initial conditions for the numerical forecasts are obtained by assimilation of suitable observations into the global prediction model in a 6-hourly data assimilation cycle. Thus for a forecast cycle commencing at 1200 hours, observations from 0900 to 1500 hours are interpolated to grid points and compared with the forecast field produced during the previous assimilation cycle at 0600 hours. The discrepancy between the two sets of values is calculated and the prediction model re-run starting at 0600 hours but this time it is allowed to relax towards the interpolated observation, thus accounting for part of the calculated discrepancy at each integration. The end result is a new initial field for the 1200 hours forecast cycle.

The vertical resolution of the model allows the boundary layer to be well represented so that forecast winds can be reliably estimated, particularly over the sea. This allows the winds to be used as a direct input to a numerical sea state model which predicts wind waves and swell (Golding 1983). The resolution of the wave models ranges from 150 km for the Global Oceans to 30 km for European Waters, including the Shelf area, and the Mediterranean and Baltic Seas.

The sea-state models are based on a discrete frequency/direction energy spectrum and include parametrizations of wave growth, non-linear interaction and decay. The European area model contains additional representations of bottom friction and refraction applicable in shallow water. The less detailed global ocean model is run twice daily for 120 hours ahead using winds from the global version of the numerical weather prediction model. The European area model is also run twice daily for 36 hours ahead but uses surface winds predicted by the more detailed fine resolution numerical weather prediction model. An example of the fine mesh wind field is shown in Figure 2.

The starting fields for a sea state forecast are calculated by means of a process erroneously called 'hindcasting', whereby the first 12 hours of the previous forecast is re-run using numerical wind fields that have been updated by means of the latest observations and analysis.

The performance of all the models was improved when new versions were introduced in late 1982 and again in 1986. Not only was resolution increased due to use of fine meshes of grid points but there was a basic improvement in accuracy. It is interesting to examine forecast wind speeds shown in Table 1. The new models have a better performance at higher wind speeds than their predecessors, indeed the present coarse-gridded version performs rather better than the earlier detailed fine-grid model. Errors calculated for the entire

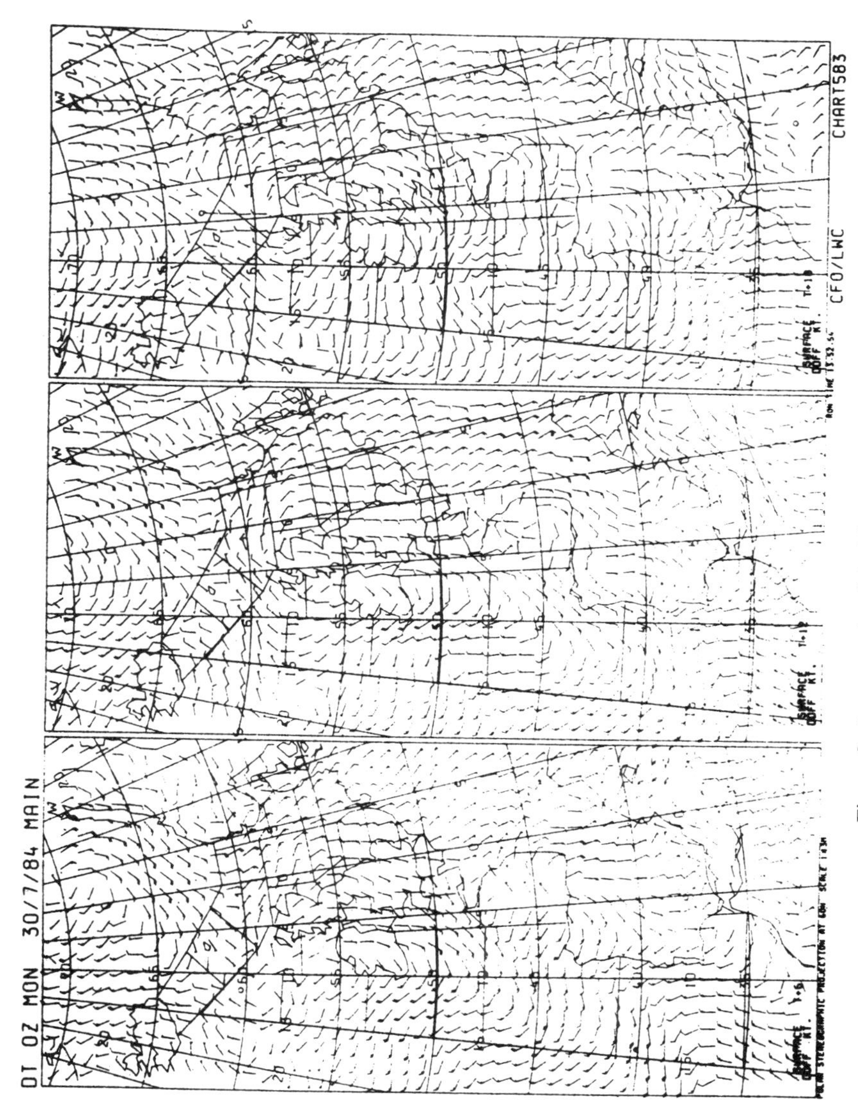

Figure 2. An example of the Fine Mesh Wind Field.

wind speed range do not show such a marked improvement because they are dominated by frequently occurring low wind speed cases. Table 2 shows annual mean and root mean square errors in forecast wave heights 24 hours ahead.

The numerical weather prediction products are fundamental tools but both short and medium range forecasts are prepared using a well tried man-machine mix. The more specialised the product, the greater the contribution from the experienced forecaster. An example of a forecast of combined sea and swell is shown in Figure 3.

Model	Year	WIND SPEED (MS^{-1})									
		0-10		10-15		15-20		> 20		ALL	
		MEAN	RMS	MEAN	RMS	MEAN	RMS	MEAN	RMS	MEAN	RMS
Rectangle 100km	1981	0.4	2.8	-2.1	3.8	-3.8	5.2	-5.8	7.5	-0.8	3.5
Coarse Mesh 150km	1983	0.3	2.8	-0.8	3.8	-1.3	4.5	-3.0	5.6	-0.3	3.4
Fine Mesh 75km	1983	0.9	3.3	0.4	4.3	0.2	4.8	-1.0	5.3	0.7	3.8
APPROXIMATE NUMBER OF OBSERVATIONS		1800		800		270		75		2945	

TABLE 1. Annual Errors (mean and root mean square) of 24 hour forecast surface winds, North Atlantic and North Sea. (Speed ranges are taken from observations, and error = observed speed - forecast speed.)

Site	Wave Height Error (Metres)	
	MEAN	RMS
OWS L	+ 0.3	1.5
Statfjord	+ 0.4	1.0
Southend North Sea (K - 13)	+ 0.1	0.5

TABLE 2. Annual Errors (Mean and root mean square) of 24 hour forecast wave height at 3 verification site

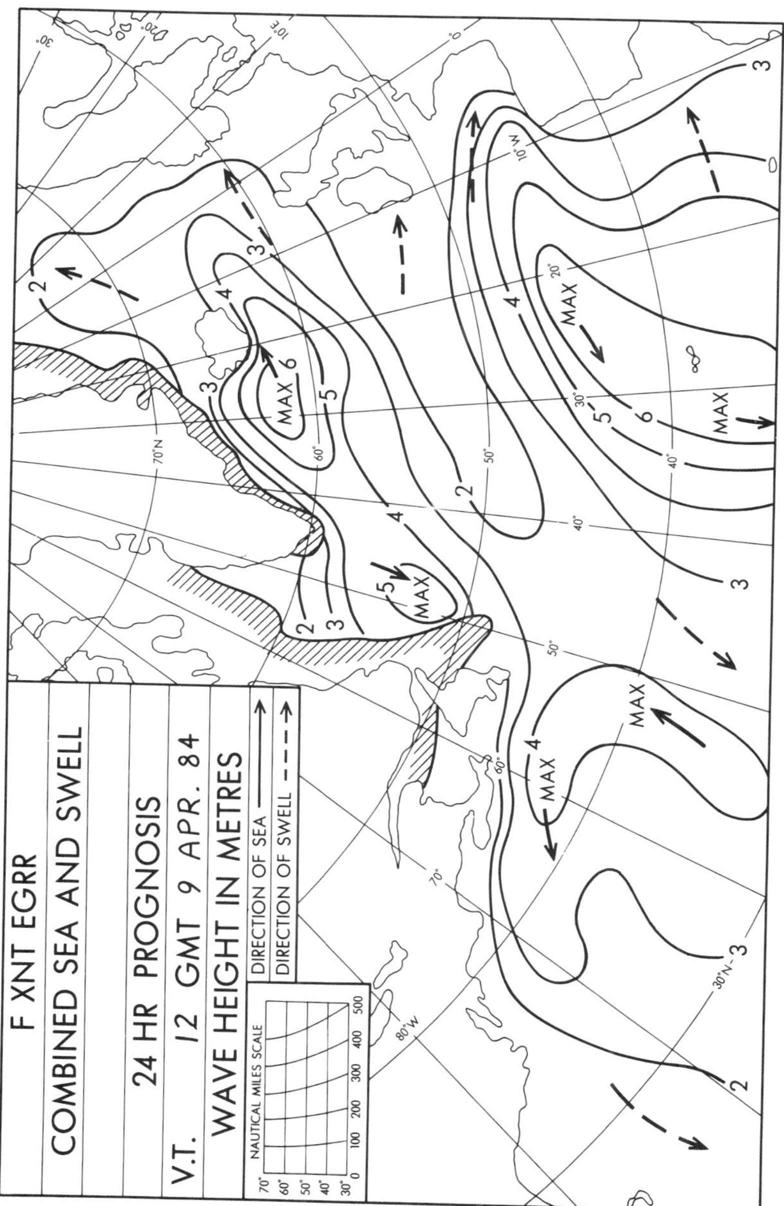

Figure 3. Combined Sea and Swell produced by man-machine mix.

2.2.2 Shipping Forecasts and Warnings

The Meteorological Office provides routine weather information for shipping operating in the 'Coastal Sea Areas', i.e. from the coasts of Western Europe to 15°W, and for a larger area of the Eastern North Atlantic extending to 40°W. The information includes synopses, forecasts and gale and storm warnings. A number of methods of communication are used including coastal radio stations, BBC broadcasts and radio facsimile transmissions.

The Atlantic weather bulletins are issued twice daily from Portishead Radio Station. Forecasts include wind speed and direction, weather, visibility and ice accretion. Storm warnings, a synopsis of weather conditions, and ship and land reports are also broadcast. Storm warnings are repeated 6 times a day.

The Coastal Sea Areas are covered by transmission from Coastal Radio Stations and also by BBC Radio 4 (200 KHz). Gale Warnings, including winds of Beaufort forces 8 to 12, are issued for these areas when necessary. They are also included in summary form at the start of the BBC Radio 4 Shipping Forecast bulletin which covers all areas. The summary is followed by a general synopsis and forecasts for the following 24 hours. In addition, BBC Radio 4 broadcasts a forecast at about midnight each day for the next 18 hours covering those inshore waters within 12 miles of the coast. Special 24-hour forecasts are also issued for fishing fleets.

It could be said that shipping forecasts are a general rather than a specialised service, provided free of charge. However, these forecasts are aimed at a particular industry, and form a quid pro quo for the observations made by merchant ships totally free of charge. The value of those observations has been assessed at between £5 and £10 M per year.

2.2.3 Sea Ice Analyses

In addition to warnings of adverse weather, daily sea ice analyses are compiled covering the Western North Atlantic Ocean, Baltic, Barents and Greenland Seas, Baffin Bay and Hudson Bay, and transmitted by radio facsimile. The charts contain isotherms of 5-day mean sea surface temperatures as well as details of ice conditions taken from a number of sources, but notably American and Canadian information. The ice edge and isotherms are reassessed and redrawn as necessary twice each week.

Copies of the sea ice charts are despatched by post as well as by radio facsimile, and the staff of the unit answer enquiries of a forecasting and climatological nature concerning sea ice. A study is being made of the possibility of using satellite data to provide advice on sea ice in the Southern Hemisphere for shipping companies with specific operational problems.

2.2.4 Ship Routeing

The forecasts and warnings issued to shipping are adequate for general safety purposes. However, they necessarily have a coarse spatial resolution. For example a gale warning will be issued for a large sea area if a wind greater than force 8 is expected somewhere in the area. A ship some distance away, but in the same area may experience much less severe conditions.

It is possible to provide more specialised meteorological guidance, and in 1968 the United Kingdom ship routeing service was created to do this, concentrating initially upon North Atlantic routes, and developing until today it has a worldwide capability. The basic aim of ship weather routeing is to give the vessel the most economical passage through the various weather systems consistent with any other factors considered by the operators to have higher priority. Some examples of such constraints are fuel economy, time on passage, avoidance of damage to ship or cargo, and also onboard operations such as hold and tank cleaning.

The staff of the ship routeing service are all ex-seagoing masters who have collectively had experience with most types of vessels. They are located in the Central Forecasting Office, and receive documentation and an oral briefing from forecasters engaged in the man-machine interaction mentioned above. The position of the vessel being routed is plotted every 6 or 12 hours, and predicted positions marked on forecast charts for up to 5 days ahead. On a daily basis a computer program is used to trace the 'least-time track' of each vessel through the forecast weather systems. In this way a vessel can be advised, well in advance, of adverse weather or sea conditions en route, and evasive action taken if necessary.

The movement of very large offshore structures leads to lengthy tows which are considerably more weather sensitive than normal ship operations. The ship routeing service caters for such projects by providing 3-day forecasts along the advised route, taking into account possible shelter ports suitable for the tow. The close liaison between experienced masters, able forecasters backed up by powerful global computer models of weather and wave systems, satellite data and global observations giving information on events such as tropical cyclones results in an efficient service giving valuable advice. Table 3 shows an assessment of the benefit to a shipping company of the ship routeing service.

	TIME LOST	TIME SAVED	POTENTIAL TIME SAVED
89 ROUTEINGS (44 OPTIMUM)	37.6	257.9	220.3
DEPARTURE FROM ADVISED ROUTE (Clients Decision)	106.8	-	106.8
TROPICAL STORMS) SHIPBOARD CONSTRAINTS)	27.8	-	-
TOTAL POTENTIAL SAVING	-	-	327.1

TABLE 3a. Analysis of Routeings of a Container Fleet for one year.

	TIME LOST	TIME SAVED	POTENTIAL TIME SAVED
41 ROUTEINGS (20 OPTIMUM)	10.0	127.8	117.8
DEPARTURE FROM ADVISED ROUTE (Clients Decision)	125.7	-	125.7
TROPICAL STORMS	-	-	-
SHIPBOARD CONSTRAINTS	7.4	-	-
TOTAL POTENTIAL SAVING	-	-	243.5

TABLE 3b. Analysis of Routeings of Bulk and Tanker Division for one year.

2.2.5 Specialised Forecasting Services for the Offshore Industry

Forecasting services for the offshore industry are provided by London Weather Centre supported by Aberdeen (Dyce) Meteorological Office. The Weather Centre is equipped with twin mini-computers linked by a high speed telecommunication line to the main computer system in Bracknell. As a result, grid point data from the new numerical weather prediction model and associated wave model are received quickly, and specialised forecast material produced locally. There is scope for more variety than originally available via facsimile, because any increase in the range of products in that system would have caused further delays. Forecasters also have visual display units linked to the system, providing instant access in a number of graphical formats, to the latest observational data, including those from rigs and platforms in the North Sea.

The forecast service most used by the offshore industry has been the twice daily site specific telex forecast for up to 5 days ahead. Such forecasts have been provided for as many as 19 individual locations at any one time. This forecast enables companies to plan resource deployments for the week ahead (for example provision of tugs, support vessels, location of drivers, etc.) and also to make decisions regarding day to day weather sensitive activities. Site specific forecasts transmitted by telex contain a brief description of the synoptic situation and how it is expected to develop. A forecast of weather, wind, and visibility is provided for 12 hours from time of issue for 12 to 24 hours, and 24 to 48 hours ahead. An outlook is given for wind and wave conditions for 48 to 72 hours and 72 to 120 hours. The forecast format is shown in Table 4. A 24-hour consultancy service is also available to provide 'up-to-the-minute' advice. A daily briefing for the marine staff is presented by forecasters from Aberdeen for local companies, and by staff from Kirkwall and Lerwick in support of tanker loading and unloading activities. All forecasts are updated during their period of validity, if the forecaster on duty considers action to be necessary, and there is also a special warning service to cope with sudden and dramatic developments in weather of the type which occurs two or three times each winter in the North Sea.

Although satisfactory for many purposes, shore-based services do not provide an adequate safeguard for some projects which require the presence of a forecaster on site. The tow-out and deployment of new structures, where the entire capital investment is at risk for a short period, is one such operation. Another example is that of complicated and sensitive diving work where a number of lives may be at risk. In these cases there is a definite advantage in locating a skilled forecaster on the structure, so that he is in contact with local conditions while still supported by his colleagues and the facilities ashore. He is also readily accessible to the local project manager, allowing discussion at any stage during the evolving operation. Considerable efforts have been made to assemble a team of skilled forecasters at London Weather Centre who have developed a knowledge and understanding of the operational problems and constraints of the offshore industry, and are used to working closely with operators in the field.

SECTION		CONTENTS
		SYNOPTIC SITUATION
1		COMPLEX LOW NEAR IRELAND WILL MOVE NORTHEAST EXPECTED NORWEGIAN SEA LATER ON TUESDAY. WEAK RIDGE OF HIGH PRESSURE WILL FOLLOW FORECAST WIND SPEED IN KNOTS WAVE HEIGHTS IN METRES PERIOD IN SECONDS
2	A	FORECAST 0800GMT 30/07/1984 TO 2000 GMT 30/07/1984
	B	WIND AT 10 METRES VARIABLE 05 BECOMING NORTHEAST 10 GUSTS 14 LATER WIND AT 50 METRES VARIABLE 07 BECOMING NORTHEAST 13 GUSTS 17 LATER
	C	TOTAL SIGNIFICANT WAVE HEIGHT 2.0 BECOMING 1.8
	D	SIGNIFICANT WAVE PERIOD 4
	E	MAXIMUM WAVE HEIGHT 3.0
	G	SWELL DIRECTION AND HEIGHT WEST 1.8 BECOMING 1.5
	H	SWELL PERIOD 07
	I	WEATHER RAIN AT TIMES ESPECIALLY LATER PERHAPS THUNDER
	J	VISIBILITY 8NM BECOMING 2NM WITH PATCHES 1000 YARDS
	K	CLOUD BECOMING 8/8 500 WITH PATCHES 200FT
3	A	FORECAST 2000GMT 30/07/1984 TO 0800GMT 31/07/1984
	B	10M NORTHEAST 10 BECOMING NORTH 28 GUSTS 30 LATER
	C	1.8 BECOMING 2.0
	D	4 BECOMING 5
	E	3.2 LATER
	G	WEST BECOMING NORTH 1.5
	H	8 OR 7
	I	RAIN AT TIMES ESPECIALLY EARLY
	J	MODERATE OR POOR BECOMING GOOD
4	A	FORECAST 0800GMT 31/07/1984 TO 0800GMT 01/08/1984
	B	10M NORTH 18 BECOMING VARIABLE 05 50M NORTH 23 BECOMING VARIABLE 07
	C	2.0 BECOMING 1.0
	D	5
	E	3.2 EARLY
	G	NORTHNORTHWEST 1.5 BECOMING 0.8
	H	08
	I	A FEW SHOWERS
	J	MAINLY GOOD
5	A	OUTLOOK 24 HRS ENDING 0800GMT 02/08/1984 10M WIND VARIABLE 05 BECOMING EAST 10 TOTAL SIG WAVE ABOUT 1.0
6	A	OUTLOOK 48 HRS ENDING 0800GMT 04/08/1984 10M WIND EAST 10 TOTAL SIG WAVE ABOUT 1.0

TABLE 4. Example of an Offshore Specialised Telex Forecast

2.2.6 Negative Storm Surges

As the introduction to the Admiralty Tide Tables points out, tidal levels can be affected by the meteorological situation, and the Storm Tide Warning Service, which was transferred from the Hydrographic Department of the Navy to the Meteorological Office on 1 April 1983, has the task of forecasting these effects. The service was established in 1953 to issue tidal flood warnings for the low-lying areas of Eastern England, but the Hydrographer, noting that the opposite effect - a depression of tidal levels - could prove a hazard to navigation, particularly for deep draught vessels, extended the unit's task to include a 'Negative Surge Warning Service' for the Southern North Sea, Thames Estuary and Dover Strait from 1973.

Making use of the observed growth of surges as the tide travels South along the East Coast of Britain, empirical equations derived in the late seventies by the Oceanographic Department of Southampton University allow 5-10 hours notice to be given of negative surges, usually on occasions of strong southerly airflow. The equations are supplemented by numerical forecasts from a surge model run as part of the operational forecasting suite.

Warnings are issued to the appropriate authorities when levels are expected to be one metre or more below astronomically predicted levels. The most notable negative surge of recent years occurred on 19 December 1981 when levels fell to 2.35 m below prediction at Sheerness.

2.2.7 Forecasting for Marine Leisure Activities

An increasing number of people are turning to the sea for leisure purposes, creating a demand for readily available but nonetheless specialised weather information for the general public taking part in marine activities. Local weather centres provide a range of specialised forecasts and documentation, in some cases making use of telecommunication devices such as DOCFAX, and can advise when selected weather parameters reach agreed limits where further deterioration may be detrimental to the safety of personnel and/or equipment. Individual briefings can also be arranged, and are particularly useful for the organisers of marine events.

However, these services may be beyond the resources of many of those now taking up marine leisure activities, due to lack of equipment of finance. Much has been done to include some details of inshore conditions in local television weather presentations, but the time available is often limited. A determined attempt has been made to provide information for the 'weekend sailor'. Several Weather Centres produce yachtsman's information packs, usually on Fridays, containing data and a set of relevant weather charts.

The greatest improvement has come with the inauguration of the 'Marinecall' telephone service. Any member of the public may dial an appropriate telephone number and obtain a pre-recorded weather forecast for local waters, specifically designed for use by amateurs engaged upon leisure activities. This service is available for inshore waters around the United Kingdom.

2.3 Legal and Post Mortem Activities

Despite careful planning and all the efforts of forecasters and shipping or offshore operators, the maritime industries remain fairly hazardous, and adverse weather inevitably takes its toll. There is a wide range of legal disputes ranging from relatively small insurance claims for damage to privately owned sailing vessels through to major disasters involving loss of a large vessel or drilling rig. There are also cases concerning breach of contract, where an operation or a voyage has taken longer than originally agreed and the aggrieved party is seeking compensation. Alternatively, production from an offshore platform may be poorer than expected and adverse weather the leading suspect.

In every case, the Meteorological Office has the responsibility of ensuring that weather is correctly identified as the reason and not merely used as an excuse. The financial implications of many of these enquiries are extremely serious, compensation of £200,000 is commonplace and some claims may be in the region of several millions of pounds. Investigations range in complexity from assessing weather at the time of the accident to expressing an opinion on what might have happened, when appearing as an expert witness in court. The expert witness has the task of expressing a technically competent opinion in entirely non-technical terms which can be understood and appreciated by the court.

The basic source of information is often an analysed surface synoptic chart retrieved from comprehensive document archives. Since these are working analyses from the Central Forecasting Office, and usually drawn on a very small scale, considerable care must be taken in interpreting the probable weather conditions at any location. In some cases it is possible to extract ship observations in the immediate vicinity from the main marine data bank, and so describe conditions with some confidence. But there are occasions particularly involving locations close inshore when available observations are misleading and a reasoned meteorological opinion must be put forward. On such occasions the practical advice of the ex-seagoing ships' masters within the nautical section of the Meteorological Office can be invaluable, as they may well have ship-handling experience in the area concerned. Software has also been developed to printout all ship observations along the track of a particular vessel, to assist in analyses where there is a dispute over the length of the voyage, and delays due to poor weather have been cited. Modern computer modelling techniques are not neglected, and archived data from the wave model in particular, is used to answer some enquiries. However, once again care must be taken in interpretation as operational model data are considerably smoothed in space and time. Often an assessment is required of weather conditions compared with what might reasonably be expected at the time of year. The main marine data bank can be accessed using one of the standard enquiry programs, to assist with this judgment, but it is often necessary to look at various published sources of climatology such as the Admiralty pilots and similar publications or climatological atlases from other countries.

3. CONCLUSION

The Meteorological Office has built up a specialised service to support marine interests. This service draws upon sophisticated meteorological techniques but always gives weight to the needs of the customer. Regular contact is maintained with industry via such groups as the Exploration and Production Forum, United Kingdom Offshore Operators Association, and the International Maritime Organisation. In addition to these formal links, there is some feedback from individual customers via meteorologists dealing with forecasts or enquiries. This information is also used to tune the service to clearly perceived requirements and is encouraged at every opportunity. Market research amongst prospective and existing customers is increasingly providing information about their special needs. The services described have evolved over a number of years, but should not be considered as a fully developed system. The United Kingdom Meteorological Office also has a very active research program. Staff involved in enquiry work often call upon the expertise of their research colleagues to develop techniques in response to particular problems, and this acts as a catalyst to both groups, resulting in further improvements.

REFERENCES

1. Shearman, R J 1983 The Meteorological Office main Marine Data Bank. Meteorological Magazine, 112, 1-10.

2. Weibull, W 1951 A statistical distribution function of wide applicability. Journal of Applied Mechanics, 18. 293-297.

3. Gamble, E J 1958 Statistics of extremes. New York, Columbia University Press.

4. Department of Energy Offshore installations: Guidance on design and construction. HMSO, London 1977.

5. Golding, B 1983 A wave prediction system for real time sea-state forecasting. Quarterly Journal of the Royal Meteorological Society, 109, 393-416.

7

The role of education and training in the use of meteorology in human affairs

C. L. Hosler, Vice President for Research and Dean of the Graduate School, The Pennsylvania State University

The Swiss Reinsurance Company's latest compilation of economic studies has listed natural disasters that claimed more than 20 lives or cost more than 10 million Swiss Francs in damages between 1970 and 1985. During this period 1.5 million people were killed, 50 million were left homeless and property damages were $700 billion in 2,305 such events. An average of three such disasters occurred each week. Earthquakes accounted for a few of the large losses but weather-related disasters are the preeminent sources of these losses. In the United States alone, 10,000 severe droughts, heat waves and severe cold typically characterize an average year. Property damages average tens of billions and hundreds of lives are lost each year in weather-related events.

These spectacular events tend to hold the attention of insurers and make newspaper headlines but I would submit that the economic losses are matched by the aggregate losses due to the billions of individual or small business or family agricultural losses incurred each day because of poor weather advice, or even more likely, failure to receive or act upon advice on day to day weather. Aside from the personal loss of life or property involved, these losses add up to a considerable loss of productivity for an entire country. Small fishing boats are lost, cut hay rots in the field, a crop of vegetables is destroyed by frost, grain drying on the ground is wet and rots, public events are rained out, concrete must be repoured, vehicles are mired in the mud, small aircraft are lost, muddy fields prevent a harvest, weather events occur which paralyze transport or commerce to cities or whole countries, which if properly anticipated and prepared for could result in much less disruption. The list of these losses, small and large, is endless.

Meteorologists must place at the top of the list of their reasons for their existence as being to increase their contribution to the productivity of national economies and to minimize weather-related losses of life and property. It is accepted that this requires high quality products using the most modern techniques and that training and retraining to assure this high quality product will always be the top priority. To a very large extent, the willingness of users to base decision-making on meteorological products will be based on their perception of their accuracy and relevance to their activities. Building credibility through accurate predictions is the best route to a high level of utilization.

In our training programs we must create a strong sense of our degree of assurance of forecast accuracy and convey that degree of confidence to the user. There is a large category of operational decisions in which weather is critical and where the activity can be postponed until a high level of assurance of favorable weather can be given by the forecaster. Too often we routinely issue forecasts giving the impression that we have equal confidence in all of them when any experienced forecaster knows that our confidence level varies from almost zero to 100 percent. Our training must dissuade forecasters from assuming a pose of great confidence in every forecast. For this reason, among others, a close association between the forecaster and the individuals and activities for which the forecast will be applied is highly desirable. For example, a week on a fishing boat will make a much better forecaster for fishing operations. He will see how weather affects operations and that real people's lives and livelihood are at stake. The forecaster may, as a result, put more emphasis upon winds and sea state than cloud cover or precipitation, for example. If the forecaster trainee cannot spend time at sea then the course should include lectures by those who have been at sea. In cold climates, ice formation on the ships' superstructure may be the greatest hazard. In this case, wind force and air temperature are crucial.

It is highly unlikely that someone who has never been on a farm and who don't know what crops are grown and the change in criticality of the various weather elements for planting, fertilizing, irrigating, applying insecticides or growth regulators, harvest and drying, transport or storage can tailor his forecasts and shift emphasis in such a way as to make his forecasts most useful and credible. If actual internships are not possible, then visits to production areas, conversations with growers or lectures from farm experts will help.

Because each regional training center has unique agricultural, forestry, fishing, hydrological, industrial and transport problems, the training program must be tailored to assure a match between forecast emphasis and presentation and local activities and how they change through the year. If aviation forecasts are being made, then the forecaster should fly the routes flown and talk with pilots. The training center should place great emphasis upon liaison between users and forecasters as part of the sociology of the profession of meteorology.

B J Maunder (Weather, January 1973) in discussing the benefit-cost ratios to be obtained from weather prediction said, "If this high benefit-cost ratio is to be realized, then the atmospheric scientist must not only work closely with experts in particular weather-sensitive but must also provide the decision-maker in the various weather-sensitive enterprises with the appropriate weather information and in the most useful form". He quotes a colleague: "The art and science of weather prediction and the application of lessons from past climates to future climatic probabilities will yield great economic and social benefits in the future, provided the communication media allow the fullest two-way contact between the producer and the consumer. It is throughout a problem in communication and marketing". Maunder also quotes a colleague as stating: "In some important instances, the greatest contribution a meteorologist can make is to find a way to establish real communication between a certain group of decision-makers and the meteorological system. This is not to say that all a national weather service has to do is to provide information to management (or to the general public), and then wait for the economic benefits to become obvious to all concerned. But it surely is true that no economic benefits will accrue to the meteorological information system from a weather-sensitive segment of the economy in which the persons making management decisions are (1) not aware of the availability of meteorological information; (2) aware of the potential value of meteorological information, but do not have any channels through which such information may be received".

We are now in a position to greatly improve small scale, short range weather prediction both due to technology and the freeing of time that used to be needed to follow the synoptic scale weather. Accurate short range forecasts and more time to relate to consumers can greatly increase the effectiveness of meteorology in contributing to the economy. A few large world forecast centers now provide the basic prognoses and computation power. Their products can be downloaded to smaller national centers who can then tailor the information to their needs and concentrate their forecasting skills and efforts on the mesoscale and in relating to users. This is already done in some regions and is the basis for the national system in very large nations. Smaller developing nations should look at the products of the larger centers in the same way they now look at the raw data and save a great deal of expense associated with trying in duplicate the large national or multinational cooperative centers already in existence.

The availability of low-priced microcomputers is gradually making it possible to make large amounts of information available to users. Further, new storage technologies make the availability of extensive climatologic and other data bases inexpensive and accessible. This information can be available when needed. However, some types of information and a sense of assurance cannot be transmitted electronically. There is a communication between humans that transcends language that can only occur when the sender and receiver are familiar with each other or in direct face-to-face contact.

In the United States more than one hundred private firms who provide specialized weather and climate information to businesses and individuals have been formed. This is not due to the inadequacy or lack of quality of the products of the National Weather Service (NWS). It is primarily due to the need for a more personalized and tailored service based upon study of the clients' needs and the timing and updating required to make the information more useful. The private forecaster is part of the decision-making process. Many corporations have employed their own meteorologists to perform this function even though the basic data and information are available from the National Weather Service. Decision-making which will affect lives and fortunes must involve an unhampered two-way flow of information. Not only does the weather change rapidly, frequently and in unexpected ways, but so also do the operations, circumstances and events upon which the weather will impact in such a way as to radically change their sensitivity to weather. In countries where the private sector is not developed there is no less need for liaison between user and forecaster and proper emphasis must be placed upon this in training.

In every country with which I am familiar, the rate of utilization of weather information is still much lower than is possible. In the industrial countries, the search for greater productivity is a major task. Information and especially, the search for greater productivity is a major task. Information and especially, weather information, is a path to greater productivity. But what is true for the developed countries is even more true for those that are developing. For it is here that the gains of bringing state-of-the-art forecasts and data to enterprises and activities - agricultural, commercial, and economic - that do not now use them at all, will be greatest. The tools are at hand, but we must train these interpreters, and recognize the unique role that they will have to learn. They must be scientifically and meteorologically sound - but much more, problem solvers, as well. They must be trained in that.

In 1980 the Select Committee on the National Weather Service of the National Research Council of the USA said "A major study is needed to define the public education required to raise the level of awareness of the excellent weather information currently available. Once additional interest and a higher level of use are generated, the feedback will enable the NWS to tailor and modify its products in accord with user needs. There is little or no evidence that this loop has been closed today. Even in those areas where efforts to improve dissemination are clearly evident it is not evident that studies have been done to assess the effectiveness of these efforts.

Productivity increases and reductions in losses will be in proportion not only to the quality of the information available, but also to how well the consumers know what information is available, how they may gain access to it, and how to use it. A series of user manuals should be produced for wide distribution to all users. These should provide clear, concise information on the types of NWS services available, procedures for obtaining them, and information on interpretation and use of the products. The design of future dissemination systems must take into account the broad spectrum of users and recognize the similarly broad spectrum of types of data and information that must be supplied. All members of society are the beneficiaries of dissemination of the information, because it results in better planning, reduced losses, and higher efficiencies and productivity for the society. Consequently, a rational basis must be provided for addressing the needs of all classes of users according to the ultimate benefits to be derived".

In summary, no forecaster should leave a training program to go into the field without being sensitized to the fact that producing good forecasts is difficult and crucial, but not of much value unless the forecasts are made in the context of the types of activities to which they will be applied and are communicated in a timely manner in terms the user will understand.

This necessitates that the forecaster be taught the economic geography of the region he will be forecasting for, visit the activities and people who will use the forecast and listen to their operational problems. This knowledge can be acquired through lectures from users themselves, forecasters experienced in the region or through an internship in the region. At the very least, a paper should be required from the trainee which discusses the economic impacts and weather related health and human safety concerns of his forecast region. Special meteorological needs, forecast frequency, derived parameters (evaporation, etc.) should be spelled out. Research for this paper could be based upon first-hand experience in the region and library research. An accumulation of these papers at the regional training centers could serve as course materials for future classes.

Accordingly, I propose that we have a new emphasis in our training programs based on the need for scientific interpreters, people who can learn what is available in the meteorological system and what the needs in the national economies are, and who can bring the former to bear on the latter. Our personnel who work with users need meteorological training, communications training, and problem-solving training. Our investments in atmospheric science will pay off only when we view communication and problem solving for users as important as being able to understand atmospheric physics and dynamics.

REFERENCES

Mason, B J *Weather*, November 1966.

Maunder, W J *Weather*, January 1973.

8

The roles of universities and meteorological training institutions in the education of users and the general public

J. R. Milford, Department of Meteorology University of Reading

INTRODUCTION

The rational planning of any educational activity must start with an analysis of the initial knowledge and skills of the target group and also of the final state which is required. When the material and skills which have to be imparted are defined, the planners can select from the available resources the best people to do the teaching, and the most appropriate methods and materials. It is at this point that the roles of universities and other higher education institutions, and of meteorological training institutions should be discussed: there are certain tasks which each of these groups are best able to carry out but there are others which can better be taken on by schools and learned societies. Such a rational and coherent approach hardly seems possible, however, when we discuss the education of users and the general public because both the starting points and the desirable ends vary so widely that educational recommendations can only refer to very specific example, even within one nation or culture. Even if we cannot generalise, a constant question in the discussions should be 'Who needs to know what about the weather?', as emphasized by Professor Lucas (1985).

Discussed here are some general principles which are relevant to the way in which we tackle the task of educating users and the general public, and illustrate them with specific examples. They are necessarily strongly biased towards my experience in the UK, or as a visitor to other parts of the world.

The two classes of people in the title, namely users and the general public obviously overlap: the need to educate the general public arises insofar as they are users of meteorological information. In fact, the most useful distinction we can make may be between those who need information about past, present or future weather and climate for use in their business or profession and those who need it for personal reasons, to enhance their convenience or enjoyment of life or reduce risk. For the first group, which I refer to as professional users there are financial gains which may be quantified, at least in principle: for the general public, assessment of the rewards of using meteorological information successfully will largely be subjective. For this reason and the consequent difference in motivation, and also because we generally expect a higher level of technical understanding from the professional users, the two groups often merit different approaches.

It might be helpful to distinguish a third group, including high-level decision makers, politicians and national leaders. It is they who must make the most onerous decisions, and make them across such a wide range of subjects that they have to rely on advice and their general knowledge rather than their own technical expertise. It is they who decide on the resources available for the development of our subject, and how it will be organised within each nation. Little can be done to educate people at this late stage of their careers, except through advice on specific problems. While it is very important that the environmental aspects of decisions, including the climatic ones, be represented properly, this will only come about if education of the leaders and their advisers has been effective while they were still members of the two classes distinguished earlier.

PROFESSIONAL USERS

We may assume that professional users of meteorology recognise a need to know a certain amount about the subject, but that they will normally hope to learn the absolute minimum. The meteorologist with whom they discuss a problem is likely to feel that substantially greater knowledge than this minimum would be useful, and would enable the user to ask the right question more often. A compromise must be reached, because the professional can afford to spend only a limited amount of time on his or her general meteorological education, and also because knowledge which is not clearly related to current activities is much less likely to be absorbed, however clearly it is presented.

Much of the education of the professional user has to be done 'on the job' during discussions of particular problems with meteorologists acting as consultants: this is undoubtedly an important role for members of the national services. The service must choose the right kind of person for the task, with a readiness to be educated in other people's fields during such discussions. Only mutual understanding will lead to effective transfer of information. Because such mutual education is not a rapid process, the meteorological service has to take a long-term view, and not insist on immediate returns from every discussion. The provision of services to the oil industry in the North Sea was an example where a long process of mutual education was obviously necessary: in other cases the case may be harder to make but no less valid.

On occasion it may be possible to identify a group with a need for a substantial block of common knowledge, and it is then appropriate to consider more formal educational events. These may be single seminars, or short courses, or series of such events. In every case the duration, content and expected background must be clearly specified and the most appropriate format discussed. It may often be most effective for such courses to be given by members of universities, or of training institutions, but their expertise in the meteorology must be balanced against the users specialist knowledge of the ways in which it can be applied. No educational event of this kind will succeed unless it is planned by a group which includes representative users. One example of this was a two-day course on Meteorology for Engineers and Builders which was run annually for a time, based on the University of Reading. The organising committee consisted of a member of the UK Meteorological Office, a senior engineer and a member of the University staff (who was, incidentally, also a Secretary of the Royal Meteorological Society). The programme aimed at a balance between presenting some basic scientific ideas, giving information on what the meteorological service could provide, and describing cases where such information had been useful both in planning and in day-to-day management. Although the courses were appreciated by those who attended, and were undoubtedly of considerable use to the industries in the long run, the market in the UK did not sustain the series which lapsed during a period of recession in the engineering and building industries: because the benefits were essentially long-term, and not quantifiable they were easy economies for firms who might have sponsored staff on the courses.

Two groups of professional users who depend critically on weather information are those involved in aviation and in nautical affairs. These groups are so large, and the knowledge they require so extensive, that they have their own qualifications in meteorology among their other subjects: meteorology tends to be but one subject among many technical ones which have to be passed before full qualification is obtained. As a result it seems that the meteorology is often taught by people with rather little background in the subject outside the professional curricula which they are teaching. There is a clear case here for better coordination. There is also a case for ensuring that the education of amateur pilots and yachtsmen is coordinated with that of the professionals since the lives of the rescue services as well as those of the amateurs may be put at risk by ignorance.

THE GENERAL PUBLIC

In countries where leisure activities have assumed a major role in the economy there are substantial numbers of people involved in weather-dependent pursuits like sailing, hill-walking or gardening. Apart from these we must assume that the general public has little wish to be educated in anything scientific. If it is to be educated, it must be largely by stealth, and in very small doses. How far is it necessary? I have always assumed that one needs to understand the background to any new information one receives to be able to make the best use of it, but I realise this is the result of indoctrination in my early years rather than scientific evidence. It does seem that interpretation of public service forecasts can be improved by knowing that rain may be associated with fronts, or isolated thunderstorms, or the difference between air and ground frosts. In the UK, presenters of televised forecasts have the most obvious opportunity to introduce such ideas, and they often respond by referring to features such as the connection between the spacing of isobars and the wind speed. The broadcasting companies naturally judge how much of this they can allow on commercial rather than educational grounds. In the UK our independent television channels appear to have introduced more new ideas, particularly on improving forms of presentation but the British Broadcasting Corporation still has a tradition inclining to favour education. The news that a 90-minute television programme on meteorology was made in Tunisia to celebrate World Meteorological Day, 1986, shows that some other countries give the matter higher priority.

In many countries academics play a prominent role in broadcasting on meteorology, but in the UK the most important way in which higher education in the UK provides education for the general public is through their activities in adult, or continuing education. Every night of the week several thousand people attend classes on some topic of personal interest, and where individuals volunteer to provide courses on weather or basic meteorology they are usually well subscribed.

The most successful educators in meteorology, as in any

other subject, are enthusiasts with expertise; individuals in this class should be given the opportunity to communicate as often as possible. It may be that the best input which employers can make to educating the general public is to allow their staff who have the ability to communicate the opportunity to do so, even if the time spent cannot be justified in immediate financial terms. Of course, the role should not be reserved to professional meteorologists, though these will be needed for the majority of the education of other professionals. The complementary pool of expertise may be found in school teaching staff, and through the learned societies, as discussed by Walker (1987).

We should accept, however, that the members of universities and professional institutes, are likely to be in the best position to prepare teaching material. They should also be given the opportunity to check the material prepared by amateur enthusiasts for accuracy. Many serious errors have been put out with the best of intentions: ".. since the moisture in the air allows it to be regarded as a fluid, the equation is called a hydrostatic one" is one of my examples, and no doubt we all have our own horror stories. Provided that such mistakes can be avoided, however, amateurs may well be the better communicators, because they are more aware of the difficulties experienced by the layman: their input is invaluable.

In addition to avoiding errors, teaching material at any level should be checked by well-qualified meteorologists to ensure that no term which has a specialist meaning is used with a contradictory meaning at any level, no matter how lowly; if this is allowed, students who make good progress may later have to unlearn ideas, which is certainly an unnecessary burden. One example of this is the statement that an overcast night will be warmer than a clear one "because the thermal radiation from the ground is reflected back by the cloud": a student who learns this will have extra difficulty in understanding that clouds are black bodies in the thermal radiation region of the spectrum. There are also opportunities for confusion in the use of the word 'humidity' where the relative humidity may be high and the absolute humidity low (in Arctic air for example): at the risk of sounding pedantic we should make this distinction as early in the educational process as possible.

CONCLUSION

In considering the most appropriate roles of universities and meteorological training institutions in the education of users of meteorology, the dominant factor is that in them we expect to find the highest level of technical knowledge, and the best facilities for demonstrating equipment and experimental techniques, such as data logging and instrument calibration. The universities should also include the highest level of expertise at conducting special investigations and interpreting climate data, and staff are therefore often used in consultancies, which include an educational as well as an advisory role. They should also contain expertise in communications skills and techniques. The staff of the national training institutions are not likely to include the foremost scientists in specialist areas, but should be able to call in the experts from their national service when required. The role of the WMO Regional Training Centres has not been discussed explicitly here, and their brief does not include general education, but they can, and occasionally do, provide knowledge and expertise which would not otherwise be available in smaller countries.

It would seem that in either case, much education will be done on the job, but occasionally groups of people with similar educational needs may be identified, and short courses set up to meet them. Those who control the finances of either type of institution will have to be convinced that the activity is likely to be profitable, and their judgement will depend on whether a short term view is taken, or whether the intangible benefits of long term developments in the use of operational or academic meteorology are allowed to count.

The direct part to be played by the universities and meteorological training institutions in the education of the general public will inevitably be small, though their indirect influence will be substantial: this will be through their role in the education of broadcasters, teachers and other communicators, whose individual enthusiasm will be the most potent factor in communicating the ideas of modern meteorology. They should also be given the opportunity to ensure the technical accuracy of courses and teaching material at all levels.

REFERENCES

Lucas, A M, 1985: Who needs to know what about the weather?. In Weather Education (Walker, J M ed), Royal Meteorological Society 271 pp.

Walker, J M, 1987: The roles of schools and learned societies in the education of users and the general public.

9

The roles of schools and learned societies in education of users and the general public

J. M. Walker, Department of Maritime Studies, University of Wales Institute of Science and Technology, Cardiff, United Kingdom

INTRODUCTION

As Battan (1983) has shown so well, weather affects us all: it affects our moods and our health; it sometimes inconveniences us, or worse; and it controls agriculture, energy consumption and many of our spare-time activities. More than that, though, the atmosphere is vital for mankind: everyone must breathe: the atmosphere shields us from cosmic matter and harmful electromagnetic radiation; and the atmosphere moves. Without atmospheric motion there would be no weather systems; and without weather systems life as we know it would be impossible on this planet. No water would fall from the sky, and there would be no agriculture. No-one can survive without water and food.

In principle, therefore, everyone in the world ought to be keen to learn about the atmosphere and its behaviour. In practice, however, most people take an interest in the weather and climate only when there is some obvious need to. They do so when disaster strikes, in the form of a tropical cyclone or storm surge, for example. They do so when there is a heavy snowfall or a prolonged heat-wave, and they do so when drought causes a water shortage. Indeed, they generally take an interest only when the weather is inclement, extreme or abnormal; otherwise, they tend to take weather for granted.

Why this should be so need not concern us here. Let us merely recognize that scope exists worldwide to improve meteorological awareness and literacy. In the words of 'Public Weather Awareness and Literacy - A Call for Global Action', the declaration agreed by those who attended the First International Conference on School and Popular Meteorological Education, held at Oxford, United Kingdom, in July 1984, "adequate awareness and understanding of weather and climate improves the quality of life and helps to mitigate the effects of weather-related hazards" (see also Green, 1957; Kraus, 1957).

The declaration specifies some courses of action by means of which general weather literacy may be improved, and it concludes with the words, "Perceived additional benefits from the actions outlined are greater interest in the practical applications of science and mathematics in schools and a broader scientific appreciation by the general public". As Eugene Bierly noted in the Foreword to Weather Education, the Proceedings of the Oxford Conference (Walker, 1985), there is growing concern over the inadequacy of science and mathematics education in countries around the world.

A further problem for meteorological educators has been put thus by Smith (1964): "[The man in the street] often has no knowledge of the scientific methods employed by meteorologists, and he is often convinced that he himself is an expert. Certainly he is liable to be more forthright in his opinions than any scientist. He is sure that the amateur can beat the professional, especially in the realm of weather forecasting. He will believe old sayings and old fallacies with a tenacity that is the envy of any politician or advertising manager. His memory of the weather is fallible and selective in the extreme, and no amount of official evidence will convince him he is mistaken. He can be difficult to contact, hard to convince and yet he must never be ignored".

The education of users and the general public therefore presents challenges. Let us now proceed to consider ways and means of tackling the problems. In so doing we must remember that climate, cultures, educational systems and other local circumstances vary from country to country. Accordingly, a teaching technique which is appropriate in one country may not be in another. Nevertheless, there is much in education that is generally applicable. In this paper there is a bias towards the British experience.

THE EDUCATION OF YOUNG CHILDREN

Education begins at birth and continues throughout life. We shall ignore the role of parents, however, and consider first the formal education of children up to the age of eleven or twelve years. This is an age group of considerable importance, for when they are in it children are passing through their most formative years. Moreover, they are not yet involved in preparations for school-leaving and college/university-entrance examinations, with all the demands, limitations (and traumas) these entail. The opportunity to mould sensible attitudes to the natural environment should certainly not be missed.

At the Oxford Conference there was a consensus that endeavours to improve public weather awareness and literacy should be aimed largely at children; and it was agreed that the emphasis should be upon exciting curiosity about the weather, developing an enquiring mind, encouraging observations and generating a sense of wonder at atmospheric behaviour. Young children are naturally inquisitive. This characteristic should be exploited. Children should be encouraged to ask questions. Why is it sunny and warm today when it was cloudy and cold yesterday? What caused the rain that stopped me going out at playtime this morning? How is it possible for lumps of ice (hailstones) to fall out of the sky? Why was there a circle of light (halo) around the moon last night? Are snowflakes really all different?

Of course, much depends upon the enthusiasm, imagination and inspiration of individuals interacting with children in the classroom and out of doors. It should not be assumed, though, that these individuals need necessarily be professional schoolteachers. As Floor (1985) has shown, in a paper presenting his approach to teaching the subject of rainbows to Dutch children aged 10-12, professional meteorologists also have a rôle to play. His approach was very simple: first he asked each child in the class to paint a rainbow; then he discussed the results with the children, correcting their misconceptions and acquainting them with the real characteristics of rainbows. In Floor's words: "The purpose of the lessons was to increase the pupils' enthusiasm for rainbows and, more generally, for the beautiful phenomena that can be observed in the open air. The lessons were also meant as an exercise in making careful observations. The physical explanation, of course, was beyond the scope of these primary school lessons".

Without saying so, Floor exploited another attribute of children: they like to be active. He did not ask the pupils to sit passively in front of him for a while to learn about rainbows. He asked them to do something they particularly liked doing. He asked them to paint a rainbow. The he discussed their work. Thus, he gained and held their interest. Similar results can be obtained by exploiting children's enthusiasm for making things. Simple equipment can be made in the classroom and used out of doors. A primitive wind vane is not difficult to make; an anemometer can be made using ping-pong balls; and the sizes of raindrops can be studied from the marks they make on blotting paper. At the primary-school level scientific rigour is not important. Arousing interest and generating enthusiasm certainly is. In any case, a competent teacher will, as a matter of course, point out shortcomings of the simple equipment.

At the Oxford Conference it was also agreed that children's studies of weather and climate should be incorporated, to a considerable extent, within broad scientifically-based programmes of environmental education. As was noted in Public Weather Literacy World-wide: Strategies for the Future (Walker, 1985, pp 248-251), interest in weather and its vagaries might be simulated by studies of relationships between agriculture and the weather, gardening and the weather, pollution and the weather, and so on. It might also be stimulated by studies of the weather to be expected in various holiday localities or the effects of weather on sporting activities (for the latter, see Thornes, 1977; Cairns, 1984).

Such studies are ideal for project work; and, judging by the work of Ilsley (1985) and the large number of enquiries which the Royal Meteorological Society receives from children seeking material, the weather is already a popular subject for such work, at least in the United Kingdom. Typically, letters are of the form: "Dear Sir, I am nine years old and I am doing a project on the weather. Please will you tell me all you know. Love from Mandy"! For answering such letters satisfactorily and assisting teachers seeking material for projects, there is a need for information sheets, booklists, posters, booklets and explanatory leaflets. The language used in this material should be as simple as possible and free from jargon, not least because many teachers know little more about meteorology than the children they are teaching.

It follows that the material must be prepared by persons who are meteorologically qualified to do so. Here again, professional meteorologists have a rôle to play. However, the assistance of teachers who are familiar with the children's capabilities is essential. At all levels in education it is important that teaching material not only meets the needs of intended users but also does not contain misconceptions or outmoded ideas.

THE EDUCATION OF OLDER CHILDREN

Studies of relationships between weather and human activities should not cease when children reach the age of 11 or 12. These relationships should prove interesting to older children (and to adults too). Indeed, some relationships are too complex or conceptually too difficult (or both) to be fully appreciated by young children. Such is the case, for example, with the rôle and influence of weather and climate in economic and social activities (see Maunder, 1971) and with portrayals of weather and atmospheric phenomena in art and literature (see Burroughs, 1981; Nyberg, 1984, Thornes, 1984).

In the education of children older than 11 or 12, though, the emphasis tends to be placed chiefly upon preparing for the examinations which must be passed to satisfy employers and college/university admissions tutors. There is no longer much scope for studying weather and climate purely for their own sake. After the age of 12 or so children's education becomes increasingly dominated and constrained by the syllabuses for public examinations.

For examination purposes meteorology has traditionally formed part of geography, and this will continue to be so for the foreseeable future. In fact, without geographers comparatively little meteorology would be taught in the world's schools, colleges and universities. Nevertheless, meteorology is not only a component of geography but also a branch of physics, and there are compelling reasons for integrating meteorology into the science curriculum. Here we turn to the pages of Weather Education once more, for a list of topics which are suitable for inclusion in the school science curriculum (see also Inspectors of Schools, 1956). The topics are suitable because they serve to stimulate observation and scientific reasoning among pupils and help bring physics and everyday lift together. In the words of Weather Education (Walker, 1985, pp 250-251), they are as follows:

(i) Observations and their analysis The use of instruments - precision and accuracy; principles of instrument design; construction of simple instruments; exposure and representativeness of the instrument site; setting up of a small weather station (see Pedgley 1980); data collection (manually, by chart recorders, by computer - see Sparks & Summer, 1984: Jenkins, 1985); simple quality control procedures; presentation of observations (graphs, wind-roses); statistics of local weather - correlations and regressions between variables; analysis of observations (e.g. by statistical techniques and graphical representation - ideal computer applications).

(ii) Laboratory demonstrations Many topics which are studied in physics courses can be shown to have relevance to everyday life when related to the 'weather machine'. Examples are convection, condensation and electromagnetic radiation. Laboratory demonstrations should be backed up with observations of atmospheric behaviour outdoors.

(iii) Experimental projects Some local meteorological phenomena can be investigated using relatively crude equipment. Examples are: raindrop size-spectra (using filter paper) and eddying around buildings and vegetation (using soap bubbles or smoke). Profiles of wind and temperature (the latter above and below ground) can also be studied, using inexpensive commercially-available instruments. Such investigations can form well-defined projects which can be completed in a limited period.

As a means of linking geography and physics through the teaching of meteorology, an innovation of the British-based Associated Examining Board (AEB) represents an important development: in 1980 the Board introduced Meteorology as a General Certificate of Education (GCE) subject, offering it at Ordinary Level (Milford, 1980, 1985; Griffiths, 1981). Examinations at this level are normally taken by pupils around the age of 16.

Using the number of candidates as a measure of success (180 in 1980, over 800 in 1986), the subject has certainly gone from strength to strength. However, those who constructed the Meteorology syllabus hoped, as Milford (1985) put it, "to combine the best from the traditions of both Geography and Physics teaching and provide an occasion for teachers of both subjects to come together, apply physical reasoning to events in our environment, and interpret the observations which have been analyzed for so long under the heading of climatology". In practice, this has proved to be too idealistic. It seems to be rare for physics and geography teachers in one school to combine to teach the GCE Meteorology. The majority teaching the subject have either been geographers in schools or instructors attached to groups of Air Cadets.

In respect of the AEB examination in Meteorology, a further development in the United Kingdom is the demise of the GCE Ordinary Level and its replacement by the General Certificate of Secondary Education (GCSE). Fortunately, Meteorology has been retained as an examinable subject and it will be offered from 1988 by a new body, the Southern Examining Group (SEG), which incorporates the AEB. Jointly with Oceanography, and intended particularly for persons interested in marine and maritime matters, Meteorology will also be offered by the London and East Anglian Examining Group.

For a number of reasons, the introduction of the SEG GCSE Meteorology is a welcome development. First, as Perry (1986) has noted, the new examination has been designed to cater for a much wider range of abilities than its predecessor and care has been taken to ensure that the papers cater not only for the most able student but also for the less able. Second, the greater emphasis on teacher-assessed course work than in the GCE will, as Perry put it, "allow the assessment of skills and qualities which traditional examinations cannot easily assess, for example the ability to make accurate observations and to relate them to the general weather situation". Third, the syllabus for the GCSE Meteorology has been designed, according to the Southern Examining Group, "to provide an introduction to a distinct body of knowledge and associated skills which will enable candidates to acquire a better understanding of the atmospheric environment in which they live". The ability of a candidate to evaluate the impact of weather and weather forecasts on society is specified as one of the assessment objectives.

Whatever the level of meteorological education, GCSE or otherwise, correct concepts must be taught. This point was made earlier, but is worth repeating because it is so important. Substandard teaching materials cannot be tolerated. Books, visual aids, computer software and so on must not contain misconceptions or outmoded ideas. At the present time far too many aids for teaching meteorology in schools are thus flawed or deficient. Moreover, teachers themselves must be acquainted with correct concepts and kept abreast of developments in understanding of atmospheric behaviour.

Again, professional meteorologists have a rôle to play, for when they review books and other teaching material, such as computer software (see Cornford & Clisby, 1983; Diver & Readings, 1986), they are duty-bound to draw attention to imperfections. In this respect, learned societies also have a rôle to play, because many of the journals which commission and publish reviews are controlled by such societies. Criticisms of teaching material should not be confined to the pages of journals, however; members of the education committees of learned societies, and, indeed, everyone else, should be vigilant and make every effort to inform authors and publishers of shortcomings of educational material.

The magnitude of the task should not be underestimated, for long-cherished concepts, even those which are readily disproven, can be extraordinarily tenacious and difficult to correct or eradicate. The traditional explanation of the foehn effect is a case in point: precipitation on the windward slopes is not essential for the occurrence of a foehn. We must be optimistic and persistent, though, and hope that words like those of Warren (1987) can never be applied to meteorology. "There are scores of fundamental concepts in physics which are commonly taught wrongly, stated incorrectly in increasing numbers of textbooks and reference books, and required to be presented wrongly in examinations. Criticism has proved useless. It is just 300 years since Newton gave the first systematic accounts of force, gravitation and waves in the Principia. In schools now these essential concepts are almost invariably taught incorrectly, as if Newton had never lived. The ideas taught about energy are quite literally incoherent nonsense. ... The situation is disastrous and is rapidly becoming even worse. This has been a generation of educational change. Either by accident or design it has been ensured that no change which would correct inaccurate teaching has been undertaken. Radical reform is urgently needed."

Learned societies can also help improve the teaching of meteorology in schools by arranging lectures for pupils and organizing seminars, workshops and short courses for teachers, student teachers and the teachers of teachers. They can help in other ways too. For example: they can exhibit at fairs and conferences; and, making use of members who are willing and suitably qualified, they can initiate, commission, supervise and co-ordinate activities such as the writing of books and articles and the preparation of leaflets, wall charts, slide sets and other teaching material. The 'Schools Supplements' which were published in the magazine Weather every month from October 1961 to September 1963 and from September 1985 to April 1986 particularly come to mind; Satpacks 1 and 2, are also, like Weather, published by the Royal Meteorological Society. These are sets of exercises concerned with the interpretation and understanding of satellite imagery in terms of surface data, upper-air ascents and synoptic analyses; and they provide a novel approach to the study of extratropical depressions in the classroom.

School Corporate Membership of the Royal Meteorological Society is also available, carrying with it a number of advantages: the nominated representative receives a copy of Weather each month; any member of the school (teacher or pupil) may attend Society meetings on the same basis as Fellows; any member of the School (teacher or pupil) may attend Society field courses at reduced rates; the Society undertakes to provide advice, as required, on matters concerning careers and education in meteorology; and the Society undertakes to supply, as required, the names of lecturers willing to speak on various meteorological topics. It is a further benefit of corporate (and personal) membership that the Society is able to grant small sums of money (normally up to about £200 per project) to support scientifically-based meteorological activities which are deserving of support but cannot attract financial sponsorship from other sources. The Society can also assist such activities by lending from its stock of instruments and apparatus.

THE EDUCATION OF ADULTS

One's initial impression may be that the number of adults needing a meteorological education of any sort is rather small. Amateur and professional mariners and aviators certainly need one; and so, perhaps, do farmers and professional cricketers, because their livelihoods depend to a large extent on the weather. It is not immediately obvious, however, why an office worker who spends all day indoors behind a desk in a centrally-heated or air-conditioned office and who takes his or her annual vacation in a resort where unbroken sunshine is almost guaranteed should bother to learn anything about meteorology. In any case, after a day at work, such a person may well be too tired or too busy with domestic commitments or leisure activities to want to exercise his or her brain unduly.

In many spare-time activities, though, weather is an important factor. It can make or mar a drive in the country, a walk in the park, a game of tennis, a session of gardening, a day at the seaside or a morning of bird-watching; and it can be a vital factor for the yachtsman, hill-walker or mountaineer. It affects domestic chores too. It is frequently on the mind of the housewife, for example, who would rather dry the washing out of doors than in a tumble-drier or in front of a fire or radiator. And there are more compelling reasons why the general public should take more than a passing interest in the atmosphere and its behaviour.

Those who live on coasts visited by tropical cyclones or storm surges, for example, need to be aware of the hazards and adequately prepared for them, as Friedman (1985) has stressed. Those fortunate enough to be able to choose a place to live should also consider weather. Is the house on a flood plain? Is the locality frequently foggy? Will my wife's catarrh, or my daughter's asthma or my father's bronchitis be alleviated or aggravated? How often would that idyllic little village be cut off by snow? And what is the weather like in winter at that seaside resort which is so picturesque and tranquil in summer?

There is clearly a need to educate the general public meteorologically. However, the educational techniques employed in schools tend to be inappropriate at the adult level. As Smith (1964) noted, "an adult person is not generally willing to be educated in an obvious fashion and education has to proceed indirectly in a form of publicity whereby the new knowledge is assimilated in a concealed and painless process". It is also an important point that no attempt should be made to teach meteorology to all and sundry. There is a need to determine precisely what the populace ought to know about atmospheric behaviour, distinguishing carefully between meteorology and meteorological awareness. The difficulties of teaching meteorological principles should not be underestimated. Meteorology is so complex that a full understanding of the subject cannot be accomplished without a thorough grounding in mathematics and physics. Such expertise is beyond all but a few members of society.

To improve weather literacy among the general public there is a need for books, pamphlets and leaflets, articles in magazines and newspapers, courses of various kinds, and programmes on radio and television. Once again, learned societies have much to contribute, commissioning books and articles, preparing pamphlets and leaflets, running courses, providing advice, and urging radio and television companies to produce programmes about weather and climate. It was mentioned earlier that simple language must be used and jargon avoided when communicating with young children. This should be a golden rule when addressing the general public too. Unfortunately, as Holford (1985) has pointed out, the ability to communicate intelligibly to laymen is a gift possessed by all too few scientists.

For civil engineers, builders, architects, offshore engineers, farmers, hydrologists and others for whom at a professional level some knowledge of meteorology, particularly applied meteorology, is beneficial, there is much scope for learned societies to arrange courses, seminars and workshops and produce training material and other literature. For members of the general public, courses involving or initiated by learned societies fall into two broad categories: field courses such as those organized by the Royal Meteorological Society (see Walker and Riddaway, 1985); and evening or weekend courses organized in association with community adult education centres or with university departments of adult education or extramural studies. Within these categories there are two kinds of course: those which concentrate upon meteorology for its own sake; and those which relate weather to outdoor activities.

To reach the general public en masse radio and television must be used. Indeed, they are already used to a considerable extent, principally through broadcasts of weather forecasts, which not only provide information about forthcoming weather but also educate the listener or viewer surreptitiously. In some countries, notably the United States of America (see Geer, 1985), the cause of popular meteorological education is quite well served by radio and television. In others, however, notably the United Kingdom, there is much to be desired.

About five years ago the Education Committee of the Royal Meteorological Society thought television companies might be interested in weather documentaries or, perhaps, 'fillers' lasting five or ten minutes on specific topics, such as 'The beauty of snow flakes' or 'Why does autumn tend to be a season of mists and fog?'. Very little interest was shown. In view of this, and the apparently indifferent attitude of most British broadcasting companies to weather bulletins on radio and television as judged by the ludicrously limited amounts of time allocated to them, particularly in comparison with the amounts of time devoted to programme trailers, the only possible conclusion was that the companies generally feel no obligation whatsoever towards popular meteorological education. This is somewhat curious, given that the atmosphere is vital to mankind and the vagaries of weather affect humans in so many ways. Furthermore, programmes about the natural environment are among the best produced and most popular on British television. The problem may be that television producers believe that meteorological programmes lack popular appeal, that they will not make "good television", because they cannot always be as colourful or spectacular as programmes about animals, birds or plants.

In the course of his work as a television weather presenter Hunt (1985) demonstrated that pupils in schools and members of the general public can be used effectively as voluntary observers for providing spatial and temporal resolution of day-to-day weather patterns. Chaplain (1985) has demonstrated this too; and in Canada more than 2,000 volunteer observers from all walks of life and all age groups assist Environment Canada by observing and recording weather twice daily (see Environment Canada, 1984). Sometimes, moreover, as Ludlam (1961) and Pedgley (1971) have shown so well, members of the general public can also be enlisted as observers in scientific research projects. Used in this way, people are likely to take an interest in whatever it is they are taking part in, especially if they also gain publicity in the local newspaper or on radio or television. Children often broadcast very effectively; and a radio or television "appearance", however brief, will almost certainly impress relatives, friends and neighbours! At the primary school level, networks of weather observers can be established, partly to kindle interest in weather and climate and partly to carry out simple studies, of for example, local climate. Of course, expert supervision of such studies will normally be required.

CONCLUSION

The worldwide concern over the inadequacy of science and

mathematics education was mentioned earlier, in the Introduction. The need for such education was put thus in Science is for Everybody (Royal Society, 1985): "Science and technology play a major rôle in most aspects of our daily lives both at home and at work. Our industry and thus our national prosperity depend on them. Almost all public policy issues have scientific and technological implications. Public decision-makers, whether Parliamentarians, civil servants, leaders of commerce or industry or voters in a democratic society, therefore need to understand the scientific basis of their decisions. So, too, do private individuals going about their daily lives. Everybody needs some understanding of science, its accomplishments and its limitations, whether or not they are themselves scientists or engineers. Improving that understanding is not a luxury: it is a vital investment in the future well-being of our society". Although the promotion of meteorological education is worthwhile for its own sake, improving public understanding of science as a whole.

As was stressed in the report of The Royal Society (1985), improving that understanding requires concerted action from many sections of society, including the scientific community itself. We must be realistic, though. Many, many things can be done, but money, time and goodwill are limiting factors. Those who serve learned societies are probably encouraged to do so by their employers and are probably allowed to attend committee meetings during working hours. Otherwise, however, they are expected to work for the societies in their own time. It is my hope that universities, national weather services and other employers or persons possessing both the enthusiasm and the necessary skills to promote school and popular meteorological education effectively will recognise the importance and potential value of such education and accordingly, adopt policies of treating their staff as generously as possible in respect of time and resources. The potential benefits far outweigh the costs.

REFERENCES

Battan, L J, 1983 Weather in your life. Freeman, San Francisco, 230 pp.

Burroughs, W J, 1981 Winter landscapes and climatic change. Weather, 36, 325-357.

Cairns, J A, 1984 The effect of weather on football attendances. Weather, 39, 87-90.

Chaplain, H R, 1985 Weather watching networks. In Weather Education (Walker, 1985), 221-232.

Cornford, S G, and Clisby, T, 1983 Weather and chips at school. Weather, 38, 111-114.

Diver, M and Readings, C J, 1986 Teaching meteorology with the aid of microcomputers. Weather, 41, 265-268.

Environment Canada, 1984 Volunteer Weather Observers. Atmospheric Environment Service Fact Sheet, Environment Canada, ISBN 0-662-12386-7, 4 pp. This is but one of a number of Fact Sheets published by Environment Canada. The Atmospheric Environment Service actively promotes school and popular meteorological education in a variety of ways. Their publications include The Atmosphere, Weather and Climate (a set of fact sheets for teachers and students, 55 pp.), Knowing Weather a book for students, 27 pp.) and Mapping Weather (a book for teachers, 31 pp).

Floor, C, 1985 The rainbow in the classroom of the primary school. In Weather Education (Walker, 1985), 134-138.

Friedman, H A, 1985 School-based and community-wide education and public information programs to increase tropical cyclone awareness and preparedness. In Weather Education (Walker, 1985), 79-85.

Geer, I W 1985 Use of the media for meteorological education. In Weather Education (Walker, 1985) 106-109. Professor Geer has also published, in 1980, A M Weather for Teachers (US Department of Commerce, National Oceanic and Atmospheric Administration, OA/W116, 25 pp) and, in 1981, Increasing Weather Awareness with NOAA Weather Radio (National Weather Project, State University of New York College at Brockport, 15 pp).

Green, S W, 1957 The place of meteorology in liberal education. Weather, 12, 301-310.

Griffiths, W A, 1981 'O-level' Meteorology in practice. Weather, 36, 260-263.

Holford, I, 1985 Communicating with the public about the weather. In Weather Education (Walker, 1985), 110-111.

Hunt, T L M, 1985 A personal account of the establishment of a regional television weather service. In Weather Education (Walker, 1985), 113-116.

Ilsley, M M, 1985 A preliminary survey of weather education in primary schools in one local education authority. In Weather Education (Walker, 1985), 61-69.

Inspectors of Schools, 1956 Meteorology in secondary schools Weather, 11, 305-318.

Jenkins, G J, 1985 Using a BBC micro in an automatic weather station. Weather, 40, 380-384.

Kraus, E B, 1957 The place of meteorology in liberal education. Weather, 12, 171-182.

Ludlam, F H, 1961 The hailstorm. Weather, 16, 152-162.

Maunder, W J, 1980 The Value of the Weather. Methuen, London, 388 pp

Milford, J R, 1980 'O-level' Meteorology. Weather, 35, 328-329.

Milford, J R, 1985 Meteorology for the General Certificate of Education. In Weather Education (Walker, 1985), 73-78.

Nyberg, A, 1984 An old painting of a halo phenomenon in Stockholm Cathedral. Weather, 39, 84-87.

Pedgley, D E, 1971 Some weather patterns in Snowdonia. Weather, 26, 412-444.

Pedgley, D E, 1980 Running a School Weather Station, 2nd edition. Royal Meteorological Society, 12 pp.

Perry, A H, 1986 GCSE Meteorology. Weather, 41, 359-360.

Royal Society, 1985 Science is for everybody. Summary (4 pp) of The Public Understanding of Science. The Royal Society, London, 41 pp.

Smith, L P, 1964 Weather and Man. World Meteorological Organization, Geneva, WMO-No.143.TP.67, 80 pp.

Sparks, L and Sumner, G, 1984 Micros in control - on-line weather data acquisition using a BBC microcomputer. Weather, 39, 212-218.

Thornes, J E, 1977 The effect of weather on sport. Weather, 32, 258-268.

Thornes, J E, 1984 Luke Howard's influence on art and literature in the early nineteenth century. Weather, 39, 252-255.

Walker, J M (Editor), 1985 Weather Education, Proceedings of the First International Conference on School and Popular Meteorological Education. Royal Meteorological Society, 271 pp.

Walker, J M and Riddaway, R W, 1985 Royal Meteorological Society field courses. In Weather Education (Walker, 1985), 98-105.

Warren, J W, 1987 Letter in Physics Bulletin (The Institute of Physics), 38, 128.

10

The introduction of agrometeorology into the curriculum of agricultural technical secondary schools

E. V. Terentev, (USSR)

No other field of man's activity is more closely connected with climatic conditions than agriculture. Weather and climate influence not only the growth and development of agricultural crops. They define the yield quality, the agricultural machines effectiveness, the usefulness of fertilizer, the conditions for the spreading of pests and diseases of crops and animals, etc. Quite often during certain periods meteorological conditions favour agricultural production but during others the same factors may cause great damage. Weather and climate have always attracted great attention from those working in the field of agriculture.

High yield of agricultural crops is possible only under the optimal conditions necessary for the matter exchange in plants beginning with photosynthesis and up to the end-product formation. Such conditions may be created by the changes in the nature and sequence of agricultural procedures. In any case, the total combination of weather conditions (rainfall, air temperature, radiation flow) plays a decisive role in the choice of agricultural technique. Therefore experts in agriculture have to have a good understanding of the available hydrometeorological data, to be able to estimate correctly the influence of meteorological conditions on the life and development of plants, to know the specific features as well as the reliability of agrometeorological forecasts and calculations and to be able to evaluate objectively the role of other meteorological factors in the process of yield formation. In addition, it must be noted that a wide introduction of automatic control systems into agricultural production greatly increases the significance of hydrometeorological information. Automatic control systems are known to incorporate the achievements of a great number of overlapping sciences in a certain field. In agricultural practice such systems are intended to make a wide use of quantitative yield formation models with the help of modern computers. They help to process rapidly a great amount of information concerning the factors influencing the growth and development of plants and to outline optimal variants of agronomical practice to get a planned yield.

It is quite evident, that the successful use of hydrometeorological information in agriculture depends on the thorough knowledge of specialists in this field.

Several decades ago agrometeorology was developed - a science treating meteorological, hydrological and climatic conditions as applied to objects and agricultural production processes. The task of agrometeorology is to make possible a rational use of these conditions in agricultural production, to develop scientific methods to overcome unfavourable climate and weather phenomena and to contribute to a high yield of crops and greater productivity in cattle-breeding.

Agrometeorological observations enable the study of agrometeorological conditions influencing the growth of crops and pasture. They also make possible the examination of the conditions necessary for farming and cattle grazing. Agrometeorological data and forecasts help to find the solution to a number of questions in agricultural production concerning both the evaluation of formed and expected agrometeorological conditions as well as the agrometeorological substantiation for the organisation and carrying out of agricultural work.

In the USSR the training of specialists in agrometeorology is organised in special educational institutions. The Odessa Hydrometeorological Institute trains engineers in the field of agrometeorology, a number of technical secondary schools train specialists in agrometeorology with a secondary education. Graduates of these educational institutions are generally employed in the system of the national hydrometeorological service and carry out the processing, control and generalisation of the compiled agrometeorological data. This is quite a complicated task. At present agrometeorological observations cover a wide spectrum of facts and phenomena. Together with the conventional measuring methods, modern non-contact measurements of different agrometeorological characteristics are widely used, such as soil humidity, crop density, leaf surface size, etc.

The generalised agrometeorological information includes agrometeorological forecasts and the descriptions of agrometeorological resources. The qualified use of such broad, specialised information obviously requires a certain level of training in the field of agrometeorology and overlapping sciences. Therefore, the curriculum in agricultural technical secondary schools contains the discipline "Agricultural Meteorology". The name of the discipline is quite conventionally used; it should not be taken for "agrometeorology" which is included together with some problems of general and special meteorology into the subject of agricultural meteorology.

Here is a representative curriculum on agricultural meteorology used in the agricultural technical secondary schools.

1. The general description of the earth's atmosphere and its characteristics.
2. Solar radiation and its role in agricultural practice.
3. Heat balance of the air and soil surface.
4. Water balance in the atmosphere and soil surface.
5. Atmospheric pressure patterns and wind.
6. The main weather phenomena unfavourable for agricultural practice and ways to overcome them.
7. Short-range weather forecasts.
8. Some problems of climatology.
9. Agrometeorological forecasts and calculations
10. The Agrometeorological Service.

The characteristic feature of the discipline "agricultural meteorology" is the sense of practical purpose. The students learn the quantitative descriptions of the weather and climate elements, as well as their combinations affecting the yield and quality of agricultural production. The stress is laid on the practical mastering of the principle agrometeorological observations. The students study thoroughly the specific features of an agrometeorological service, including the preparation and compiling of the simplest agrometeorological forecasts (available soil moisture storage, phenological forecasts, etc.). They also learn to understand the probability forecasts, that is, the agrometeorological calculations (probability frost forecasts).

The general system of education in the USSR also comprises a wide network of vocational schools. Some of them train specialists for agriculture. The students of such vocational schools master the main concepts of general meteorology and agrometeorology through the discipline "nature study" or some other courses connected with their future profession.

In 1984 the reconstruction of the educational system was implemented in the USSR. First, the reforms of the comprehensive and vocational schools were started. Two years later the reconstruction embraced higher and special secondary education. The main task of the reconstruction was the creation of a whole uninterrupted educational system to meet the requirements of modern production in a period of scientific and technological revolution.

11

Contribution of the Ecole Nationale de La Météorologie towards the training of users of road meteorology

F. Lalaurette, Ecole Nationale de La Météorologie (France)

1. OVERALL SCHEME FOR ROAD METEOROLOGY (SEMER)

The road meteorology plan was drawn up jointly in 1985 by the Department of Roads (DR), the Department of Road Traffic and Safety (DSCR) and the National Meteorological Office (DMN), and was the result of a realization of the impact that meteorological conditions were having on road traffic in the following areas in particular (see Figure 1):

- Road accidents;
- Measures taken to combat snow and ice;
- Imposition of restrictions on heavy goods traffic on thawing roads;
- Roadworks (particularly involving earthmoving);
- Fog (accidents - traffic);
- Traffic news.

The economic stakes involved in taking high-quality meteorological forecasts into account are high: (necessarily arbitrary) reductions of 1% in accidents due to fog and ice, 1% in the cost of winter road clearance services and 1% in the cost of thaw damage would mean savings of the order of 20 million French francs per year nationally. This realization went hand-in-hand with the realization that the meteorological service, with the following means at its disposal, had the capacity to meet some of the needs expressed:

- Fine-mesh numerical forecasting models (PERIDOT 31 km);
- The METEOTEL system for displaying and processing satellite and radar images and data from the PERIDOT model;
- The ability of regional forecasters to make local adaptations, possibly backed by statistical adaptation models or expert systems.

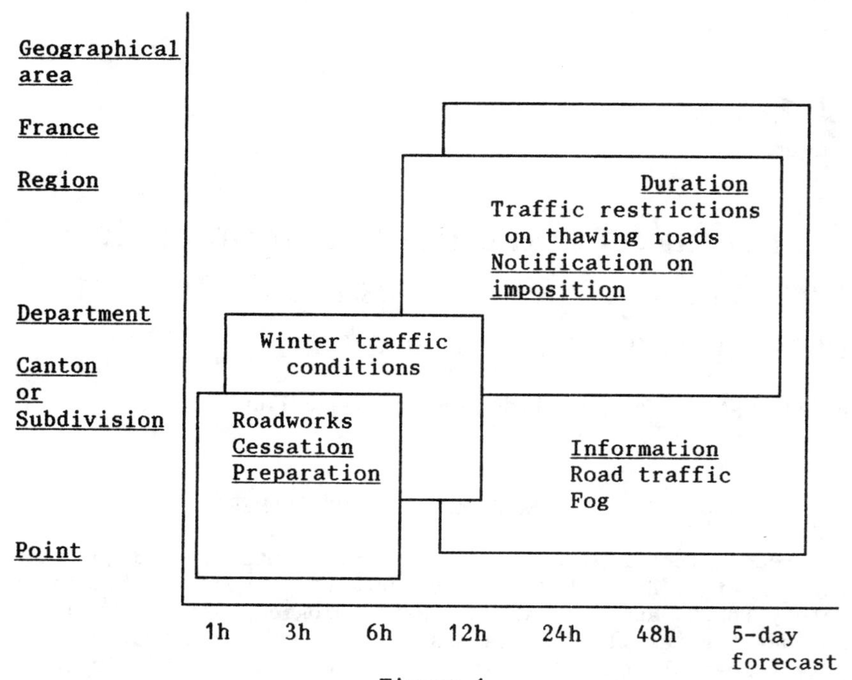

Figure 1
Roads and meteorological forecasts
Summary Table of requirements

The purpose of the SEMER plan has therefore been:

(1) To optimize the provision of weather information for use by the road services (compilation of customized bulletins, provision of satellite and radar images, a fresh integration of the PERIDOT on the basis of the 1200 UTC network);

(2) To draw up a programme of directions in research and development towards providing better coverage of short and long-term needs.

Training of road service personnel to make optimal use of

the new meteorological data made available to them (images, "professional"-type bulletins) has, as one might expect, been found to be an important part of the plan.

2. THE TRAINING SECTION OF THE SEMER PLAN

2.1 The four levels of training

An examination of the training needs arising out of the SEMER plan showed that there were four levels:

1: Meteorology for beginners;
2: Refresher level;
3: Revision level;
4: Basic training (vocational colleges).

Level 1 is aimed at a large audience of users with an involvement in roads (equipment operators, motorway companies, road traffic news centres and so on). The objective of level 1 is to provide a general knowledge of French meteorology (methods, organization and vocabulary) and develop the ability to read and interpret basic meteorological documents (surface maps and bulletins) in synoptic situation terms and also in terms of approximately what the weather implications are. This level is run by the regional meteorology services (there is therefore a direct link with those responsible for regional forecasting) and lasts about two days.

Level 2 is run by the Ecole Nationale de la Météorologie. Details are given below.

Levels 3 and 4 are still in the planning stage.

2.2 Level 2 (refresher) at the Ecole Nationale de la Météorologie (ENM)

Only one training course, for 12 trainees, has been run so far.

2.3 The training course

The initial aim of the course was to train one or two members of staff from each centre (regional road traffic information centre, road services offices in the departments) who would be called on to make decisions on the basis of meteorological information of a traditional or more recent kind (bulletins or satellite and radar images). These people would have taken a beginners' course (level 1) and would be selected by their degree of interest in meteorology. The teaching level would be at the school leaving stage. The objective set for the first course was to give the trainees a basic knowledge of the techniques of analysis, forecasting and local adaptation used by forecasters along with techniques for interpreting satellite and radar images on the bases of regional bulletins, making it possible for them to monitor changes in the situation in the very short term. They would also be made aware of road traffic problems affected by weather conditions: this was to be done with the assistance of experts in winter road conditions from the road research centres (Centres d'Etudes Techniques de l'Equipment). On the basis of this information, the ENM put together a two-week training course as follows.

METEOROLOGICAL REFRESHER COURSE

(DSCR)

EMM/Toulouse, 6-17 April 1987

I Large-scale atmospheric motion

1. General circulation concepts
2. The scales of atmospheric movement
3. Standard characteristics of the atmosphere
4. Air masses: origins, evolution, conflicts

II Effects of atmospheric movements on weather type

1. Ground observation charts
2. A particular situation: 11 to 15 December 1986
   - The various types of weather observed
   - Their relationship to upper air movements
   - Fronts: positioning criteria (types of weather, physical parameters)

III The major atmospheric equilibria: application to interpreting meteorological charts

1. Geostrophic equilibrium
2. Hydrostatic equilibrium. Thicknesses

IV Vertical structure of the atmosphere - variability - interpretation

1. The emagram (presentation)
2. Effects of rising air masses. Föhn effects
3. The situation from 11 to 15 December 1986: interpretation of radiosonde data:
   - Superposition of air masses
   - Stable and unstable layers:
     - Relationship with clouds observed;
     - Fog zones

4. Study of a storm situation: 16 September 1986
   - Synoptic situation
   - Generalization: types of favourable weather
   - Forecasting criteria. Limits
5. Study of a fog situation: 2 December 1986
   - Synoptic situation
   - Formation - dissipation: physical causes. Criteria for determination

V Numerical models for meteorological forecasting - interpretation
Illustration: situation from 14 to 17 January 1986

1. Characteristics of the models
2. Products available
3. Production of forecast charts and general guidelines
4. Adaptation by region and department
   - Use of fine-mesh model (Péridot)
   - Forecasting of local phenomena: Mistral, Autan, relief effects
   - Contribution of Meteotel
   - Regional guidelines

Four teachers took part in the training course. The resources used, apart from the standard environment of a forecasting centre (observed and predicted charts, model outputs), were two Meteotel terminals used by the trainees on selected situations.

2.4 Evaluation of the training course

Two snags were encountered during this experimental training course:

- A lack of motivation on the part of the personnel at whom the course was aimed (three applicants for 12 places), probably due to the length of the course. This shortfall was alleviated at the last minute by personnel from the road research services, whose motive was the impact of meteorological conditions on their areas of research;

- Objectives which were badly suited to the needs of road officials, who would be more motivated towards a refresher course in road meteorology (a study of the optimal use of meteorological information in the decisions they would be called upon to take) than in a refresher course in meteorology.

At the cost of adapting the teaching to the needs of the group of trainees, the course can be considered to have achieved its objectives. Moreover, it made possible a meeting between meteorology teachers and experts in winter traffic conditions (who were both trainees and participants in increasing awareness of road problems sensitive to weather conditions). These experts felt a need to meet for two days' work on skill sharing and to put forward some proposals concerning road meteorology (content of a bulletin of forecasts for traffic needs, impact of certain meteorological situations on the road traffic situation and so on). This should make it possible to give the various training courses a more practical content.

3. PROSPECTS FOR DEVELOPMENT

Despite the relatively major steps taken, road meteorology does not yet have a sufficiently solid content for it to stand alone as a subject for training courses. The implementation of the SEMER plan, and its training section in particular, had made it possible for meteorologists and road officials to set their knowledge against their requirements, and it is to be hoped that it will soon be possible to give road meteorology at least a part of the content.

The beginners' course now has quite a solid content, which is simply going to be strengthened from one to two days of basic training in interpreting Meteotel images.

The refresher course on the other hand will have to develop to move further into the forefront of road user needs.

- An initial version could take place as first planned, for road officials called upon to take decisions related to meteorological conditions using meteorological bulletins and radar and satellite images put at their disposal: this stage would have its "practical use of weather forecasts" part reinforced.

- A second version of this course would be aimed at the road research services, whose area of research is very affected by weather conditions. The course would have its theoretical and meteorological part strengthened, with the "road-meteorology" interface aspect coming down to the skill of the trainee.

4. CONCLUSION

The French SEMER plan is a global and original approach to the optimal use of meteorological information by a given sector. The implementation of the training section of the plan has given the Ecole National de la Météorologie problems of adapting to a new audience which is very different in its level of training and motivation to student meteorologists. The content of the courses offered is thus still developing so as to better reflect the users' expectations, and to provide them with more precise vocational elements.

12

Education of potential users of meteorological information in the field of water resources

L. Oyebande, College of Science and Technology, University of Lagos, Abeokuta

INTRODUCTION

Education and training of personnel at various levels is a prerequisite to any real progress in the national hydrological and water resources management activities. And just as water resources are indispensable for the development of a nation or group of nations, human resources are also important for such development. Human resources are often subdivided into manpower, information resources, (that is the body of knowledge available to all mankind), and the cultural resources which include the overall cultural background as well as the institutional structure which renders the human resources effective. All these three aspects of human resources should be developed through the education process.

Some factors have induced rapid changes and trends in water resources education. One is a growing awareness that water is a finite resource, vital to every function of an increasing and ever more demanding population. Another is the advancement of technology. The introduction of computer techniques has improved the speed, scope and depth of analysis the scientist may perform. He needs to be prepared to use this tool however, hence it has to be properly incorporated into the education process. It is necessary not only to master computer applications, but also to see the relationship of the computer to the whole methodology of problem-solving in the field of water resources. Also, water resources planning has changed with respect to objectives, criteria and manner of decision making as affecting civil projects. The changes in planning require that water resources education be broadened to include treatment of social, environmental, political and economic issues and concepts.

One of the basic principles of imparting knowledge is consideration of the psychological peculiarities and background of the students. The background of the students is known to have significant influence on their traits and reactions with respect to educational programmes in water resources. In the case of a civil engineering student, for instance, teaching should proceed from the particular to the general at each phase of the water cycle. This is because the student focuses on design and construction of various structures without being much aware of the roles played by the natural forces of wind, rain, ice and water in their design (Clausen, 1975). Thus, in order to retain his interest, it helps him to know that data on precipitation, evaporation, streamflow and groundwater are available and numbers can be used. This is in line with his interest in problem solving.

On the other hand, the science student has earlier been involved with the study of natural systems or phenomena. Thus in his hydrology and water resources education the student expects to know the how and why of the hydro-meteorological elements such as the wind, rainfall, ice and water as well as their physical and chemical inter-relationships. For these students, introductory courses in hydrology are helpful in retaining their interest, focusing on the occurrence of water, water balance of the world, the hydrological cycle and the regional and world water balances which show mean values of precipitation, soil moisture and runoff as it occurs as a function of climate.

NATURE AND OBJECTIVES OF EDUCATION IN THE FIELD OF WATER RESOURCES

Water resources development requires the involvement of persons of diverse backgrounds. Education to achieve such involvement in an effective way is no casual undertaking. If too much emphasis is placed on the interdisciplinary facets of the educational process, specialized education in the major field of water resources may be too shallow for competent professionals. On the other hand, specialisation within a narrow field can lead to non-utilization of professional talents in the solution of water resources problems.

Larin (1980) opined that modern educational methodology demands that the teacher convey knowledge in a comprehensible form and a certain orderly system, forming in the students a correct understanding of phenomena and firm practical skills.

Much of the practical skills will, of course, be acquired from out of classroom experiences in the form of laboratory studies, excursions, practical field training, individual projects or assignments. Laboratory study is one of the principal means of establishing a connection between theoretical education and practice. In our particular case, it provides the opportunity to train students in the processing, analysis and generalising of meteorological and hydrological data. Excursions to operating facilities also enable students to see structures and water resource processes in real life conditions.

In order to educate meaningfully the potential users of meteorological information, such education needs to be orderly, comprehensible and based on some clear knowledge of the phenomena involved. Some knowledge of some hydrological concepts and methods are also indispensable since hydrology is the science of water.

Moore (1975, p 87) reported how shocked a professor was to find mimeographed handouts containing objectives of the course lying on the table. The professor exclaimed, "You don't let the students see these, do you?" The person who had prepared the handout however, not only confirmed the professor's fears, but also added that the student would indeed learn all the objectives. It is believed that a clear written statement of objectives - what is expected - will surely help the student to learn.

Table 1 shows some purposes of water-resources development. These range from flood control, sediment and salinity control and drainage irrigation, navigation, recreation, watershed management, fish and wildlife preservation, and pollution abatement to the all-important domestic and industrial water supply as well as hydro-electricity. The description of the elements and the types of structures and the works and measures needed to accomplish such purposes indicate the type of meteorological and hydrological information required for the planning, design and operation of the specific water projects. Increasing demand of the listed purposes are making it necessary for more and more countries to undertake comprehensive water management measures.

METEOROLOGICAL AND HYDROLOGICAL INFORMATION

What are elements of the meteorological information? According to Andrejanov (1975, p 9), meteorological information used in water resources planning and development can be classified into two types. The first consists of those needed for a general climatic description of the conditions of the construction and the operation of structures. Such data are usually arranged in calendar months, and include data such as:

air temperature (average, maximum and minimum annual values); relative and absolute humidity (average and extreme annual values); cloud amount; liquid and solid precipitation; snow cover and water equivalent; wind speed and direction; number of rain days (with over 0.1 mm of precipitation); number of days with temperature below freezing point; duration of sunshine; temperature of the soil at surface and various depths, etc; and diurnal variations of temperature and relative humidity, depth of freezing, glazed frost and perma frost, if any.

The second type consists of some special data required for the calculation of hydrological characteristics and include:

long-term mean annual values of solar radiation, albedo, effective radiation, the radiation balance for water and heat-balance calculations for the purpose of determining evaporation from a land and water surface; frequency of totals of precipitation of various intensities and durations for calculating flood flow; and frequency of calm and windy periods of different speeds and direction for calculation of airflow over watersheds.

Data on river flow are of basic importance, however, in planning the use of water resources. For this reason, it has been suggested that the greatest emphasis should be placed on river flow and less attention devoted to other meteorological and hydrological elements when considering the methods of data analysis. We know however that there are, more often than not, cases where such basic data are not available and have to be estimated from meteorological elements especially in developing countries.

Column (4) of Table I lists meteorological and hydrological data required for various purposes of water-resource development. This knowledge which relates the data to the types of works or structure and finally to the purpose of the water project, should make the process of learning more meaningful and goal-oriented. The student may thus be better motivated.

PERSONNEL AND CURRICULUM

We have already indicated that the planning, design and operation phases of water management all require more than a joint consideration of meteorological and hydrological events. This realisation highlights the need for trained meteorologists and other specialists in the atmospheric phase of the hydrological cycle who are also competent in water management applications of meteorology. The idea is, of course, not to make them qualify as hydrologists or specialists in water resources engineering, but to equip them to use available information efficiently and effectively.

WMO (1984, p 191) makes a distinction among four categories of such personnel as follows:

Class I: Hydrometeorologists (specialists with a University degree). A minimum of 12 years of schooling plus university training of at least 4 years.

Class II: Hydrometeorologists (specialists up to Senior Technician level). A minimum of 12 years of schooling plus meteorological training of over 2 years.

Class III: Hydrometeorologists (specialists technician level). A minimum of 12 years of schooling plus 8-10 months of meteorological training.

Class IV: Hydrometeorologists (specially trained observers). A minimum of 9 years of schooling plus formal training in meteorology for at least 4 months.

For each of the above, education and training should be supplemented by adequate practical on-the-job training of 6-9 months.

The curriculum is the main document upon which education and training of all the four categories are based. It should consist of a list of subjects or courses with clear indications of their workload or volume and sequence of presentation as well as the ratio between the theoretical instruction and practical training. The curriculum that will adequately equip potential users of meteorological information in the field of water resources must include courses in general hydrology and specific topics.

Classes I and II meteorologists must have already had a course in general hydrology (WMO, 1984, section 3.3.6) as listed in Table II(a). Those categories are, therefore, expected to be familiar with the basic elements of hydrology and aspects of its application to water resources. In teaching general hydrology to these categories of personnel, it may only be necessary to organise revisional work with greater emphasis on practical training.

However, in the case of Classes III and IV, meteorology personnel and Class I scientists who are neither hydrologists nor meteorologists, a course in general hydrology is a prerequisite. The course content listed in Table II(a) is appropriate for Class II scientists, but for the Class III personnel, a course syllabus for Elementary Hydrology and special Field Techniques (Table III(a)) is adequate to provide the necessary background before specific topics in the use of meteorological information in water resources are taught.

The rest of Table II (ie, b, c, d and e) contains the course syllabi for education and training of Classes I and II personnel in the use of meteorological and hydrological data in the field of water resources (see also WMO, 1984, Chapter 13). This is to be accomplished through teaching of analysis of meteorological and hydrological data for design purposes as well as meteorological and hydrological aspects of water management operations.

For the Class II personnel, it should be noted that greater emphasis should be placed on practical aspects of the programme. The field in which the student will be eventually employed should be given due consideration.

The remaining sections of the Table III (ie, b and c) are intended to accomplish the same purpose for personnel of the Class III category.

Table IV contains the curricula recommended for the Class IV personnel (see also WMO, 1984), Chapter 13.4).

METHODOLOGY

Comprehensive and optimum curricula and syllabi are of great importance, as may be seen above. However, methods of teaching and teaching aids are also significant. The problem of teaching aids and establishment of appropriate and adequate laboratory facilities together with provision of suitable textbooks should be given proper attention. It is helpful that rapid technological advances have made an array of communication media available and within reach of many, even in developing nations. The film media: the film strips, slide shows and the motion pictures have been used as teaching aids for upwards of 50 years in water resources education. Now, closed-circuit TV, video cassette recorders of several types and modes, programmed instruction, teaching machines, and computer-assisted instruction are in popular use not only in centres for educational technology, but by many teachers who appreciate the importance of harnessing the modern communication methods to design and implement the learning systems in water resources education.

We have earlier noted the significance of computer applications. Actual water resources problems can be solved through the applications of computer software. But, in addition, it can be used in the education of water resources staff, particularly in the teaching of specific techniques of water resources. This, according to Chidley (1975, p 290) can be done by using the standard software and some specially devised learning programmes.

Given appropriate hardware facilities - a small computer system - a number of hydrological and water management software packages are available dealing with extreme value analysis, storm estimation for drainage design, backwater curves, irrigation network design, etc. In fact, the Hydrological Operational Multipurpose Sub-programme (HOMS) of the WMO meets this vital need. HOMS is designed to facilitate the transfer of technology used by hydrologists and water resources specialists. The technology takes the form of computer programmes, technical manuals, or hydrological instruments. The technology, available through HOMS (from originator countries), is provided in the form of separate HOMS components and published in the HOMS Reference Manual, the basic document of HOMS. The document give the addresses of the HOMS National Reference Centres through whom the components may be obtained. The WMO Secretariat in Geneva will, however, readily provide additional information on aspects of HOMS.

Sections H, J, K and X of HOMS, for instance, deal with the following aspects:

H - Primary data processing (including H25 - General meteorological data used in hydrology).

J - Hydrological models for forecasting and design.

K - Analysis of data for planning, design and operation of water resources systems.

X - Mathematical and statistical computations.

The components included in the above sections can be applied directly to, or used for training in, the use of meteorological and hydrological data in the field of water resources (ie, the contents of most of the curricula and syllabi listed in Table II and IV).

TRAINING METHOD

The different methods of training hydrological technicians include the service-sponsored institutes, technical colleges, polytechnic institutes, short courses and in-house on-the-job training. The categories of senior and junior hydrological technicians correspond to those of Classes II and III meteorological personnel respectively (WMO 1984, pp 269-70). The last two of the methods (short courses and on-the-job training) can be adapted to the training of potential users of meteorological information in water resources management. Oyebande (1983) has however proposed some modifications which could make the methods more effective. He proposed a modified on-the-job-training scheme through establishment of Research and Training Divisions in the Hydrological Water Resources Hydrometeorological Services to plan and coordinate such training and liaise with appropriate national and international bodies. It is also proposed that a properly structured on-the-job training programme be introduced for all major medium and long-term water resource projects awarded to national or international contractors and consultants.

Short training courses are conducted by presenting formal lectures on certain themes which incorporate basic knowledge designed to upgrade the knowledge of the participants who have no previous experience of the particular subject. The course can also provide a valuable refresher course for others. Courses may include practical work and have a duration of a few weeks to a few months. Numerous courses of this kind have been mounted by WMO, UNESCO and other international agencies. This training method is to be strongly recommended, but care should be taken to ensure that the background of the students in both basic education and experience in the field are not too divergent. Furthermore, practical and field work should be given greater emphasis than is done at present.

Two relatively new training methods now being used in training hydrology personnel are also to be recommended. They are roving seminars and correspondence courses. While students usually travel out to attend short training courses, roving seminars bring consultants or experts to students in the familiar environment of their own countries, or to a group of countries in the same region. Oyebande (op cit) observed that these seminars can be most useful and cost-effective as a training method if properly streamlined such that problems relating to the optimum number of participants, selection of target trainees, curriculum and course syllabi are successfully resolved.

Correspondence courses have great attractions and have played a vital role in the education system of developing countries. There, a vast majority could not afford the cost of full-time post-primary education. Even many of those going their earning power on which so many members of the extended family depend. These courses could be adapted successfully to the more practical training of personnel in the field of water resources. To accomplish such a task however, the setting of examinations and issuing of certificates and diplomas, promotion and job prospects, course writing, practical and field work as well as aspects of funding require close and careful study and attention (Oyebande, 1983).

SUMMARY AND CONCLUSION

It is helpful to recapitulate that water resources management requires joint consideration of meteorological and hydrological phenomena, and entails undertakings which require information on those same phenomena for their design and operation. Education of science-oriented potential users of meteorological information should therefore proceed from teaching of properties and phases of the hydrological cycle and water balance concept and the magnitudes of the forces of the components and how such forces affect man-made structures associated with water projects. The need for practical laboratory and field work cannot be overemphasized.

It is gratifying to note that through the education and training activities of the WMO and the agencies with which it cooperates, the National Meteorological and Hydrological Services of many developing countries have succeeded in providing much of the required trained personnel. Such training has enabled them to carry out their functions and to adapt to some of the new methodologies and technologies. There is, of course, much room for improvement and it is hardly time to relent on the efforts of international cooperation.

REFERENCES

Andrejanov, V G, 1975 Meteorological and hydrological data required in planning the development of water resources. WMO Operational Hydrology Report No 5, WMO No 419. Geneva.

Chidley, T R E, 1975 Requirements for education in computer techniques in developing countries. Water Resources Education, UNESCO/IWRA, 285-294.

Chow, V T, 1979 Water as a World Resource, Water International, 4(1), 3-6.

Clausen, G S, 1975 Teaching hydrology and water resources to civil engineering students. Water Resources Education, UNESCO/IWRA 94-95.

Larin, V V, 1980 Education of teachers in technician training courses ...., Paper presented to the international workshop on hydrological education (Smolenice, Czechoslovakia), 16-19 September.

Moore, Walter L, 1975 Teaching methods - philosophy and approaches. Water Resources Education, UNESCO/IWRA.

Oyebande, L, 1983 Proposals for training hydrological technicians in developing countries. Technical report to the Commission for Hydrology, World Meteorological Organization, Geneva.

WMO, 1983 Guide to hydrological practices, volume II - analysis, forecasting and other applications. WMO, No 168. World Meteorological Organization, Geneva.

WMO, 1984 Guidelines for the education and training of personnel in meteorological and operational hydrology, third edition, World Meteorological Organization No 258. Geneva.

Table I: Some Purposes of Water-Resources Development

Purpose	Description	Type of works and measures	Reqd. Data
Flood control	Flood-damage abalement or reduction, protection of economic development, conservation storage, river regulation, recharging of groundwater, water supply, development of power, protection of life	Dams, storage reservoirs, levees, floodwalls, channel improvements, floodways, pumping stations, floodplain zoning, flood forecasting	4,5,6,8, 10,12,15.
Irrigation	Agricultural production	Dams, reservoirs, wells, canals, pumps and pumping plants, weed-control and desilting works, distribution systems, drainage facilities, farm land grading	1,2,3,4,5, 6,7,8,9, 10,11,14.
Hydroelectricity	Provision of power for economic development and improved living standards	Dams, reservoirs, penstocks, power plants, transmission lines	1,2,3,4,5, 6,7,8,10, 11.
Navigation	Transportation of goods and passengers	Dams, reservoirs, canals, locks, open-channel improvements, harbor improvements	2,3,4,6,7, 8,12,13.
Domestic and industrial water supply	Provision of water for domestic, industrial, commercial, municipal, and other uses	Dams, reservoirs, wells, conduits, pumping plants, treatment plants, saline-water conversion, distribution systems	1,2,4,5,6, 7,8,9, 10,11.
Watershed management	Conservation and improvement of the soil, sediment abatement, runoff retardation, forests and grassland improvement, and protection of water supply	Soil-conservation practices, forest and range management practices, headwater-control structures, debris-detention dams, small reservoirs, and farm ponds	See Flood Control & Sediment Control
Recreational use of water	Increased well-being and health of the people	Reservoirs, facilities for recreational use, works for pollution control, reservation of scenic and wilderness areas	1,2,3,4,5, 6,7,8,9, 10,11, 16,17.
Fish and wildlife	Improvement of habitat for fish and wildlife, reduction or prevention of fish or wildlife losses associated with man's works, enhancement of sports opportunities, provision for expansion of commercial fishing	Wildlife refuges, fish hatcheries, fish ladders and screens, reservoir storage, regulation of streamflows, stocking of streams and reservoirs with fish, pollution control, and land management	2,3,4,6, 7,8,9.
Pollution abatement	Protection or improvemnt of water supplies for municipal, domestic, industrial and agricultural uses and for aquatic life and recreation	Treatment facilities, reservoir storage for augmenting low flows, sewage-collection systems, legal control measures	1,2,3,4, 5,6,7,8,9, 10,11.
Groundwater recharge	Involves use of reservoirs, ponds and bank infiltration wells	Dams, reservoirs, ponds, wells	1,2,3,4,5, 6,7,8,9, 10,11,14.
Drainage	Agricultural production, urban development and protection of public health	Ditches, tile drains, levees, pumping stations, soil treatment	2,3,4,6,9, 10,12.
Sediment control	Reduction or control of silt load in streams and protection of reservoirs	Soil conservation, sound forest practices, proper highway and railroad construction, desilting	1,2,3,4,5, 6,7,8,9, 10,11.
Salinity control	Abatement or prevention of salt-water contamination of agricultural, industrial, and municipal water supplies	Reservoirs for augmenting low streamflow, barriers, groundwater recharge, coastal jetties	2,3,4,5,6, 7,8,9, 10,11.

1. Series of Monthly and Annual Volume of Streamflow
2. Mean Daily Discharge Series
3. Low-flow Frequency Distribution
4. Frequency Distribution of High Discharges
5. Frequency Distribution of Large-Volume Floods
6. Shapes of Typical Flood Hydrographs
7. Ice Cover Information
8. Sediment Transportation
9. Quality of Water
10. Precipitation Distribution in Space and Time
11. Evaporation Distribution in Space and Time
12. Stage Discharge Rating Curves/Tables
13. Snowmelt Distribution
14. Groundwater Level
15. Bivariate Data on Heavy Rainfalls & Floods
16. Air Temperature Distribution
17. Wind Distribution

SOURCES: CHOW [1979], TABLE 1 W M O [1983], TABLE 7.5

Table II: Curriculum for Classes I and II Personnel

(a) General Hydrology

Introductory material: definitions and relations of hydrology to other sciences; the hydrological cycle; physical characteristics of the watershed; variability and randomness of hydrological phenomena.

Precipitation: determination of amounts, intensity and duration, spatial and temporal distribution; measurement and accuracy; estimation of missing data.

Evaporation and evapotranspiration: determination by measurement and by computations; the energy and mass transfer approaches; evaporation from water surfaces; soil, snow and ice; transpiration, total evaporation and total losses; evaporation control.

Infiltration: soil moisture; laws governing infiltration; measurement; empirical formulae.

Groundwater: origins and occurrence of groundwater; types of aquifer; interrelation between ground water and surface water; depletion; water-table fluctuations; movement of groundwater.

Surface run-off: elementary hydrograph; types of run-off; depression storage, overland flow, surface detention; unit hydrograph.

Water balance; calculation of yield; short-term and long-term variations; water balance of lakes, swamps, watersheds and regions.

Hydrometry: various methods of measuring water-levels, velocities and, solid and liquid discharges; storage and processing of data.

(b) Analysis of Meteorological Data for Design Purposes

Computation of mean annual basin precipitation taking into account physiographic factors; rainfall intensity-frequency data for a point; relationship between point and areal rainfall intensities of specified frequency; depth-area-duration data for storm precipitation; probable maximum precipitation (PMP); snowmelt computations; maximization of snowmelt in conjunction with PMP; estimate of free-water evaporation and actual evapotranspiration; conjuctive design of climatological and hydrometric networks; time trends in meteorological data.

(c) Analysis of Hydrological Data for Design Purposes

Methods of computing runoff hydrographs: mass curves of runoff; flow-duration curves; design floods; probable maximum flood, minimum flow computations; flood frequency analysis; empirical flood formulae, flood plain mapping; co-ordinated mapping of precipitation, evaporation and runoff; time trends inflow data.

(d) Meteorological Aspects of Water Management Operations

Meteorological effects on the runoff cycle; precipitation types and related intensities, calculation of mean basin storm precipitation; radar measurement of precipitation; water management requirements for quantitative precipitation forecasts; meteorological basis for flash flood warnings; causes and predictability of droughts; irrigation requirements.

(e) Hydrological Aspects of Water Management Operations

Hydrometric measurements; relationship between discharge and stage; rainfall-runoff relations and conceptual models for predicting the discharge hydrograph; hydrological basis for flash flood warnings; long-range forecasts of seasonal flow volume and ice phenomena, such as formation and break-up.

(Source: WMO (1984), Sections 3 and 13)

Table III: Curriculum for Class III Personnel

(a) Course Syllabus for Elementary Hydrology and Special Field Techniques

The Hydrological cycle; Water balance and water balance equations; Erosion, transportation and deposition processes; Drainage basin and stream networks, lakes and swamps; features of stream valleys and channels; Flood and the flood hydrograph; Low, flow, drought; Important features of hydrology of watersheds and regions; Hydrometry: Various methods of measuring and recording water levels, velocities, solid and liquid discharges; storage and processing of field data; Rating curves, quality control (recognising errors); Elementary analysis of data - mean daily, monthly, annual flow, water yield, runoff; Runoff coefficient, flow duration curves, flood frequency curves.

(b) Analysis of Meteorological Data for Design Purposes

Computation of mean annual basin precipitation taking into account physiographic factors; rainfall intensity-frequency data for a point; relationship between point and areal rainfall intensities of specified frequency; depth-area-duration for storm precipitation; probable maximum precipitation (PMP); snowmelt computation; maximization of snowmelt in conjunction with PMP; estimates of free water evaporation and actual evapotranspiration; conjuctive design of climatological and hydrometric networks.

(c) Meteorological Aspects of Water Management Operations

Meteorological effects of the runoff cycle; precipitation types and related intensities; calculation of mean basin storm precipitation; water management requirements for quantitative precipitation forecasts, meteorological basis for flash flood warnings; causes and predictability of droughts, irrigation requirements.

Table IV: Curriculum for Class IV Personnel

The following syllabus covers only the rudiments, without mathematical development:

Rudiments of hydrology, and hydrological measurements; the hydrological cycle;

Intensity and frequency of precipitation at a given point;

Estimating evaporation from a free surface and actual evapotranspiration;

Snow; evaluating snow storage; snowmelt;

Influence of meteorological elements on the hydrological cycle;

Relationships between climatological and hydrometric networks;

"Height amount of precipitation" relationships;

Calculation of the average annual depth of water falling as precipitation over a basin;

Depth area duration relationships;

Probable maximum precipitation (PMP);

Calculation of average depth of water falling as precipitation, homosensity of series;

Forecasting of precipitation, meteorological elements in flood forecasting;

Causes and characteristics of droughts and excess precipitation;

Water and its users; requirements (in particular for irrigation).

(Source: WMO (1984), Section 13.4)

13

The education and training of "applied meteorologists"

J. F. Griffiths, Professor of Meteorology and of Geography, Texas State Climatologist, Texas A&M University, USA

Climatologist, Texas A&M University, U.S.A.

1. INTRODUCTION

Most of the institutions of higher learning that offer the equivalent of a degree or degrees in meteorology require the student to concentrate on certain aspects of physics and mathematics. This is, of course, a logical approach for a geoscientific subject.

However, most curricula do not permit the student much flexibility or time to take courses in other disciplines. This is a marked disadvantage to the student wishing to specialize in applied meteorology, that branch dealing with interrelationships between atmospheric conditions and organisms, structures, water bodies, etc.

During the recent decade there has been a definite increase in employment opportunities in the field of applied meteorology and it is becoming progressively more difficult to fill these posts with well qualified personnel. I understand that this is a situation that applies also to experts in the subject required by WMO in its excellent Technical Cooperation Programme, in which there is a need also for experience in addition to knowledge.

The main body of this paper will be concerned with methods used during past years to institute or enhance this specialized education and training.

2. AMERICAN ASSOCIATION OF STATE CLIMATOLOGISTS

The state of affairs noted above was discussed at meetings of the American Association of State Climatologists (AASC) and during my period as President a committee on Education was formed to address the matter. There was a degree of self-serving in this undertaking because a number of the State Climatologists are reaching the age of possible retirement and suitable replacements are needed.

The Final Report of this Committee was as follows:

Proposed Resolution to the AASC from the Committee on Education

Be it resolved, that the American Association of State Climatologists establish the following recommended curriculum guideline for students aspiring to become State Climatologists.

Because state climatologists work with a wide variety of climate data and information users, and because State Climatologist must be innovative in providing products which relate to the specific needs of the users, and because State Climatologists must be adept in verbal and written communications of a scientific nature, and must be able to interpret scientific commentary, we recommend that aspirants to State Climatologist positions hold at least a Masters degree, and complete the following curriculum.

Since climatology largely rests upon meteorology, statistics and data handling, we suggest the following courses as necessary to provide the basic training for State Climatologists. The courses are listed without order of preference.

Introductory Meteorology
Synoptic Meteorology
Dynamic Meteorology
Meteorological Instruments
Regional Climatology
Programming, emphasis on micro computers, data base handling
Climatological Methods, or Applied Climatology
Physical Meteorology or Physical Climatology
Physics and Mathematics through differential equations (should be prerequisite to Meteorology and Dynamic Meteorology)
Statistics through analysis of variance, time series analysis and multivariate analysis

Because State Climatologists need to converse and correspond knowledgeably with a wide variety of climatic data users, a student should complete at least four of the following courses:

Agricultural Climatology
Air Pollution
Bioclimatology
Oceanography
Hydrology
Applications of Climatology to Architecture
Climatology Hazards, Consulting Climatology
Marine Climatology

We further recommend that graduate students aspiring to a State Climatologist position commit themselves to at least one semester of an internship to be spent at one of the State or Regional Climate Centres (other than that which may be associated with the university of their residence), or the National Climatic Data Center. This practical experience will acquaint the student with common inquiries made to State Climatologists as well as available service products, and research products means of "independent study" under the supervision of their major professor in cooperation with the scientist and the place of internship. Prior to an internship, the student would write a proposal to the intern facility and his/her major professor for acceptance. This "contact" would serve as the basis for evaluation at the end of the internship.

It is appreciated that, within the structure of most degree programs, it may be impossible to cover all the suggested topics. The consensus was that to develop the necessary skills a State Climatologist should hold at least the Master's degree.

The last paragraph suggests an important adjunct to the regular meteorological education - an internship. This arrangement ensures the trainee gets some practical experience in the appreciable breadth of topics met with in applied meteorology. This facet will be considered later in this talk.

3. THE CIC BIOMETEOROLOGY GRADUATE PROGRAM

After a planning meeting in 1960, a group of 11 midwestern universities of the USA (CIC = Committee on Institution Cooperation) a program was funded by the Air Pollution Division of the US Public Health Service to train graduate students in interdisciplinary studies. The program lasted for 7 years and involved 30 students and many faculties in various disciplines. Students could take training from two universities; workshops and field work were organized and in the later years visiting foreign lecturers were brought in. Of the students, 5 resigned, but the fields of specialization of the others were

Agricultural Engineering (2)	Plant Ecology
Animal Science (2)	Plant Physiology
Botany (3)	Physiology (2)
Civil Engineering	Sanitary Engineering
Forestry	Soil Physics
Geography (2)	Soil Science (3)
Mechanical Engineering (2)	Zoology (2)
Meteorology	

Those who were trained under this Program speak very positively of their experiences, feeling that they have gained much by their in breadth exposure to aspects of other disciplines.

4. SPECIALIZED CURRICULUM IN HUMAN BIOMETEOROLOGY

A curriculum in human biometeorology was prepared by Dr D M Driscoll. This was presented at the CCAM, VIIIth Session in Washington, April, 1982. Abbreviated details of this are given below. It is to be noted that similar curricula can be prepared for other fields of specialization, such as engineering, veterinary sciences, hydrometeorology, etc. These would doubtless prove of use to investigators in many regions of the world.

CURRICULUM FOR INSTRUCTION IN HUMAN BIOMETEOROLOGY FOR PERSONNEL IN NATIONAL METEOROLOGICAL SERVICES

a. INTRODUCTION

Definitions: biometeorology, human biometeorology.

Historical development.
Distinction of various scales e.g., micro-, meso-, macro-scale; importance of the meteorology of the boundary layer; bio-sphere.

b. THERMOREGULATION

Basic physiology; meteorological conditions affecting: comfort/discomfort indices, maintenance of core temperature, effects of work and exercise, metabolism, man-thermal environment mathematical models.

c. RADIATION
Effects of radiation.
Erethyma-skin cancer-pigmentation, genetic differences;

ozone, its space-time variations and their consequences.

d. RESPIRATION

Partial pressure of O_2 - variation with altitude; man's adaptive mechanisms; temporary and long-term (acclimatization).

Air pollution: sources, gases vs particulates; meteorological conditions affecting; epidemiological considerations.

e. CLIMATE, WEATHER AND DISEASE

Diseases so influenced; direct vs indirect (e.g., vectors) effects. Patterns in developed and developing countries, deseasonalization. Inferential human biometeorology; statistical associations, incomplete etiology.

f. METHODOLOGIES IN HUMAN BIOMETEOROLOGY

Investigative methods; statistical associations to the evidence of clinics and climate chambers. Morbidity and mortality records, variations in national practices.

Meteorological data needs of human biometeorology.

It is anticipated that the subjects given here should comprise about ten to twenty hours of classroom instruction. Text material could be the new Technical Note, or other suitable text.

5. PERSONAL OBSERVATION

During the past four decades it has been my good fortune and privilege to be associated with problems in applied meteorology ranging over many disciplines and countries. This experience has helped me to appreciate some of the basic concepts and approaches that are necessary to tackle the practical questions most often asked. These are given here so that others may become aware of some of the fundamental imports.

The first step in answering the questions is the acquisition of quality data, both meteorological and of the discipline to which it is to be applied. The World Climate Data Programme is helping take steps in the right direction, through INFOCLIMA and CLICOM, and the setting up of data banks. However, the relevance and reliability of the observations can only be judged fully if station and instrument histories are archived, preferably in digitized form. This makes essential a back-up programme of station inspection. With the improvement of computers it is possible to process vast amounts of data but all this must be tempered with understanding of the techniques to be used. One must not be caught in the embarrassing situation that arose with regard to the 1982-83 El Nino-Southern Oscillation event and the Antarctic ozone-ring phenomenon. In both these cases the computer had been programmed to disregard (not enter into the data bank, but only retain the data) readings outside a predetermined domain. Such readings were considered aberrant or wrong. It was not until individual observers working with the raw (undigitized) data identified the singularly unusual conditions that it was realized that "the baby had been thrown away with the bath water".

The next step in the investigation is to realize the limitations of the instruments used. This way the measurements will not be used beyond their limit of accuracy.

The third aspect is to decide the form of mathematical (statistical) analysis. These are detailed in Table I. It must be noted that all six stages are important but the last is of major concern for the interpretation of the result is the end product, that which is communicated to the user.

TABLE I. STEPS IN STATISTICAL ANALYSIS

In most statistical climatological studies the analysis can and should have the following sequence of steps:

1. Numbers - obtain reliable, relevant data in numerical form (be aware of the many ways in which the data may contain errors).

2. Problem - postulate the problem to be studied and answered (paying special attention to the null hypothesis to be tested).

3. Techniques - decide upon the statistical techniques to be applied, examine them to discover if they are valid and allowable in this problem; note particularly the assumptions, if any, inherent in the use of the technique.

4. Arithmetic - carry out the calculations dictated by (3); include all checks possible.

5. Arithmetic-Answers - extract from (4) all the required numbers needed to answer (2); if approximations and groupings are carried out only at this stage they are less likely to cause misleading or spurious results.

6. Interpretation - use (5) to phrase the practical answers and interpret them in terms of the discipline to which they are applied; either statistical or meteorological, or both.

However, before presenting the findings it is desirable to check the susceptibility of the result to natural inaccuracies. It may be found that the answer is unstable and should not be given great credence. Additionally the answer must be couched in terminology familiar to the user's discipline.

It is realized that most meteorologists would welcome some guidance in the tackling of interdisciplinary problems and I shall now describe an undertaking with which I have been associated since 1979. This unique and most interesting project is funded by the Association of South East Asian Nations (ASEAN) and the United Nations Development Programme (UNDP), and guided by WMO. This has been the production of a Climatological Atlas and Compendium of Climatological Statistics for the ASEAN region. In the process of preparation of these publications data were subjected to numerous quality checks, not as rigorous and ideal as perfection demands but as strict as was possible in the circumstances.

After the completion of the project it was decided that a Manual for the benefit of users of the Atlas and Compendium was necessary. This new project is now underway. The Manual will include a section on relevant statistical techniques, their strengths and limitations, plus other sections dealing with the application of climatological data to problems in agriculture, water use, energy, engineering, architecture, transport, energy and tourism.

This Manual will illustrate all the techniques cited with examples taken, when relevant, from data given in the ASEAN Atlas or Compendium. Such an approach will allow users to apply similar methods in their own specific problems. The Manual will include numerous references so that the user can pursue the subject to the depth required.

I believe that such a publication can be of great use to applied meteorologists in many countries since the techniques are independent of region. Unfortunately it is intended to print only 1400 copies due to financial restrictions, and most of these will be distributed to workers in ASEAN region.

Many countries are in the stage of development in which careful and correct use of climatological data could assist improvements in their economy. It would be beneficial therefore to hold regional workshops at which the current techniques in applied meteorology could be explained and illustrated, using the Manual as a starting point, but retaining the flexibility to address the problems of most importance to the particular area. It is likely that a period of 3-4 weeks would suffice for each Workshop and, to reduce travel expenses, one could be held in each of Africa, Asia and Central/South America.

14

The weather package: distance learning techniques in meteorology

A. Cooper, Open University (UK)
D. B. Shaw, Meteorological Office College (UK)
G. Turner, BBC (UK)

Teaching at all levels is changing its character. A wide range of styles is now in use, from pure class teaching to totally independent learning from specially prepared material, using a mix of media. Different styles are appropriate to different situations. Perhaps the greatest changes have occurred in the University sector, through the successful operation of the Open University, and in training material within (or for) individual firms and organisations. School education is changing more slowly, through the introduction of computers and video players, commonly using tape, but also disc.

The teaching and learning of meteorology will surely follow these trends, so that the subject expertise must be merged with experience in various types of distance learning. With this in mind, the Meteorological Office College, the Open University and the BBC have, over the last 18 months, developed a proposal for a set of three packages. Why three?

We feel that the whole audience of people interested in meteorology can be split into three parts, whose requirements are very different. First, there is professional training, for people whose work is, or depends closely on, meteorology. Secondly, there is an audience of people who take a serious interest in the weather through work such as horticulture or the construction industry, or perhaps through a hobby such as sailing. Thirdly, there is a very wide general audience, including the whole area of schools. These audiences are so different that they need different mixes of media, so we take them in turn.

We assume that professional training will always need a teacher and here the aim is to maximise the efficiency of the teaching. This means making powerful teaching materials readily available. We believe that videodisc is the best available method. It also means freeing the teacher from routine items of teaching, since this can be done by the computer, leaving the teacher free to deal with particular, individual problems (and personal problems), for which a computer is of little use. BBC/OUPC has already produced several such videodiscs.

In considering the videodisc concept it should be noted that meteorology lends itself readily to this very visual form of presentation. In teaching the subject, the need to refer to pictorial material is almost always present and many examples spring to mind. Common to most training of professional meteorologists of all classes, throughout the world, is the recognition of cloud types. Here the student would benefit from reinforcement learning, carried out without supervision. Animation of cloud sequences would also help to distinguish the different processes of formation for the different types. A second example of visual material in common use by the professional is satellite imagery. Again there is an important training requirement which is increasing as both the range and the resolution of the products increase. Training in interpretation of satellite products would benefit from an ability to readily relate satellite imagery to other meteorological information - overlaying of satellite pictures and synoptic analyses, for example, in both static and animated sequences would be very helpful. The ideal is where such a capability is coupled to suitable computer-aided learning software so that the student's learning is directed, but without the need for tutor supervision.

In modern forecasting offices it is increasingly the case that satellite imagery can be supplemented by rainfall displays based on weather radar networks. Effective interpretation of such displays is a skill that has to be taught. Videodisc would present such training in precisely the same medium as is used operationally i.e. a colour monitor.

These are just a few examples of how videodisc could be used as a teaching aid in meteorology be it for class or self-instruction. Many other examples could be readily identified and indeed its uses are not limited to its training role. The storage capacity of video disc is sufficiently extensive to enable it to be treated as a data base for reference material. Sequences of major meteorological events - such as the onset of the SW monsoon or El Nino could be stored, providing finger tip control over an enormous range of charts, climatological data and model simulations. There is no better means of bringing this range of visual material together, under direct keyboard control, than videodisc.

For the second audience, with a serious but not professional interest, the approach must be different. The Open University Continuing Education programme produces material in the form of packs combining text and video cassettes. For a weather pack we would seek to lead the reader to an understanding of the physical basis of atmospheric behaviour, and of the place of local weather in large scale, even global, weather patterns. The level would be non-mathematical, but graphics and animations can convey the essence of these areas which in a more advanced course would be explored through calculus. Indeed one of the major changes in meteorology is, of course, the availability of beautiful and very revealing, satellite imagery. Clearly video material can exploit this very efficiently.

The second audience being targeted in this Open University package of text and video cassette is the educated non-meteorologist - an increasingly large body which often includes the user of meteorological products. As non-specialist users they need a proper appreciation of meteorology. Some training establishments have induction courses which partly meet that demand but there are many constraints which prevent it - cost, the remoteness of the training centre to the trainee, the inability to release the trainee at the time a course is scheduled, failure of the training centre to cope with the demand, and so on. What is required is a means of overcoming these problems; the Open University package seeks to do just that, providing a format and a syllabus which are within the scope of the educated user of meteorological products. Everyone has some knowledge of meteorology but a good understanding can only come from a carefully designed syllabus which takes the student through basic principles in a well ordered and interesting way. The package seeks to do this starting from a familiar image - the atmosphere as seen from space. This global view provides a framework for identifying the basic mechanisms at work in the atmosphere. The dynamics of the atmosphere, and the physical relationships of temperature, wind and pressure, are explained in non-mathematical terms leading into accounts of many of the concepts and features of the science which the student will readily relate to. The objectives of this package are to help the student to fully understand weather maps as they are normally published, to learn the use of the most common meteorological instruments, to be able to relate his local conditions to the general weather pattern and to understand the relationships between the weather and various human activities.

Through television, the public has come to enjoy, and to expect, a very high standard of science documentary, fully satisfying the BBC's aim "entertain and inform". Many excellent series on astronomy, natural history and earth science have been produced. Strangely, the weather, which almost everyone has some interest in, has not featured prominently among such series. We have developed a proposal for six television programmes whose theme would be people and weather, but would build into the story some of the most visual aspects of weather processes. There would, in fact, be an overlap between the visual material for the pack and for the TV programmes. The importance of the video component of all three parts is a feature we wish to stress.

In drawing up these proposals we have had in mind a content for a British audience, with possible extension to European and American audiences. However the techniques and general philosophy could be applied elsewhere in the world.

The needs of Meteorological education and training worldwide are enormous, no-one here needs to be told of its significance to the lives and prospects of the world's population, particularly in those vulnerable areas of the Third World. However, such a big task needs proper guidance and advice. We recommend that the WMO set up a working group to consider the approach we have adopted, to advise us on our next steps, and to look to an extension to wider audiences.

The implementation of current and future plans obviously depends on funding. As far as our current plans are concerned, each of the three items will cost in the region of £100,000, rather less for the pack and a little more for the TV series. They cannot be fully self-funding - no serious educational project can. They can, however, bring in quite a lot of money, perhaps as much as three quarters of the total cost.

SUMMARY

We have set up a working consortium of specialists in meteorology, video techniques and distance learning and propose a package of materials to satisfy a wide range of needs and interests in the weather. We seek advice and support on the implementation of the scheme, and its extension to a wider, global audience.

15

Experiences with meteorological in-service training of school teachers in England

R. Reynolds, Department of Meteorology, University of Reading

INTRODUCTION

Meteorology is taught formally in schools in the United Kingdom as part of the upper secondary level Geography syllabus, and since 1980 has been offered as a distinct subject at G.C.E. Ordinary level. This 'O' level will be replaced with the first examination for the General Certificate of Secondary Education (GCSE) in the summer of 1988.

As a specific 'O' level subject, Meteorology has grown healthily, both in terms of candidates and of examination centres (Milford, 1985). The combination of this development with the widely taught 'O' and 'A' level Geography syllabuses forms the basis for a potentially large demand for teacher training.

The subject is perhaps one of the few for which the provision of training, particularly at the more 'technical' secondary level cannot easily come from the school teaching community itself; or even from tertiary level Education Departments. Indeed, even at primary level the demand for project ideas and more general weather-related classroom material (Ilsley, 1985) needs, ideally, to be satisfied by someone quite well-versed in the science.

So who is to provide the training? One obvious solution is

that professional meteorologists who possess the enthusiasm and application to develop short courses could. In fact this theme of education in meteorology has long been a role of the Royal Meteorological Society, although reaching out to nonprofessional meteorologists has developed relatively recently in the Society's history (Walker, 1985).

Some of the earliest courses for secondary level teachers were organised for the Society by James Milford and involved one-day meetings at Imperial College, London. These were developed in response to the foundation of 'O' level Meteorology and covered selected aspects of the syllabus intensively. It soon became apparent that there was a good deal of demand for such courses and partially in response to this, a new Subcommittee (of the Royal Meteorological Society's Education Committee) on Courses was established in the spring of 1985, chaired by the author.

IN-SERVICE TRAINING

Initiation

One of the first acts of the Subcommittee was to write an introductory letter to Geography/Environmental Studies Advisers in seven County Education Departments in South-East England. The idea was simply to bring to their attention the existence of the Society, and its interest in the provision of educational expertise in Meteorology at school level. All county authorities own both residential field centres and non-residential teachers' centres which could act as venues for teacher training.

A positive response was received from four authorities, and no reply whatsoever from the remaining three. The first reply, from Berkshire's Education Department, was enthusiastic and set out a number of possibilities which were condensed into a short course after personal discussion with Mr Lawrence Taylor, the Adviser in question. An interest in developing such courses was expressed also by the Inner London Education Authority and both Hampshire and Dorset Education Departments.

ORGANISATION

Weather Projects for Middle Schools

Funds are available to support In-service courses not only from County Authorities but also from University Education Departments. The money is of course limited so that the funding authority will normally quite naturally want to stipulate some of the cost-related parts of a proposed course.

Thus the training provided for Berkshire was aimed primarily at ideas for projects for Middle School (8-13 year olds) teachers on a non-residential 1½-day course held in the Department of Meteorology at the University of Reading. Given the constraints affecting both tutors and participants, a Friday afternoon and all day Saturday in early July 1986 were determined to be the best period. The advice of Reading University's In-service coordinator was invaluable, particularly with regard to when Berkshire's (and adjacent Counties') teachers would be most likely to be able to take a half day off.

The non-residential nature of the event kept costs to a reasonable level, and meant that all twenty-seven teachers were sponsored by their Authorities. All of the administrative tasks were carried out by the University's Education Department which has a regular mailshot to all Berkshire schools and to those in adjacent areas of surrounding counties. Seventeen teachers attended from Berkshire, four from Oxfordshire, three from Hampshire and one from each of Buckinghamshire and Inner London.

A Weekend on Weather

Positive replies to the Subcommittee's original letter also came from the adjacent counties of Dorset and Hampshire, both of which offered a residential centre as a venue for an In-service course. A decision was made to use Leeson House Field Studies Centre, Langton Matravers near Swanage, Dorset but to advertise the course in both counties.

As with the Reading event, the funding was provided by the Authorities concerned and all the administrative chores were undertaken by the Principal of the Field Centre, David Kemp, and his staff. The only early work required of the tutors was to decide, liaising with the 'users', what level the content would be pitched at, and to provide a brief outline and tentative schedule of the weekend - which was 30 January to 1 February 1987.

The course was advertised as being aimed at teachers involved with the 10-16 year age range, and as being of particular relevance to those involved with science, technology and geography.

Execution

Although the training was specifically stated to be of interest to teachers of children within restricted age ranges, both classes in fact involved staff from primary to senior secondary level. This problem proved to be not too severe, and the tendency was generally to aim the content at those involved with older pupils. In this way the participants interested in such a level received directly useful expert instruction, while those concerned with the lower age range generally gain a reasonably firm grounding in selected aspects of the subject.

Instruments

The use of weather instruments is relevant at virtually all levels, and forms a prime part of course content. A wide range of equipment is presented to illustrate what can be measured easily and reasonably cheaply. The demonstration, given by Dr Geoff Jenkins at these two courses, thus provides an opportunity to see and handle items ranging from the very basic, home-made type up to quite sophisticated sensors. Teachers find the opportunity to ask advice of someone so well versed in 'cheaper' as well as standard instruments invaluable. Knowing what items will suit their ideas for projects, given financial and educational constraints, is clearly of great practical use. With the development of GCSE and its project component, there is considerable pressure on teachers to develop suggestions that can be accommodated by using a relatively small stock of good equipment.

Observations

A common problem in the classroom is what to do with the observations once they have been logged - or how to acquire observations from other sources for project work. Both of these tasks were addressed by running through a simple exercise of processing selected daily observations, either in real time using a school's weather station or by illustrating the value of using one-day-late observations published in the UK's quality press. The availability and scope of weather data for schools has been discussed by Singleton (1985a,b), while the use of a BBC Micro linked to an automatic weather station was outlined by Jenkins (1985).

We stress the way weather data can be used to develop skills in numeracy and communication, by suggesting the use of up to 60 British stations' observations from the previous day published widely in the UK in two or three good daily news papers. Each student can be assigned one or more stations and collect daily values of maximum temperature, sunshine hours and rainfall total - and then think about how to process and present a time series for a project period of say one week. Should they average a sequence of sunshine or rainfall totals? What does the result mean? How does their station fare in a rainfall 'league table' which may be updated daily using all students' values? Can a given site's values be used to assess regional forecasts issued in the press? Are the newspapers' daily weather maps useful for a project display to provide background for class discussion? These ideas have been developed in more depth and with more formalism with reference to differentiation in a GCSE Meteorology class (Reynolds, 1986).

Recurring Themes

Experience of previous training courses highlighted the frequent appearance of problems relating to specific topics within school syllabuses. These are normally covered selectively at each course and include pressure and wind, airmasses and fronts, jetstreams, static stability and convection, clouds and water in the atmosphere, and satellites and the weather.

Open House

A final but important ingredient of short courses is the provision of an open session at which the participants are free to ask and discuss any question relating to Meteorology that they wish. Indeed the style throughout the course is one of informality which engenders a favourable environment for discussion. With an open session however, it is advisable to have say three tutors present since none of us possesses individually the compendious knowledge required to satisfy all the queries!

Conclusion

Within the school system in Britain there exists a large and wide-ranging demand for the provision of meteorological education. There also exist, to a reasonable degree, the funds and the machinery for implementing planned courses, along with the expertise of advice for potential tutors regarding likely educational markets. Both the courses outlined here were successful, judging by the participants' remarks, and the potential problem of primary to senior level teachers sharing proved to be mainly minor. However, future courses will be more explicit in their advertising in an attempt to ensure a better sorting of attendance.

Acknowledgements

I would like to thank my colleagues Geoff Jenkins, Bob Riddaway and Dave Shaw for acting so enthusiastically and effectively as fellow tutors.

References

Ilsley, M.M., 1985: A preliminary survey of weather education in primary schools in one local education authority. Weather Education, Royal Meteorological Society, 61-69.

Jenkins, G.J., 1985: Using a BBC micro in an automatic weather station. Weather, Royal Meteorological Society 73-78.

Milford, J.R., 1985: Meteorology for the General Certificate of Education. Weather Education, Royal Meteorological Society 73-78.

Reynolds, R., 1986: GCSE Meteorology, A Teachers' Guide. Southern Examining Group, Tunbridge Wells, 36pp.

Singleton, F., 1985a: Weather data for schools - part one. Weather, 40, 267-271.

1985b: Weather data for schools - part two. Weather, 40, 310-313.

Walker, J.M., 1985: The educational activities of the Royal Meteorological Society. Weather Education, Royal Meteorological Society 10-19.

16

Computer aided teaching in meteorology

C. N. Duncan, Department of Meteorology, University of Edinburgh

INTRODUCTION

The possible uses of computers in meteorology have now extended beyond data analyses and numerical forecasting to include education and training. The ways in which computers can be used in training are numerous, but it is reasonable to divide them into three classes; rule learning and memory testing; tutoring systems; and explanatory methods. Perhaps these divisions are too rigid but they will serve to illustrate three very different styles of education. Rule learning and memory testing programs were among the first to be produced for computer teaching. They followed on from the use of multiple choice questions in conventional training. While these methods may still have their place in certain subjects they are now accepted as too constraining and inappropriate in any field in which reasoning and understanding are to be taught and assessed. Tutoring systems attempt to simulate the teaching of a human tutor. As it is unlikely that every student can have an individual human tutor these systems are a cost-effective way of providing training, provided the tutoring system emulate some of the basic functions of a human. Exploratory methods are usually used in conjunction with other teaching media. Students may be presented with various hypotheses, either by a human tutor, book or any other medium, and are presented with a computer system which can enable them to test the hypothesis. This may be done by providing a database which can be tested for certain conditions, or through a simulation in which the student controls certain parameters.

The case for and against using computers to teach meteorology is given below with examples of both tutoring and exploratory systems. It is suggested that when there are large numbers of students computer aided teaching is very economical, both in terms of absolute cost and accessibility by students.

THE CASE AGAINST USING COMPUTERS

The arguments against using computers are separated into those which can be answered and those which cannot. Firstly, those which cannot be answered:

- Very good computer tutors will never be as good as humans, probably not even as good as mediocre human tutors. To justify this statement consider some of the things students often say in tutorials, "I don't understand", "Why can't I do this?", "What would happen if ...", "Show me". Humans understand these comments instantly and use their knowledge of the student, and of the subject to make decisions about how to reply to that individual in a way which may be unique. Computers may try to do this but will not succeed as well as humans.

- In any subject area the amount of information available in a computer will never equal that of an experienced teacher.

- The range of techniques available for human communication is far greater than can be produced by a computer. Teachers can use voice tone, gesture, and pointing and drawing movements very effectively. They can also use pauses in speech for effect.

Now for some of the arguments which are often used against computers which can be answered and are answered in the next section.

- Learning how to use a computer takes too much time. Even when one computer system has been mastered its commands are no use on another system.

- Using a keyboard is difficult and is not a necessary part of learning the subject.

- It is difficult to get help when you run into trouble on a computer. It is especially difficult for a novice to be able to differentiate between the main problem and side effects which might arise.

THE CASE FOR USING COMPUTERS

Before considering the positive advantages of using computers the answers to the last three points above against computers are as follows:

- It should not be necessary to learn how to use a computer in order to use a teaching program, although it is still necessary to learn how to use the program. It is highly desirable, and not unattainable, that all programs should have the same appearance (i.e. the same man-machine interface) so that when a student has learned how to use one program the same skills are used in any other program.

- Learning how to use a keyboard is now hardly necessary. Computers which use WIMP (Window, Icon, Menus, pointer) systems are easy to control using only one or two buttons.

- Getting help from a computer is difficult but it is getting easier. It is possible to make help available on a particular topic by pressing a single button (see tephigram example below). Also, help can be provided in different ways depending on the context of the program. These improvements to help systems generally require much more text to be held on-line.

The answers to previous criticisms against using computers for teaching are now being removed as a result of advances in technology. Increased memory and new 'standardised' interfaces make learning to use a program relatively painless and mean that the knowledge, once gained, will be of use again and again.

Now we can consider the positive advantages of using computers.

- Ability to make mistakes, or take risks, without being accountable.

- Objectivity of assessment.

- Speed of progress through material is dictated by the student.

- The impression of continuous feedback is encouraging.

- Access at any time of day or night.

An example of a tutoring system

Aim

To explain the structure of a tephigram, outline its uses, and give many exercises to enable the student to become familiar and confident in its use.

Expected Users

University undergraduates in a first year Meteorology course.

Approach

The uses to which the tephigram is put have been identified and sorted into order to determine what previous knowledge of the diagram is required for each task. This led to the establishment of eight 'layers'. Each layer contains material which depends on knowledge of the previous layers.

1. Structure of the tephigram, potential temperature lapse rates, plotting environment curves, height and thickness.

2. Water vapour, saturated adiabats, relative humidity, vapour pressure.

3. Path curves, condensed water, Föhn effect.

4. Normand's theorem, use of wet bulb potential temperature for air mass analysis.

5. Lifting condensation level, convective condensation level, convective cloud development and surface temperatures.

6. Vertical static stability and instability.

7. Level of free convection, latent instability, the effect of diurnal variations.

8. Potential instability.

Since the first level does not include moist processes the diagram is simplified by removing the saturated mixing ratio lines and the saturated adiabats. This presents the students with a slightly simpler diagram at the beginning.

At each level there are a number of questions for students to answer which test their knowledge and also show how that knowledge may be applied. For each question a set of the most common errors made by students has been identified. (This was achieved by calling on the expertise of someone with thirty years experience in teaching students). Each time a student answers a question by identifying a point on the screen it is possible to check whether he/she is confusing dry and saturated adiabats, is assuming that pressures greater than a certain value lie above rather than below that level on the diagram, or any one of a number of other common mistakes. The feedback provided to the student is therefore not simply 'wrong' or 'correct' but an explanation of why the answer is wrong and what the flaw in the student's reasoning appears to be. Since the knowledge required is in an earlier layer of the program it is easy for the program at this stage to suggest appropriate revision. The student may then revise the appropriate section and return to try the question again.

The program tries to emulate a good human tutor. It provides questions which are varied by randomisation and gives feedback and advice at all stages. Other assistance is provided for the student through a sophisticated 'help' system, an interactive notepad and the ability to ask for worked examples. These are described in the section on special features below.

Hardware/System

The hardware used is a Torch Triple X computer (68010 processor) with 2 Mbytes of memory and a 20 Mbyte hard disk. The operating system is Unix and the man-machine interface is a WIMP environment produced by Torch. An inkjet printer is included to allow students to make notes which incorporate sections of the tephigram in colour.

Special Features

The software for this system is intended to be 'state-of-the-art' in the way it provides assistance for the user. Three of the features which deserve special mention are the help system, the notepad and the worked examples.

The help information is stored as a set of short descriptions which cover items such as the definition of lifting condensation level, or how to equalise areas when doing a thickness calculation. The help information is graphical as well as textual. Without the use of sketches it is impossible for a teacher to show how to use a tephigram. The most common time for a student to need help is when they do not understand a question which has been asked. This is often because they are unsure of the meaning of one or two words in the question. By using the mouse the student merely needs to point at a word in a question and a definition of that word will be provided.

The notepad may be called on by the student at any time. The ability to decide to pause at any point and use another of the facilities provided is a feature of WIMP environments. When using the notepad the student may copy any part of the tephigram, and the constructions on it, into his/her own 'notepad'. The question may also be copied and the student can add any notes he/she wishes. It is possible to store these notes on-line for future recall or to have them printed. The student may then take away parts of the tephigram, the question, the response to the question, and any additional notes, on a single sheet of paper.

The worked example is a replication on the computer of the common teaching technique where a student who doesn't understand how to solve a problem is shown by the tutor how to solve an analogous problem and then asked to try the original problem again. The computer chooses the worked example randomly and explains how and why each stage of the solution is obtained. The student may ask for extended explanations or keep them brief and may alter the speed at which the demonstration runs.

Feedback

At the time of writing the system has been used by one third of the class as a trial in order to refine the software before full scale use in the next academic year. The feedback from the students has been very positive. The use of a WIMP environment enabled each student to use the program without difficulty after a period of five minutes instruction. Several students said they preferred this method of instruction because they felt more willing to keep trying something when they got it wrong but would not have been willing to keep asking for help from a human tutor. Others commented that they enjoyed getting rapid feedback on their progress compared with in a tutorial when they often had to wait until a tutor was free.

An example of an exploratory system

Aim

To provide a data set of surface synoptic observations and programs for manipulating the observations. These can then be used to test hypotheses about synoptic scale weather systems.

Expected Users

Children in the later years of secondary school.

Approach

The package is designed to stimulate the user to pose questions about synoptic scale systems and to use the data provided to try to answer those questions. As well as data files and programs on disk the package includes synoptic charts, satellite images, booklets and worksheets. The worksheets are divided into five categories: observations and plotting; air masses; fronts; depressions; and anticyclones. Exercises are suggested in which the user has to examine all the available material to answer questions such as 'What is the weather like as a warm front approaches?' or 'How might you distinguish a polar continental air mass from a tropical maritime one?'

The meteorological data are synoptic observations for four periods of three days each. The case studies have been selected to illustrate a developing depression, a cold air outbreak, a mature depression, and a blocking anticyclone. Mean sea level charts are provided at six-hourly intervals and pairs of infra-red and visible NOAA9 satellite images are provided for every day.

The four programs allow the data to be displayed in different ways: a full synoptic station plot for one station at a time; one weather element at up to six stations at a time; a graph of the variation of two weather elements with time at up to three stations; a horizontal distribution of one weather element at all stations at one time. The programs which display data at only one synoptic time have the ability to step forwards or backwards in time with a single key stroke.

Hardware/System

An Acorn BBC micro or Master with a colour monitor and disk drive is essential. An optional Epson compatible printer can be added which will allow hard copy of any screen display to be produced. This configuration is very common in UK schools where a recent survey found that, on average, there are fourteen microcomputers in every secondary school.

Special Features

This system provides schools with a data set of almost 2500 full synoptic observations. Although worksheets are provided suggesting some exercises there is enormous potential for teachers to develop their own worksheets and use the data and programs to allow students to explore features not detailed in the package, such as diurnal variations, see breeze fronts, or the relationship between wind and pressure gradient.

The system is entirely controlled by menus. Choosing from up to twenty-five stations leads to large menus, so it is possible to move from left to right as well as up and down through the menus. The user then only needs to use cursor keys. This leads to many fewer errors in the program's interpretation of what the user wants to do and provides an interface to the program which is identical for all twelve menus in the system.

A window at the bottom of the screen is reserved for 'help' information using the 'help' option which exists on every menu.

Feedback

At the time of writing, the system had not been released for trial.

THE FUTURE

Producing educational software is an exceedingly time-consuming task. Some estimates suggest that the ratio of the time taken to prepare a program for use, to the time taken for a student to work through the program, is about 300:1. It is obvious that the major item of cost in using computers to teach is the cost of the skilled manpower to produce the system. By comparison, the cost of hardware on which to run the programs is not great. Of course it is possible to reduce the relative enormity of the development costs if the system is used by large numbers of people.

In meteorology we are fortunate to have a worldwide community with common goals in education and training. We also have a need to provide high quality training to many more people than there are skilled teachers to teach. We should use computer aided learning as a means of providing relatively cheap and reliable instruction to many. A room containing ten microcomputers can give tuition equivalent to four or five skilled tutors. What is more, the microcomputers can work endless shifts which could double the number of tutors they represent. The cost of a WMO instructor at present is at least US$ 30,000 per year. The cost of ten microcomputers is about the same but they would have an expected lifetime of between five and ten years. Neglecting the development cost, a computer aided teaching laboratory of ten machines, might deliver the equivalent of at least 20 skilled tutor-years (or at best 50 tutor-years) for the cost of one human tutor-year. However, we cannot neglect development costs!

An interesting intellectual exercise is to estimate the cost of providing enough teaching software to enable meteorological institutions throughout the world to run computer aided teaching programs for about two hours a day during a teaching year of, say, 200 working days. This does not mean the computers would be idle the rest of the time as they could be used for data analysis when not being used for teaching. If the ratio of 300:1 is correct this would need 120,000 hours of development time to produce the necessary 400 hours of teaching software. The development effort would be about 60 person-years, or six years for a team of 10 people. We might estimate the cost of developing the software by considering that a person-year costs about US$ 30,000 which would give a total development cost of US$ 1.8M. The cost is high, but the provision of 400 contact hours of teaching could enhance education and training considerably. Of course, the greatest benefit would be gained if a large number of institutions used the software. For example, if 60 institutions found it useful the cost per institution would be US$ 30,000; the same as the cost of the hardware for ten machines. What would an institution get for this one-off cost of US$ 60,000? Ten machines working 400 hours a year with a lifetime of between five and ten years. The total would be between 20,000 and 40,000 hours of contact teaching for US$ 60,000. To get 20,000 hours of contact teaching from a human tutor, at present, rates, would cost about US$ 300,000. The cost of computer based teaching would be about one tenth of human teaching.

These figures are just examples and are based on present day costs. It is worth noting that the cost of a human tutor would undoubtedly rise during the lifetime of the computer based systems. There is a trade-off in the economies of scale. If the number of institutions using the software is small the development cost becomes disproportionately large. On the other hand, the initial cost of equipping the institutions should not be too large. A gradual approach would obviously be sensible but the cost-benefit analysis still shows that there are considerable advantages to be gained by using the advances in technology to increase the use of computer aided teaching in meteorology worldwide.

17

The role of education and training in increasing the use of meteorological and environmental science in Saudi Arabia

A. Henaidi, Meteorology & Environmental Protection Administration, Jeddah, Saudi Arabia

RECENT HISTORY OF EDUCATION AND TRAINING IN MEPA

Our operations in meteorology date back to the very early '50s and, up until around 1980, my organization was known as the General Directorate of Meteorology. Then, on 21.4.1401 (AH) (which corresponds approximately to early 1980) a Royal Decree assigned as additional duties and responsibilities in the area of environmental protection and pollution control; our name was changed to Meteorology and Environmental Protection Administration (MEPA) to reflect those additional responsibilities.

In those days, (the 1970s), Saudi graduates and technical staff in Meteorology and this new field of Environmental Protection were essentially not available and we had already realized, even prior to the formalization of the General Directorate's added responsibilities, that something would have to be done quickly. A number of actions were taken that have already had far-reaching effects and, we trust, will continue to do so. These are:

(i) In 1974, a Royal accord was granted to establish a specialized Institute of Meteorology and Arid Lands Studies. This Institute was to be a co-operative venture between the General Directorate of Meteorology and the King Abdul Aziz University in Jeddah. The Institute began its activities in the academic year of 1975-76. Later (mid 1980's) the Institute became the Faculty of Meteorology, Environment and Arid Land Agriculture, within the University.

(ii) Mainly on the initiative of MEPA's Senior Executives, the Institute was upgraded into a Faculty of Meteorology and Environmental Studies within King Abdul Aziz University. As a result, MEPA has benefited from the influx of young Saudi nationals trained in the Departments of Meteorology and Environmental Sciences, two of four Departments within this Faculty.

(iii) Between 1977-81, 230 MEPA staff received technical training at the Bailbrook College in England under the Operations, Maintenance, Enhancement and Training (OMET) Contract awarded by MEPA to International Air-Radio Ltd. of the UK.

(iv) An agreement between the Saudi and Australian Governments aimed at providing 20 highly experienced staff to assist MEPA to upgrade the effectiveness and efficiency of its meteorological and environment protection services. The 20 expert staff comprised 15 officers from the Australian Bureau of Meteorology and 5 staff from other Government and private sector sources experienced in environmental law, environmental planning, pollution control and standards, marine and earth pollution control and standards, marine and earth sciences. These people have worked closely with our Saudi Executives and their staff to give the latter the benefit of their experience both in Australian Government Departments and in the consulting field.

EDUCATION AND TRAINING OBJECTIVES IN MEPA

The objectives of our Education and Training program are:

To ensure that the best possible training and development opportunities are provided to MEPA personnel.

To maintain and operate a centre of expertise in the fields of Meteorological and Environmental education and training for the benefit of MEPA and its personnel.

To meet the special training needs of MEPA covering technical fields such as electronics, systems analysis and programming, communications, weather observing and forecasting environmental protection and climatology. Also to provide basic training in administration and financial management is an important facet.

INTERNATIONAL RECOGNITION OF NEED FOR SPECIALIZED TRAINING

(i) On 6 November 1985, the Heads of State of the Gulf Cooperation Council (GCC) countries signed a set of 13 policies and General Principles for Environmental Protection. The 11th of these states:

> "Work on securing the manpower responsible for environmental affairs by supporting qualification and training plans at the local and regional levels by utilizing the benefits of training courses, symposia and meetings pertaining to the environment which are held in member States, and also encourage universities and relevant establishments to initiate qualifying and training programs for these purposes".

(ii) On 14 October 1986, the First Ministerial Conference on Environmental Considerations in Development pronounced the Arab Declaration on Environment and Development. There were five main decisions, the third of which emphasised the importance of introducing and upgrading training programs.

(iii) Saudi Arabia is an active participant in the programs and projects coordinated by the Regional Organisation for the Protection of the Marine Environment (ROPME). The Organization, based in Kuwait, was formed through Article XVI of the Kuwait Regional Convention for Cooperation on the Protection of the Marine Environment from Pollution, signed by the eight member States in April, 1978.

The Action plan, as it is called, emphasizes the need for training in all aspects of pollution control and environmental protection, particularly in the marine science areas. Many of our graduate staff have attended workshops arranged by ROPME over the past years and there has been great benefit to all.

ROPME is essentially a coordinating agency but it does have one "operating" arm - the Marine Emergency Mutual Aid Centre (MEMAC), based in Bahrain. This central agency was set up as a result of the member States signing, together with the above convention, a Protocol Concerning Regional Cooperation in Combating Pollution by Oil and Other Harmful Substances in Cases of Emergency.

Although most member States, including Saudi Arabia, have established their own operations in this important field, MEMAC has played an important coordinating role, particularly in the field of training. Again, many of our staff have benefited greatly from attendance at seminars and workshops designed to impart knowledge of recent developments in oil spill combat, surveillance of oil slicks, computerised modelling of the movement of oil slicks in the Arabian Gulf and the like. An example was the attendance of a group of MEPA personnel at a recent training course organized by the Bahrain Petroleum Company (BAPCO).

The point of all this is that international recognition is now being given to the importance of proper training in environmental protection. These international accords will provide a firmer basis for funding of training schemes within the signatory States.

OVERSEAS TRAINING

Over the next two years, through the Australian - Saudi Government to Government Agreement, we hope to second twelve young Saudi graduates to a number of different Government Departments in Australia; each graduate will spend six months there. The fields covered will include pollution control (air, water, noise, wastes and chemicals), environmental impact assessment, earth sciences, geographic information systems (GIS), monitoring and analysis of coral reef and coastal zone management. We are very enthusiastic about the potential for this on-the-job training scheme.

In addition, MEPA may soon enter into an agreement with a specialized contractor to assist MEPA in its day-to-day operations, maintenance and training; both meteorological and environmental protection services will be covered. In drafting the specifications for this agreement, we paid close attention to our perceived training needs. The agreement would require the contractor to undertake two important tasks in the training of personnel in the meteorological and environmental protection areas.

Firstly, he will be required to provide a specified number of expert staff covering the whole, wide field of pollution control, conservation and environmental protection. In the first year of the agreement, about ten percent of these staff are required to be Saudi Nationals. In the second and third years of any such contract, those percentages must be increased to 20 and 30 per cent, respectively. Clearly, the contractor will need to arrange appropriate training programs if he is to meet this tight specification.

Secondly, in recognition of the fact that environmental protection in the Kingdom will succeed only if there is awareness of its need across the board, the contractor is required to select 38 Saudi Nationals educated to secondary school level and to place them in appropriate institutions (either in the Kingdom or overseas) and to have them trained to technician level in specified areas. He is also obliged to take 14 university graduates and train them even further in specific areas of pollution control and environmental protection. The contractor will also be required, over the 3 years of the contract, to select and train 95 Saudi Nationals as, Observers (80), (WMO Class 3), Forecasters (12), (WMO Class 2): and Observer Training Instructors (3). MEPA sees this arrangement as an important departure from traditional training methods.

TECHNOLOGICAL DEVELOPMENT TRAINING

One of the key features of MEPA's Education and Training program is recognizing the need to train staff on new technological equipment and systems as they are progressively introduced into MEPA's operations. It is our view that training in this area can best be undertaken in the contractor's country of origin or he can provide "in-house" courses in MEPA.

An example of a new major training initiative in this area is the agreement between MEPA and the United States National Oceanic and Atmospheric Administration (NOAA) for the development and implementation in MEPA of a major meteorological and communications modernization program over the next 8 to 10 years. Under this agreement, training for three groups of MEPA staff is proposed; viz,

- Forecasters
- System management and system operations staff
- System maintenance staff

The first three staff to be trained in the new systems at the Florida State University, USA, have been selected and will take up their appointments shortly.

Other examples of contractor training of MEPA staff in new technologies include training operators in the use of Local User Terminals as part of the satellite-based environmental data collection system called "ARGOS".

The participation of MEPA staff at WMO Seminars and Workshops on topics such as Scientific Developments and Systems Development is also an important part of our Education and Training program. We believe it is important for our personnel to be given the opportunity to meet with professional staff from sister administrations to stimulate interest and discussion on ways to solve common problems and make the best use of new developments as they come along.

MEPA'S PRODUCT USERS

The importance of utilising to best advantage the products produced by MEPA, as the National Meteorological and Environmental Centre, throughout the Kingdom, cannot be overstated. As with most similar agencies, our principal product users are the Aviation Industry, the Armed Forces and the General Public for the Public Weather Service. These areas are traditionally our most important clients in terms of workload and priority. However, MEPA has for some years now been fostering the development and tailoring of new services to meet the needs of a wide range of new and potentially new users.

These customers include industries involved in agriculture, transportation, construction, insurance, environment protection, the oil industry and agencies responsible for long-term infrastructure planning in the Kingdom. This work is growing in importance and we see in the near future that it will require the dedication and training of specialized staff to meet the needs of these organizations. I believe our long-term future lies in recognizing and nurturing these growing demands and in that we must also tailor our products to meet their needs.

18

Designing meteorological services to meet the need of users

P. G. Aber, Environment Canada, Atmospheric Environment Service

INTRODUCTION

Aim

The intent of this paper is to discuss an approach to designing meteorological services that meet the needs of users. A model is presented with this aim, and its elements are described using examples from the Canadian experience. The conclusions are generalized to be applicable to other national and international meteorological organizations.

The viewpoint is principally from the government or public sector perspective, but the role of the meteorological private sector is considered. The scope of this presentation is compendious, touching on the significant features of the design process.

Outline

The lecture is presented in four parts. The first section describes a model for designing meteorological services that accommodate the needs of the user communities, and discusses its application, benefits and constraints. The second section looks at the influences on users' requirements for meteorological services in Canada as conveyed through its geography, people, climate and politics. The third section discusses some of the details of this model from a Canadian perspective, while the last section summarizes the significant conclusions.

A MODEL FOR DESIGNING METEOROLOGICAL SERVICES

The Model

Weather affects everyone. It is pervasive, influencing, virtually all aspects of society and the economy. Today, people are placing an increased value on information about the changing behaviour and character of the atmosphere as it applies to their specific meteorological-sensitive social and economic activities.

National Meteorological Services are constantly being pressed by people, to provide tailored, meteorological information to support their activities. Responding to these pressures by introducing or improving services is a challenge, especially, when operating from within fixed budgets.

An approach to designing meteorological services must attempt to accommodate the needs of the user community, yet also, must be sensitive to the political environment and the keen competition for resources from other social and economic programs of government. The approach of targeting services to specific users conjures up ideas relevant to modern marketing, such as market research, promotion, service design and client follow-up. On the other hand, political realities and competition for limited resources address mandate and policy issues.

A model for designing meteorological services is formulated around this approach and is presented for discussion. The model is depicted in this first figure.

STEPS FOR DESIGNING METEOROLOGICAL SERVICES

IDENTIFY TARGET SECTOR

CHECK MANDATE AND POLICY

ASSESS USER NEEDS
CLIENT SATISFACTION

CONSIDER COSTS

DESIGN SERVICE

IMPLEMENT SERVICE

MONITOR PERFORMANCE

This service design model provides a framework for the designing of meteorological services that meet the needs of users. It comprises seven major areas of activities.

The process begins with selecting the target sector. This may be chosen by assigning priorities or by political directive. The second activity area considers the political requirements; the mandate of the government or the National Meteorological Service to provide services to this sector, and the limits or restrictions that might apply to the level of service dictated by policy. The third activity area investigates the nature of the user's requirements. Researching of user needs and client satisfaction of the particular sector provide the information base on which to design customized meteorological services. Once the scope of the service requirements are known, the fourth activity area assesses the costs of the service and identifies a source for the necessary resources. The fifth activity area of the model creates the blueprint that specifies the functional aspects of the specific service, and transforms this blueprint into a working prototype. Also, the service design activity involves: acquiring the infrastructure support; testing and delivering of the service; soliciting user feedback; and, assessing the prototype service and modifying appropriately. The final two activity areas of the model focus on the implementation process and performance monitoring and assessment.

Applications and Benefits

This model provides a framework with several applications. It can be used for the design of services for particular sectors, or it can be generalized to the development of a National Meteorological Service.

This model combines a rational decision making framework with the innovation process of designing services that meet the needs of the organization's clients. By applying this model, Service managers acquire the necessary information to test the validity of continuing existing meteorological services and to determine when services need to be added or improved.

Such an approach encourages the organization and the government to take an outward looking focus on its activities and service. In particular, the organization's goals and priorities can be stated as responding and meeting the needs of the various user communities, measured in terms of quality and performance. The value and effectiveness of the National Meteorological Service programs can be evaluated by its clients, while the resource expenditures of the organization can be linked directly to servicing the taxpayers. Their political representatives can evaluate the fairness and appropriateness of their legislation and policies that govern of the provision of meteorological services to their constituents.

The model provides an integrated and rational approach for deciding: who are the organization's clients; what is the scope and dimensions of their requirements; and, what services are to be provided by the Meteorological Service versus those that may be provided by the meteorological private sector. Customer follow-up, monitoring the clients' satisfaction and measuring performance which also are integral parts of this model.

Constraints

The model is subject to and constrained by the environment in which it operates. There are principally three external aspects of the environment that influence the National Meteorological Service's ability to accommodate the requirements of the user community. They are: the political environment, science and technology, and the demands for service.

Political Environment

National Meteorological Services are supported and maintained by government, and therefore, they are influenced and to some extent, controlled by political motives, priorities and objectives, rather than purely by client needs. This is not to say, that the Government Service is disadvantaged by this reality. Government's have made substantial investments in establishing and operating the meteorological infrastructure that is necessary to provide any service.

A Government controls its National Meteorological Service by establishing and approving its mandate, missions and policies, and by the allocation of resources. Mandates, mission and policy statements evolve, primarily, based on the public's expectations of government, such as providing services and providing employment. From the Meteorological Service's perspective, expectations could include services that support: national security and the safety of its people; spur the growth of an efficient, internationally competitive economy, by providing, for example an adequate information base for the effective planning and management of activities; and, safeguard the quality of the environment and maintain its citizens' quality of life. These expectations are interpreted and translated by politicians into laws and priorities which direct and guide the decision making in the National Meteorological Service.

Today, fiscal restraint and budget reduction policies of government act as a severe constraint to the provision of services. Supplementary funding alternatives such as recovering costs or generating revenues from users or sponsors are being considered as a means of maintaining or improving meteorological services.

Science and Technology

The capability and, to some extent, the capacity of the Meteorological Organization to satisfy user requirements also are curtailed by the current state in science and technology. This limitation can be appreciated by considering the impact of the advances over the past decade.

Technological improvements have accounted for much of the productivity improvements in Meteorological Organizations in recent years. Improvements in efficiency in their infrastructure, in many cases have permitted the diversion of monies to improve services to users. Improvements have been realized in areas of service production and delivery which have benefited the user. For example, the use of automatic telephone answering devices has improved the public's access to weather information without placing additional demands on resources. Scientific advances have introduced better understanding and modelling of atmospheric processes. These advances have broadened the scope of meteorological services; being able to respond to environmental emergencies and identifying impacts of acid precipitation and climate change.

Demands for Services

Demands for services are influenced by social and economic issues, such as: adequacy in the supply of food, fresh water and energy; wars; threats to quality of the environment; and, expanding economic activities. The extent to which the awareness and understanding of the benefits of meteorological information is changing, influences demand. Increasing demand for services coupled with a fixed operating resource base constrains the National Meteorological Service's ability to maintain its existing level of services. Increased demands from one sector, like aviation, can divert efforts away from others.

THE CANADIAN ENVIRONMENT

Canada is a vast and diverse country; a developed democratic nation, rich in natural resources, subject to notable fluctuations in weather and climate, yet small in population.

Canada's land area encompasses 10 million square kilometres, and is engulfed by three oceans. All Great Britain could be contained in almost any one of Canada's provinces. Canada is characterized by rugged terrain, an extensive cordillera region, frozen arctic tundra, numerous lakes and rivers, and 60 thousand kilometres of coast line. These features, as you would expect, greatly influence the weather and climate in Canada, with extremes between winter and summer common place. Situated in the zone of the westerlies, much of southern Canada is affected by synoptic and meso-scale storms bringing, on occasion, adverse and sudden changes in weather.

Canada's population is dispersive with fewer than three Canadians for every square kilometre. Almost all of the population live in southern Canada, concentrated predominantly in urban centres (75%), with the majority living near the Lower Great Lakes. Thus, transportation is an important social and economic activity in Canada.

The Atmospheric Environment Service (AES) is operated by the federal government, providing weather, climate, ice and air quality services in support of all Canada's activities, including the military. It employs 2400 people and spends $210 million annually in the provision of services. Compared with other countries, like Great Britain, the level of expenditure per square kilometre by Canada on meteorological services is significantly less.

The demand for services in Canada have been sharply rising in recent years with the number of requests for services doubling since 1982. The federal government's policies of fiscal restraint and deficit reduction in recent years has required the Atmospheric Environment Service to look to other ways of meeting the growing demand for services. Encouraging the growth of the meteorological private sector is one means of meeting these demands. Capitalizing on the productivity improvements realized from new technological methods, the Atmospheric Environment Service has been able to redirect the savings into new priority areas in air quality and climate services without reducing services elsewhere.

Designing Meteorological Services in Canada

Returning to the model for designing meteorological services, I now plan to discuss the elements of the model, using examples from the Canadian experience.

TARGETING THE SECTOR

Identifying Sectors

Identifying a target sector involves: identifying the sectors or client groups that have requirements for meteorological information; and, setting a priority to each of these sectors as means of selecting a particular client group. In Canada, the identification of client groups has been done by analyzing user requests for information, joining user associations, attending user meetings and conferences, conducting surveys and commissioning sectorial studies.

Selecting Sectors

The selection of a target sector or client group depends upon the political or management objectives. In particular, government policy makers or politicians may direct the National Meteorological Service to improve or provide new services to a particular sector. Managers of the organization may have their own agenda, and may select a sector based on a growing problem area or anticipated changing government priority. Selecting a sector because of its priority is another means of targeting. This last method supports decision making in the areas of: validating the need for existing services, introducing new services and improving existing services.

Priorities can be set for servicing these client groups by testing the hypothesis of providing service against a set of decision making criteria based on consistency with political objectives (mandate/sensitivity), and socio-economic (value), marketing (demand), financial (costs) and technical (capability) considerations. Other criteria that influence the priority setting involve the extent to which the characteristics of the sector are known, whether the sector has representatives or national associations, and whether the sector is already being serviced. Anticipated changes in political direction, priorities and motives, also alter priority setting. For example, with the decline in the value of agricultural yields, the government may give high priority to services, like weather, which would benefit the farmers by improving the efficiencies of their operations.

MANDATE AND POLICY CONSIDERATIONS

Mandate and Mission

The second element of the design model, checks to determine if servicing the target sector is consistent with the mandate, mission and policies of the National Meteorological Service. In Canada, the mandate of Atmospheric Environment Service is legislated as part of the Environment Act giving its Minister national responsibility for meteorology. However, the Environment Act does not restrict other federal departments from having their own meteorological service.

The mission or the business of the Atmospheric Environment

Service is to ensure that adequate information on the past, current and future conditions of the atmosphere, ice and sea state is available. This has been revised from earlier mission statements that conferred that its business was to provide adequate information. This reflects the change in the Service's strategy from doing it all to sharing the provision of services with the meteorological private sector.

From the perspective of mandate and mission, the Atmospheric Environment Service's business is in meteorology, and in the provision of services to users. However, not all meteorological services may be provided. Some are provided by the private sector.

Policy

Policies are generally more restrictive than mission and mandate statements. For example, the Atmospheric Environment Service's basic services policy states that meteorological services are provided to support public safety, security of their property, the maintenance and enhancement of environmental quality and the greater efficiency of economic activity. All other services should be provided by the private sector. In the absence of the private sector, these services may be provided by the Atmospheric Environment Service on a cost recovery basis.

The mandate, mission and policies need to be checked to decide whether a particular sector can be served. In Canada, meteorological services are provided to specific user groups that are consistent with the government's policy. If the user's requirements for services are not consistent, some thought must be given to decide if the policy should be changed. If not, the client group is referred to the private sector.

RESEARCHING USER NEEDS AND SATISFACTION

Assessing User Needs

The third element of the model researches user needs and monitors client satisfaction of the target sector. The latter activity is important as it provides information that complements all aspects of the design process. However, the first is necessary to flesh out a statement of their requirements on which to base the design.

The statement of requirements must document the scope and dimensions of the user's needs. Scope refers to the scale, timeliness, detail and importance. Dimensions are the user's specifications of the spatial and temporal domain, lead time, nature of the variable or event, form of the information, content and format of the information packages, and the means of delivering the service. For example, mariners on the west coast want actual and forecast information on wind, precipitation, temperature and sea state for the 200 mile offshore economic zone, at least 4 times a day, with at least 1 day's lead time for forecasts of gale and storm force winds.

Researching user needs is done by: evaluating existing information on users; questioning users directly about their meteorological services through surveys, interviews and sectorial studies; and, by using a decision analysis approach where the user's decisions are modelled and discussed in relation to his information requirements. The first method generally yields incomplete results. The second method requires care in developing the questions and a somewhat narrow focus to acquire any useful information. The decision analysis method is considered the best, where the user's decisions are linked with the value of the meteorological information he requires.

Monitoring Client Satisfaction

The second aspect of the research activity is monitoring client satisfaction. It provides information on the value, utility and appropriateness of current services which feed into the refinement of existing services and the design of new ones. It also builds an information base on which Service managers can make financial, investment, policy and program decisions. In particular, the client's satisfaction provides the feedback in the assessment of the current service. If the clients are satisfied, there is no requirement to improve service. Another priority sector can be selected.

Cost Considerations

The fourth elements of the model addresses the cost considerations. An estimate of costs can be made once the extent of the user's requirements are known. A source for the resources is necessary before proceeding with the design phase. New resources could be sought from other government agencies or reallocated internally. If resources cannot be found, offering the service on cost recovery basis could be considered. If revenues are not available, another provider should be encouraged to provide the service.

SERVICE DESIGN

Developing the Blueprint

The fifth element of model addresses the design characteristics of the service. The service design involves: developing the blueprint or layout, developing a prototype service, acquiring the infrastructure support; testing; and, evaluating and refining the service. The blueprint depicts the working specifications of the service. It incorporates the scope and dimensions of the user's requirements, the infrastructure and service system requirements, service procedures, and the production and performance standards. From this, a prototype service is developed.

The service system and infrastructure requirements details the additional needs for data acquisition, the demands on the forecast production system, and the level of support for additional communications, training, and research and development activities.

An action plan is developed for the testing and evaluation phases to indicate when, where and for how long the prototype will be tested and how its performance will be measured. During the testing phase, the service is offered in a particular location or area. The procedures, standards and delivery are measured and assessed. Users are consulted during this process and their views are assimilated during the testing phase. Their feedback and performance information are used to make modifications to the service, before it is implemented on a national basis.

Importance of Delivery

Delivery is an important aspect of the service. It is the last step in the production process before the service is consumed by the user. This is the time when the user assesses the value and quality of the service provided. If delivery is poor, the user may retain this view of the entire service.

The means of communicating or disseminating information to the user should be carefully considered. Some methods include: providing access to the data banks; communicating directly to the users through a multiplier point (call out service); or, broadcasting the service to users. Services could be distributed through the private sector where value added information could be added to the service at the time of delivery. This is a popular choice in Canada, where public and marine weather is broadcast by the commercial media.

Service Implementation

The implementation phase of the design model finalizes the procedures and standards for the production and delivery of the service to the users.

Part of the implementation phase involves promotion to broaden the users awareness and understanding of the availability, utility and value of the new or improved service. Promotion expands the client base for a particular service increasing the benefit to the country as a whole. It also serves as a vehicle for educating the users on the benefits of the meteorological service. Promotion can establish new contracts, and encourage existing and new users to provide information on the utility and appropriateness of the service.

Monitoring Performance

Monitoring client satisfaction and assessing service performance provides: a profile of the users, information on the utility and appropriateness of specific services, performance data and information on changing user requirements and user sensitivity to meteorological conditions. Performance information and user feedback provides a means of comparing the demand and the quality of similar services. It motivates the Meteorological Service to strive for further improvements. It also provides information for Service's managers to indicate the quality of the service, to justify its expenditures and to demonstrate the effectiveness of the organization's programs.

SUMMARY AND CONCLUSIONS

This lecture presented a service design model aimed at meeting the requirements of users. The model provides a structured approach to introduce new services or improve existing ones. The model is practical and applicable to any Meteorological Service organization in the design of specific meteorological services. The first four facets of the model, collect and assimilate information. Who are the clients? Should the organization be servicing them? What meteorological information do they require? What are the costs and where does the organization acquire the resources? The last three facets of the model develop and test the service prototype, modify the service based on user feedback, implement and establish an ongoing program of monitoring performance and the level of client satisfaction.

The impetus of this model for designing services is that the organization's focus is on the users of its services. Consultation with the user is a key aspect of the success of this approach. From the user's perspective, the important aspects are the applicability of the meteorological information to his activity, its accuracy and credibility, and service delivery. From the perspective of a National Meteorological Service, the important aspects are the effectiveness of its programs and the accommodation of political policies, priorities and motives.

19

Operational aspects of using meteorology for energy purposes

Emeritus Professor J. K. Page, former Head of the Department of Building Science, University of Sheffield

INTRODUCTION

Meteorological information is critical to the assessment of the energy performance of many different types of system, especially those systems where indoor environmental control is important like buildings. Rather more than half the national energy consumption in most countries is consumed to aid indoor environmental regulation, and the cost to the national economy is consequently very large.

The weather affects both energy flows from the natural environment to the energy consuming system, and also the energy flows from the energy consuming system to the external environment. The most desirable energy state to achieve is the comfortable free running state in which an environmentally acceptable balance is established by suitable system design between natural energy supplies and natural energy demands. Frequently the gap cannot be spanned. The consequent fossil and nuclear fuel demands needed to operate the system at satisfactory levels are then determined by the differences between energy demand, which is strongly influenced by external environmental factors like air temperature, wind velocity and long wave radiation and natural energy supply, especially solar radiation which is also very weather dependent.

THE DYNAMIC NATURE OF THE PROBLEM AND THE IMPORTANCE OF THERMAL STORAGE

The problems encountered in this field have a strong diurnal dynamic component, both in supply and in demand, which frequently operate in antiphase, for example nights are typically colder than the days, and there is most solar radiation available in the middle of the day. The dynamic variations produced by the various weather systems passing through are superimposed on top of the diurnal variations. These may have a longer or shorter frequency than one day. Periods of sustained hot weather or sustained cold weather create the maximum energy demands, and are of special importance in basic design, both of individual systems and of supply utilities, which must have the capacity to meet peak energy demands. The thermal capacity associated with enclosures operating in interaction with the exterior environment, is very important in determining the internal dynamic thermal responses to the varying external environmental factors, so the assessment of dynamic thermal storage effects forms a key aspect of the design of thermal energy systems, and performance is closely linked with weather variations. Such problems can only be handled by examining the dynamic characteristics of the natural energy supply system in relation to the dynamics of the external natural environmental factors creating the various internal demands, impacting on the dynamic thermal response characteristics of the basic system, be it a building, car, ship, aeroplane, refrigeration plant. The implications are that external data are needed at the hourly level rather than at the daily level, if these dynamic interactions are to be properly studied.

KEY INTERACTIONS BETWEEN ENERGY SYSTEMS AND CLIMATE

The impact of climate on energy systems and their operation can be analysed in five interrelated modes:

1. Climate as an energy supply source
2. Climate as a key influence on energy demands
3. Climate as a risk factor to supply utilities
4. Climate as a risk factor in the design of the user energy system
5. Climate as a critical resource for the safe dispersal of pollutants produced in the supply of energy
6. Pollution from energy production impacting on the global meteorological system

The impact of energy systems on future world climate will not be addressed here, though obviously one of the ways of reducing the adverse long term impacts, through acid rain, carbon dioxide etc. is by making better use of natural energy resources, like solar energy and wind, and through conservation, so less fossil fuel needs to be burned per annum (Page 1980). Important though it is, the issue of safe pollutant disposal will not be discussed further in this presentation. Obviously knowledge of the vertical structure of the atmosphere is important for tall chimneys. For small chimneys local aerodynamic effects within the boundary layer are more critical.

CLIMATIC NEEDS FOR RENEWABLE ENERGY DESIGN

The author made a systematic review of the problems faced in supplying meteorological data for applied solar energy applications as an invited paper to the World Solar Congress in Perth, Australia (Page 1984). WMO has also prepared a very useful Technical Note on the subject (WMO, 1981a). WMO has also prepared a Technical Note on Meteorological aspects of the utilization of wind as an energy source (WMO, 1981b). Wind energy is very site specific, and special local studies may be required of wind flow over specific hills etc. (Lamming, 1985). The concentration in this paper will not be on alternative energy systems, but rather on data needs for the system design and assessment of energy consumption in buildings. This includes, of course, an important passive solar energy contribution, which can be enhanced by appropriate weather sensitive building design. Natural energy exchanges impinge strongly on all buildings. The preparation of climatic data to support passive solar heating and natural cooling design are discussed in detail in a manual being prepared for UNCHS (habitat) by the author (Page, 1987).

BUILDING ENERGY BALANCES

As about half the world energy consumption is used to secure a satisfactory energy balance for man indoors, and as sometimes the external environment is too cold and sometimes too hot, the positive natural energy resources like the sun are only sometimes useful. In hot climate situations, the energy losing aspects of the external environment like outgoing long wave radiation and air movement over heated surfaces, become beneficial. Table 1 outlines the natural climatic resources that may be used given appropriate preplanning in suitable situations to supply some of the energy needs of engineered systems. Table 2 indicates the conditions favouring their different uses. When a climatic factor is acting to increase demand, the principle of shelter from that factor has to be invoked, for example the use of overhangs to keep solar radiation from entering buildings, which requires knowledge of the geometry of the solar movements and the associated energy of the beam.

Basically the analysis has to proceed in terms of energy balances about some desired balance point, normally defined in terms of human need, but sometimes in terms of plant need as in horticulture, or animal needs as in the case of farm buildings. This demands defining physically concepts like thermal comfort.

Essentially one is forced into two analyses, a cold weather analysis where the energy demands are strongly influenced by air temperature, wind speed and long wave radiation, and a hot weather analysis where the energy demands in cooled buildings are strongly influenced by the solar radiation, with the amelioration coming through air movement, evaporation, and long wave radiation exchanges. Design objectives for cold weather are given in Table 3. Design objectives for hot humid weather are given in Table 4, and for hot dry weather in Table 5.

THE ROLE OF WIND IN BUILDING ENERGETICS

Wind is important both as a potential alternative source of energy supply, but also as an agent for promoting cooling by air movement. The wind also impinges on the demand side, by increasing surface convection and ventilation in cold weather, especially in buildings that are aerodynamically loose. Microclimatic factors are very important in determining the actual impacts on buildings, and design for wind shelter is helpful in reducing cold weather energy demands. Attention has to be paid to the directional nature of the wind in association with the air temperatures typically encountered with winds from different directions.

DATA NEEDED FOR ASSESSING THE PERFORMANCE AND DESIGN OF ENERGY SYSTEMS

Data of different complexity are needed at different stages of the design process. Bioclimatic analysis is an important tool used in the early stages of indoor environmental design to establish climatic priorities in design (UNCHS Habitat, 1984, Milne & Givoni, 1979, Page 1987) Monthly mean climatic data on maximum and minimum dry bulb temperature, and associated humidity are plotted onto suitable charts relating to human comfort, so the building design solution space can be more clearly identified in terms of principle. The charts themselves make it possible to perceive the effects of indoor air movement. Account is taken of typical dynamic thermal performance by distinguishing heavy weight buildings from light weight building, and by identifying the period of the day during which natural ventilation for cooling is introduced. Overlays on the chart indicate design solution areas, at which it is sensible to aim.

The next level of complexity is to study the mean vector characteristics of climate, identifying favourable and unfavourable orientations from the point of view of energy supply and energy demand. Daily data may be used, but in directional association with other variables. Such data is particularly valuable in deciding layout in relation to external shelter. The work of Reidat (1960) in this field provides a good example of clear presentation of such information.

Vector information about prevailing wind direction, and strength is very important in assessing ventilation design, especially in places with hot humid seasons. In areas with very cold winds, effective design for shelter requires knowledge about the typical directions of cold winds. There are underlying statistical implications, but, if overcomplex presentations are made, they will not be used by designers. The aim should be to present the minimum data needed to aid a good design decision. It is worth remembering the English proverb 'The best is the enemy of the good'.

The third level of complexity is to present data at the monthly mean hourly level for different types of day, like overcast days, sunny days or the average hourly values. The advantage of hourly data is that a dynamic analysis can be attempted using the assumption of a run of similar days. This type of approach was adopted, for example for diurnal temperature variations, in producing design temperature data for a Handbook of climatic data for the UK (Page & Lebens, 1986). The overcast day data were based on days with the daily mean cloud amount as 8/8, while the clear sky data were based on the mean cloud amount less than 2/8. Such approaches enable the characteristics of the different types of day to be studied in relation to energy demand patterns. The important meteorological decision is how to define the selection criteria. In the UK there is no shortage of overcast days, but there are very few totally cloudless days, and hence, in order to get a reasonable sample, the criterion for sunny days had to be set at less than 2/8 cloud, and not zero cloud. In a desert area, the situation would be just the opposite. In the UK manual, the necessary associated data for thermal assessment, namely solar radiation and long wave radiation was then estimated by theoretical modelling taking account of cloud amount. The models are based on observed data.

The final level of complexity is the supply of full hourly data tapes for simulation, containing hourly values of a wide range of relevant observed variables. As full simulation is a long process, usually shortened series are used, often one year long, and selected by various criteria to be statistically representative of long term data, using appropriate techniques. For example considerable work has been carried out in Europe on test reference years. (Lund 1986). The simulation process is thrown by data gaps, and methods for filling observational gaps have to be applied. It should be noted the tapes in themselves do not provide all the necessary meteorological information to model energy exchanges with slopes, and essential complementary algorithms have to be available to deal with the effects of slope and orientation, and with the effects of microclimate, like vertical wind gradients. A practical difficulty with the preparation of such data tapes is that radiation data is often compiled on an hourly basis using solar time to take advantage of the symmetry of the solar movements about the north south line, while the rest of the data is recorded in standard climatological time. Special interpolation procedures may have to be developed to produce simulation tapes. These interpolated tapes can produce unacceptable slope predictions at low altitude angles, as the consequence of averaging procedure used to estimate the very low solar altitude associated with the period immediately after sunrise and immediately before sunset. There are other difficulties due to the change of daylength across the month, when working with monthly summaries.

ALGORITHM DEVELOPMENT

It is clear that more and more work on applied energy studies is going to be carried out using main-frame and microcomputer based techniques. Considerable effort on a world scale has been made to develop reliable algorithms in order to link standard meteorological data into the building design process. Standard meteorological observations have important limitations for building design, because buildings consist of a set of mainly inclined planes of relatively random orientation, set at a range of heights above the ground, usually set in an urban pattern that substantially modifies the rural boundary layer. It simply is not possible for meteorological services to make, on a routine basis, observations to cover such a wide range of factors. Modelling is the only route. The role of special observations, say measurement of short wave radiation on inclined planes, is to provide data by which the accuracy of the modelling process can be checked in a specific climate.

Systematic attempts have been made in many centres to assemble algorithms, for example, just before I retired from the University, we assembled a Handbook of Algorithms for Building Climatology Applications. (Page, Thomson & Simmie, 1984). A new version is now in preparation, which will probably provide computer subroutines as well in a suitable language, probably Basic or Pascal.

The prediction of urban climate and of the impact of building microclimate on energy use presents severe practical difficulties (WMO Technical Note No 652, 1984), but there is not space to explore the issues further here.

SOLAR RADIATION ALGORITHMS

The author has always dedicated considerable effort to the study of the prediction of solar radiation on inclined surfaces. Some of the earlier work, which emphasized the critical importance of understanding diffuse radiation, has stood the test of time against more recent models, (Page, 1964) and has recently been incorporated into International Energy Agency recommendations, and ISO standards. However, over the last ten years, considerable new progress has been made working on a team basis in the European Community, combining a wide range of observational data, with an extensive modelling effort. One key goal was to produce an inclined surface radiation Atlas for Europe (Palz, 1984). This Atlas is finding an increasingly wide use, and is beginning to be adopted as the basis for new European wide codes for glazing design, and for assessing energy performance. It is also finding use in agricultural applications. Full details of inclined surface algorithms for solar radiation prediction on slopes used to produce the CEC European Solar Radiation Atlas, Vol 2, Inclined Surfaces may be found in the book the author edited for the European Economic Community (Page, 1986a). The basic scientific overview emerged from collaboration between radiation specialists in a number of centres. As well as describing the underlying science, this book gives the full computational methodology developed by the European team. This can be applied any where in the world, and can be easily implemented at the microcomputer level, using as inputs observed monthly mean daily sunshine and observed monthly mean daily short wave irradiation on horizontal surfaces. Subsequent improvements by the author have allowed the methodology to be extended to tropical areas with less detailed observations using a development of the methodology proposed in WMO Technical Note No 172 (WMO 1981). This WMO publication is very important in this area. The clarity of the atmosphere in the absence of observations is established from the station height, vapour pressure, and the colour of the sky. This data enables the clear sky impacts, important for sizing plant, and predicting overheating, to be assessed. Monthly mean daily sunshine is used in addition to establish mean levels of monthly radiation on slopes. This new work has yet to be published in detail, but will be demonstrated at the meeting.

Additional algorithms may be found in the UK Department of Energy's Climate in the United Kingdom, a Handbook of solar radiation, temperature and other data for thirteen principal cities and towns (Page & Lebens, ed 1986b). The assessment of site wind from standard wind velocity observations and the prediction of long wave radiation exchanges with slopes under different conditions of cloudiness is discussed in detail in this publication.

Work is nearing completion by the author for UNCHS Habitat, which concentrates specifically on lower latitude developing country situations, where the observational data base is less detailed. The issue of daylighting is also being addressed as part of the Habitat effort. The CEC daylighting model described in Chapter in the CEC book (Page, 1986a) has been checked against observation and good agreement found. This model makes use of the CEC radiation model, using the concept of luminous efficacy, however unlike most other models, the luminous efficacy is turbidity and solar altitude dependent, and the illuminance characteristics of the diffuse radiation are assessed separately from those of the direct beam. All this represents quite a number of years of manwork. In addition the problem of predicting the effect of obstructions on radiation availability has been taken forward to a more sophisticated level. There is a demand for such data for town planning purposes, and for the assessment of the predicted energy consumption of buildings. For example, the problem is to be addressed in a new Eurocode for predicting the use of energy in building, so a new procedure has now to be developed to allow the CEC Atlas inclined surface radiation data for any unobstructed site in Europe to be modified to allow for actual site obstruction.

MICROCOMPUTER MODELS FOR ESTIMATING SOLAR RADIATION

In order to make it possible to use all this work easily, a considerable effort has been made to develop microcomputer based models programmed interactively for easy user use. Outputs may be presented at the monthly mean daily level as in the CEC European Solar Radiation Atlas (Palz ed, 1984) or at the monthly mean hourly level as in the book, 'Climate in the United Kingdom' (Page & Lebens, 1986b).

In the short wave radiation model for slopes developed by the author for world wide applications, a series of interlocking programs are used. The first program develops the necessary horizontal data at an hourly level and stores the data in a reference file from monthly mean daily sunshine data, complemented where available by observed values of monthly mean daily global radiation. It is necessary for the program to prepare hourly estimates of horizontal surface diffuse radiation as well as of beam radiation in order to handle subsequent hourly slope calculations. Both monthly mean and cloudless day irradiance estimates are prepared. An hourly illuminance file for daylighting studies is also generated simultaneously. If observed daily data is available, the daily sum of the computed data is compared with the observed data, and the turbidity found by reiteration. The program has the capacity to deal with situations of varying horizontal data availability, sunshine alone, sunshine plus global radiation, sunshine plus global radiation plus shade ring corrected diffuse radiation. It is also possible to compare satellite sites with only sunshine observations to a reference station with more complete observations. The horizontal file is computed only once and becomes the basis of all subsequent inclined plane computations. The slope predictions are made using a non isotropic clear sky model, in conjunction with the CIE Moon & Spencer overcast sky model, the blueness of the sky being determined by the relative sunshine duration. The diffuse model is not linear between overcast and clear, but takes account of the fact that most diffuse radiation is associated with partially overcast skies. By inputting the albedo, the ground reflected component can be estimated. Full computational details are given in Page (1986a). Once a slope file has been prepared, a subsequent calculation enables the impact of glazing in different months to be assessed, taking full account of differential transmission effects. A final calculation enables impacts of horizontal overhangs in different months to be predicted.

For Europe there is a simple input data base, from which the horizontal file for that site is developed. One can then breed out from that horizontal file, as many inclined surface files as are required, which may represent quite a considerable volume of data. Essentially one only keeps the input monthly mean daily input data, and the algorithms on the program disc, and generates all additional data as needed.

Work is also progressing on a solar radiation data project sponsored by the Commonwealth Science Council (Page, 1986c). This project, which has a special African emphasis, is directed towards helping Meteorological Services make more effective use of their data bases to provide design data for solar energy applications, as well as helping improve knowledge of the transmission properties of the tropical atmosphere. Especial emphasis has been laid on improving knowledge of the diffuse radiation climate, as diffuse radiation forms such a significant part of the short wave radiation income on slopes in low latitude hot humid climates.

THE SIGNIFICANCE OF COMPUTER GRAPHICS

The applications situation is being revolutionised by advances in information technology, especially the capacity it offers to provide both visual and numeric information within a single computational structure. A large part of the human cortex is dedicated to processing visual information, and many complex design problems can be better appreciated visually than numerically. Deciding the best forms of visual presentation is important. WMO has already published one study in this area (Lofness, 1982). A new suite of programs has been prepared for the UNCHS (Habitat) project (Page, 1987) to provide microcomputer graphic aids for energy layout studies. The geometrical movements of the sun are output dynamically in terms of the chosen geometry of the building projection, so the impacts of adjacent buildings, trees and landscape can be rationally assessed.

Increasing importance is being attached to layout studies, and the assessment of the quantitative effects of obstructions on energy availability. Graphic techniques can be very valuable in this field. For example, my former team has assessed the systematic use of trees for shading purposes (Sattler, Page & Sharples, 1987), using graphic output techniques.

ALGORITHMS FOR THE ASSESSMENT OF LONG WAVE RADIATION ON SLOPES

Rather less attention has so far been given to long wave radiation studies for slopes, but a model for computing slope long wave radiation has been developed. (Page & Lebens, 1986). (Refer also Page, Thomson & Simmie, 1984). The model was developed from the work of Monteith and Unsworth (1975) and Unsworth (1975). These author's made extensive use of the USSR work of Kondratiev and his collaborators. Proper attention has to be given to the non-isotropic nature of the long wave radiance from the sky. A critical current difficulty in modelling long wave radiation on slopes is the difficulty in estimating the long wave radiation from the ground surface, which is especially important in the case of vertical surfaces. There is very little systematic information available about ground surface temperatures.

DEGREE DAYS

As temperature is such an important factor in determining energy demand, the availability of long term monthly degree day temperature data is essential in design for assessing the future energy consumption of buildings, while recent month degree day data is an important tool in systematic energy management to enable actual performance to be compared with estimated performance, taking proper account of fluctuations in year to year climate. A high priority must therefore be established in the preparation of energy design data to the working up of such degree day data for the various sites in a territory. Fortunately this is not too difficult a task, especially if the observed temperature data is held on magnetic tape. As the weather may vary substantially from year to year, as long a period of temperature data as possible should be used for the design series. This data should have been observed using standard meteorological practices.

DEFINITION OF MONTHLY HEATING DEGREE DAYS (HDD_m)

The monthly heat losses from buildings are estimated using the concept of monthly heating degree days, (HDD_m). An important concept is the balance temperature which is used to select the base temperature adopted for the estimation of heating degree days. The balance temperature of a building in a particular month is the temperature at which the mean gains, internal and external are exactly balanced by the mean monthly losses. The balance temperature is building design specific. The degree day base temperature, T_b, is set equal to the balance temperature of the occupied building. Implicit in the definition of the balance temperature is the indoor winter comfort temperature level considered acceptable in any region, which is used to establish the balance temperature in conjunction with the internal incidental heat gains. In the heating season these incidental gains are of positive benefit.

The number of heating degree days (HDD_d) for any one day in this paper is defined as the difference between the base temperature T_b and the mean daily ambient temperature, T_m, counting only positive differences.

Some countries, for example the UK, have a special computational procedure for days when the maximum and minimum temperatures straddle the base temperature. Hitchin has described the procedures used in the UK (Hitchin, 1981). Two straddling conditions are identified, one when the daily mean is above the balance temperature. The other when the daily mean is below the balance temperature. Fuller information may be found in Hitchin's paper, which provides a good historical review of the development of the degree day concept, and of the practical difficulties involved in assessing degree days to different bases in different situations.

There are two ways of making heating degree day estimates:

1. Using the actual observed daily values of T_{min} and T_{max}, and working on a day by day basis, forming the daily values from the base temperature, and hence finding the monthly values, and finally the long term mean monthly values.

2. Using statistical methods based on assumptions about the observed distribution of the daily mean temperature about its mean value.

THE CALCULATION OF DEGREE DAYS USING A LONG TERM OBSERVED SERIES OF VALUES OF DAILY MEAN AIR TEMPERATURE, T_M

As degree days for a number of different bases are required, it is worth giving some consideration to computational effort, when there are a lot of observations to handle. It is always important to use as long a reliable record as possible.

The first stage in preparing the degree day estimates is to check all the monthly daily records for completeness of data, for every day in each month in the record. If a monthly record in a specific year is to be used, any gaps will need first to be filled with suitable estimates, otherwise the integral values for that month will be too low. Any months with a large number of observational gaps should be rejected.

The usual computation method in the past, was to choose a base temperature, subtract the mean daily temperatures below the base temperature, and then add up the resulting degree days for all days with temperatures below the base for different periods of integration. This process was repeated for each base temperature. This required many passes through the data. A more economic way of computing degree days to a number of base temperatures from actual observations is given by Bushnell (1979), who showed it is possible to avoid this repetition, and that, in fact, only two computation passes are needed, one to order the temperatures, and another to form the sums. The chosen base temperatures are input after the basic processing has been done. The original paper should be consulted for details.

Another advantage of this method is that it also provides the data to establish the winter basic design temperature for sizing heating equipment, at any required frequency level of occurrence, using the lower end of the temperature distribution range.

STATISTICAL METHODS FOR ESTIMATING DEGREE DAYS

The most widespread currently used statistical model is the model developed by Thom (Thom, 1954a, Thom, 1954b) Thom's model which was based on pragmatic corrections to a Gaussian normal distribution model, has the important advantage of using two quantities which are fairly easily calculated, namely the long term monthly mean screen air temperature T_m, and the standard deviation of its monthly mean value from year to year. The method uses standard deviation of the monthly mean about its long term monthly mean value as a means of estimating the standard deviation of the daily data, using the theory of the standard deviation of the means. This value has then to be pragmatically corrected as explained later. As this method does not require a full detailed calculation of the standard deviation using daily data for each month in the data set, the amount of data to be handled is reduced by a factor of 30, 31 or 28 (29 in leap years), according to month. It is therefore particularly suitable for hand held calculator estimates of degree days.

Thom made the starting assumption that the daily average temperatures for a particular data of the year were normally distributed. It was then assumed that the daily standard deviation could be estimated from the value of the standard deviation of the monthly mean, obtained using year to year data for that month, as (sigma monthly means for month $m.n^{-2}$), where n is the number of days in the month. This result follows directly from the standard statistics of a normal distribution, and the theory of standard error of the mean of a data set. A modifying factor I was then developed using observed US data from 30 sites (Thom, 1966). This factor is applied to correct the predicted results. The modifying factor is needed because the autocorrelation of ambient daily temperature causes the value of (sigma monthly mean . n^{-2}) to be substantially larger than the value of sigma found from daily observed values in a specific month. Erbs (1984) has shown for the USA that the true value of the standard deviation, $sigma_m$, found from the daily data is typically half that estimated from the standard deviations of the year to year monthly means.

In the Thom method, the distribution of ambient temperature is symmetric about the monthly average temperature. Thom never demonstrated that the monthly ambient temperature is normally distributed in the US. Erbs (1984) among others have studied the actual form of the ambient temperature distribution function.

It should be stressed that Thom's pragmatic corrections are based on US data. The method has been successfully applied in other latitudes, for example by Shellard for the UK (Shellard, 1959), but it is not currently known whether the method is fully suitable for all sites in the world. It would therefore be wise to process the data for at least one site in each territory using the full method, and to used these results to check the validity of the Thom statistical method in that region, before applying the Thom method more extensively in the territory. If there are significant differences, it is possible that these could be accommodated by changing the form of the Thom correction function I, to match the data better.

General experience with Thom's method is that it is least satisfactory when the base temperature is close to the monthly mean temperature, but this is not the situation in which the economic costs of heating are particularly significant, so any errors due to this cause will not make much impact on the overall economic assessment.

HITCHIN'S STATISTICAL METHOD

Hitchin (1983) has suggested another statistical approach, which is somewhat simpler than Thom's method, but related in principle to it. Using twenty-year mean monthly degree day values for base temperatures between 5 deg C and 20 deg C for five sites in the UK, Hitchin found using his model for all five sites 93% of the estimates were within ± 5.5 degree days, and 60% within ±1.5 degree days. This compared with 66% and 33% using Thom's procedure for London Heathrow. Hitchin's method has been less widely tested than Thom's method, but as the methodology is simpler, and the results accurate for the UK, it is a method that deserves wider testing against observed data.

It must however be pointed out, if standard deviation values are not already available, it is more logical, more economic of computing effort and more accurate to adopt Bushnell's method, than to work indirectly through the standard deviation values.

COOLING DEGREE DAYS

The severity of the overheated period in any area can be estimated from the number of cooling degree days.

Cooling degree days are important in estimating the energy performance of air conditioned buildings, which necessarily, in some temperature conditions but not in others, operate with a restricted air flow rate. The base temperature for cooling can be defined in the same manner as for heating. Again it is logical to select it using the concept of the balance temperature. In this case the selected indoor comfort level, $T_{set\ cooling}$, has to be set to match summer requirements. Normally people will wear much lighter clothing in hot weather and the comfort temperature can be set at a higher level than for heating. Typical values are around 24 deg Celsius.

The incidental internal heat gains in this situation are of negative value, as are the solar gains, which need to be included with the incidental gains in the estimation of the balance point. It will be found, as a consequence of this, the balance temperature for cooling, in the presence of restricted ventilation, may be surprisingly low, in any building whose glazing is not adequately shaded in hot weather. As the solar gains will vary from month to month, the balance temperature for cooling will also vary.

A number of cooling degree days CDD_d for one day is defined as the difference between the base temperature for cooling T_{bc} and the mean daily ambient temperature T_m, counting only negative differences. The monthly cooling degree days for month m, CDD_m, are the sum of the daily values over the month. When the outdoor temperature lies between the cooling balance temperature and the selected indoor comfort temperature, the cooling load can be satisfied, at least in part, by bringing in additional outside air. This leads on to the concept of ventilation cooling degree days, which accumulate differently than conventional cooling degree days (Erbs, 1984).

THE ESTIMATION OF MONTHLY COOLING DEGREE DAYS

The procedure for estimating monthly cooling degree days are related closely to the procedures for estimating heating degree days. Two basic approaches are again possible, the observational data base approach and the statistical approach.

In Bushnell's method for monthly cooling degree days, CDD_m, the first stage of the procedure is identical with the procedure for heating degree days, namely the ordering of the monthly data into an incremental series of temperature observations in that month for a period of years. For cooling degree days, in the second stage of the process, the summation is started at the high temperature end of the data set working down in temperature. This produces an analogous table to the table used to estimate heating degree days. This Table may then be used to estimate cooling degree days to any base.

Thom's statistical method can also be used for estimating cooling degree days.

A much fuller discussion of all these degree day issues may be found in a review by the author, which forms part of the UNCHS Habitat project (Page, 1987). (The material on degree days in this paper is summarized from that report).

THE NEED TO DEVELOP ENERGY MODULES WITHIN THE CLICOM PROGRAM

There is considerable work yet to be done, under the CLICOM program, to provide internationally suitable procedures for processing observational data for energy applications. Thanks to work already completed, the algorithmic basis exists. The essential task is to get key facets of the work completed into the CLICOM system. The appropriate processing and presentation of observed data for practical decision making is the key to success. This places a great emphasis on the need for user collaboration. Meteorological Services must provide that applications information in forms that are useful and appropriate to the various groups that need it.

The issue is, therefore, how are the necessary funds to be raised to ensure climatic aspects of energy design can be more effectively handled across the world? This is especially important in the case of developing countries, with limited resources to develop such systems. Graphics packages will be important to aid practical applications.

Unfortunately the author's present developed interactive

system is Apple Computer based, and not IBM PC compatible. However once the development work is complete, it will not be a difficult task to reorient the software.

OPERATIONAL FAILURE RISKS AND OPERATIONAL SAFETY

Risk analysis is another aspect of energy design needing climatological data. A heating plant or a cooling plant is expected to give acceptable performance in extremes of weather. Overdesign gives low risks at high design costs, while underdesign gives low design costs at high risk. The statistical problems of risk assessment have to be approached with some care, especially if more than one variable is involved. The consequences of failure have to be explored in deciding acceptable risks. The sudden failure of an electricity supply utility distribution system, due say to overhead line failure caused by heavy icing, is obviously far more serious than the failure of the energy plant of a single building to provide a comfortable indoor environment, especially if, in the former case, the distribution system is connected to a nuclear energy generation system. The risks have to be evaluated accordingly.

There are many operational issues related to climate concerning the environmentally safe generation and operationally reliable distribution of energy, including the threats to distribution systems presented by extreme value situations like high winds, heavy icing, severe thunderstorms. The extreme value threats often occur at the same time as the energy demands on the system are highest, when the supply system will be operating near to maximum capacity, and so least able to cope with any severe disruptions.

The problems may be presented to meteorologists as problems in statistical time involving estimates of appropriate extreme values and their statistical frequency of reoccurrence, or as real time operational problems involving forecasting of special operational risks, so that special operational steps can be taken to contain the anticipated risks. If a hurricane or severe line icing is forecast, do you shut down a nuclear power station?

In the case of statistical time problems, the safest path for the meteorologist is to provide the range of extreme values encountered at various frequencies of recurrence using extreme value analysis, and then let the specialist decide the risk he believes acceptable, on a probablistic basis. It is obviously wise to use as long a time series as possible. Sound statistical processing of the data is essential, otherwise false conclusions may be drawn. It is also important to stress the probablistic basis of the prediction. The author has personal experience of the highest wind in the century occurring twice in one week in Sheffield. This simple strategy works best when a single variable is involved. When a combination of climatological variables is involved, the process becomes more difficult. This is one of the fields where simulation is particularly valuable for establishing the relative importance of the different climatological variables for a specific design solution. The extreme design load in one type of air conditioned building, for example an office, may be particularly sensitive to solar radiation gains through windows, while the extreme load in another building type, say a theatre, may be particularly sensitive to high vapour pressure in association with high dry bulb temperature, and the wet bulb temperature may become the key variable.

For real time problems, like hurricane forecasts, the special forecasting structure has to be set up in advance, and the communication structure established to ensure the appropriate messages travel quickly to the action points concerned with energy safety. There has to be a standing organizational structure, fortunately seldom needed, but always operational. The system needs to be tested by occasional training exercises to ensure a smooth efficient service has been set up to reduce risks by appropriate precautions because the consequences of the failure of energy systems, often occurring at a time of maximum demand, are so serious. Wind generating systems are particularly at risk.

CONCLUSIONS

In many countries, buildings account for about 50% of the energy demand, and the majority of this demand is weather sensitive. Heating and cooling design are especially important, and there is a growing interest in the rational user of daylight to reduce lighting energy demands. Air temperature is the critical variable on the demand side. The presentation of appropriate temperature data is therefore especially important. Heating and cooling degree days are very useful, both in design, for forecasting probable monthly energy use, and, for energy management, for following the current efficiency of operation of plant in near statistical time. This places an onus on Meteorological Services to publish monthly degree day data quickly, so it becomes rapidly available for energy management purposes.

Radiative exchanges with buildings are especially important. The short wave radiation in times of underheating is of considerable value, but in times of overheating produces many problems, so sometimes one is trying to make the building short wave radiant energy accepting and sometimes short wave energy rejecting. In hot weather, the long wave radiation exchange with the sky is beneficial, and in cold weather not so. The environmental radiant energy exchanges with buildings take place mainly on inclined and vertical surfaces, so the capacity to provide information about such inclined surface exchanges is especially important.

There is an increasing demand for good data to explore potential alternative energy systems, and the meteorological data needs have become very much better defined. Wind energy is very site specific, and special studies may be required of wind flow over hills etc. Solar energy availability is very slope and orientation dependent, so it is essential that good slope/orientation data can be made available.

The role of computing, both main-frame and micro is stressed, and the need to develop the CLICOM modules associated with energy design and alternative energy supply is emphasized. It is hoped that the computer demonstrations to be given after the paper, will convince meteorological services of the value of approaches, which marry conventional meteorological data with applications orientated programs, preferably containing a graphics element. The critical issue for WMO would seem to be how is all this achieved progress going to incorporated into the CLICOM project of the World Data Bank Programme, to make it much easier for developing country Meteorological Services to provide an effective energy service to the potential users.

Meteorologists also need to take note of the increasing use of quite sophisticated modelling techniques for predicting building energy consumption, which demand meteorological data tapes based on hourly observations, with no data gaps that will otherwise wreck the simulation process.

Urban climate which is critically important will be discussed elsewhere in this Conference, and is therefore not addressed here.

Brief mention has been made of current projects underway by the author for various international organizations. The UNCHS Habitat study contains considerable detail on how to approach building energy design problems in developing countries. All these projects make extensive use of microcomputing techniques. The economic significance of making progress in the field of climate applications in relation to energy policy is substantial, as energy use forms such a significant proportion of national budgets.

Above all, Meteorological Services should remember a proper service can only be provided by good user collaboration, and advisory bodies of an inter-professional nature are needed in the field of energy to ensure the data communicated is the data actually needed by users. More and more data communication will be computer to computer, and main-frame systems will need to be able to communicate effectively to microcomputer systems, sometimes by disc and sometimes by modem. Meteorological Services need to become less data defensive in the modern world of information technology. The big main-frame computer boys must swallow some of their pride, and accept the implications of the information technology revolution. In this context, the CLICOM project could be just as important to developed countries as to developing countries.

It is the service that counts for users.

TABLE 1 Natural climatic resources that may be used with appropriate preplanning to supply some of the energy needs of urban complexes

Resource	Use	Energetic economics
Solar energy	Daylight	Most economic use
	Passive heating	
	Active heating	
	Active cooling & dehumidification	Least economic use
Wind energy	Natural ventilation	Most economic use
	External surface cooling	
	Wind power water pumping	
	Wind power electricity	Least economic use
Nighttime outgoing long wave	Cooling by air drainage from adjacent higher surfaces	Most economic use
	Surface cooling of courtyards and open spaces sheltered from wind	
	Radiant cooling panels	Least economic use
Precipitation	Latent heat cooling through fountains & water walls	Most economic use
	Cooling by plant evapotranspiration	
	Desert coolers	
	Cooling by creation of large bodies of water	
	External building surface sprays	Least economic use

TABLE 2 Factors promoting the use of various natural climatic resources

Solar energy	Low cloudiness Clean atmosphere, low pollution Favourable orientation of buildings & streets Terrain slope favourable Reflective ground cover Absence of terrain obstruction
Wind energy	High basic windiness Open planning Absence of excessive wind barriers & trees Exposed terrain Height, both above ground level & above general terrain level
Night time long wave radiation	Clear night-time skies Low dewpoints Terrain suitable form for useful air drainage Absence of obstructions blocking flow Absence of obstructions blocking view of cold sky Low wind speeds
Precipitation for evaporative cooling	Adequate rainfall Suitable aquifers, and underground stores Controlled wind circulation to conserve cooled air Relative humidities not too high

TABLE 3 Urban planning in relation to the control of energy demands - cold seasons

1. Maximize passive solar gains by appropriate orientation.

2. Provide shelter from cold winds, especially to reduce excessive ventilation

3. Avoid or block adverse air drainage from higher colder terrain

4. Provide good drainage of surface water to avoid dampness

5. Maximize advantage of urban heat island effect by controlling wind flow by aerodynamic drag

6. Pedestrian movement in sun and out of wind, whenever possible

TABLE 4 Urban climatic planning in relation to demands of energy: Hot seasons dry climates

1. Minimize solar impacts on building surfaces by appropriate site orientation.

2. Use high ceilings with small windows to provide good daylight with small openings.

3. Facilitate external shading by appropriate site orientation

4. Where water resources allow, provide trees for shading and fountains in wind controlled micro-environments.

5. Make good use of lower nighttime temperatures and high outgoing long wave radiation by promoting nighttime cold air drainage from cooler adjacent terrain

6. Reflect incoming heat at roof level and shade streets to reduce heat island effects

7. Use nighttime air drainage to break up urban heat island

TABLE 5 Urban climatic planning in relation to control of demands for energy: hot seasons humid climates

1. Minimize solar impacts on building surfaces by appropriate site orientation

2. Facilitate the provision of external shading by providing adequate space for tree planting

3. Maximize air flow characteristics of urban areas using appropriate building form

4. Avoid high densities liable to block air flow

5. Reduce urban heat island effects by dispersing solar energy at the tree top canopy level, and encouraging good air movement below

6. Remove by good drainage excessive soil water to control relative humidity

---

Source of tables: Page, J K, (1986), Energy related issues, in Urban climatology and its applications with special regard to tropical areas, Proc. of the Technical Conf, Mexico D F, 1984, T R Oke (ed), WMO Technical Note No 652, WMO, Geneva.

REFERENCES

Bushnell, R H, 1979, Climatic Kelvin degree days below any base, Monthly Weather Review, Vol 107, 1083-1086

Erbs, D G, 1984, Models and applications for weather statistics related to building heating and cooling loads, Ph D Thesis, Engineering Experiment Station, College of Engineering, University of Wisconsin-Madison Refer also Erbs, D J, Klein, S A and Beckman, W A (1983), Estimation of degree days and ambient temperature Bin data from monthly average temperatures, ASHRAE Journal, June.

Hitchin, E R, 1981, Degree-days in Britain, Building Services Engineering Research & Technology, Vol 2, 73-82.

Hitchin, E R, 1983, Estimating monthly degree days, Building Services Engineering Research and Technology, Vol 4, 159-162.

Lamming, S D, 1985, Investigations of wind energy potential in the Eastern Caribbean, Proc Commonwealth Science Council Workshop on Meteorological data for Solar and Wind Energy Applications, Commonwealth Science Council, London.

Lofness, V, 1982, Climate/energy graphics, Climate data applications in architecture, WMO MCP-30, WMO, Geneva.

Lund, H, 1986, EC work on short reference years (SRY), in Proc Conf Advances in European Solar Radiation Climatology, Conf C43, UKISES, 19, Albemarle Street, London, WIX 3HA.

Milne, M and Givoni, B, 1979, Architectural design based climate, in Energy conservation through building design, Watson , D, (ed), Mcgraw-Hill, New York, 96-113.

Page, J K, 1964, The estimation of monthly mean values of daily total short wave radiation on vertical and inclined surfaces from sunshine records for latitudes 40N to 40S, Working paper no. E/CONF. 35/5/18, Proc UN Conf New Sources of Energy, Rome, May, 1961, Volume 4, 378-396.

Page, J K, 1980, Climatic considerations and energy conservation, in Interactions of Energy and climate, Bach, W, Pankrath, J and Williams, J, (eds), D Reidel Publishing Co, Dordrecht, Holland, 73-88.

Page, J K, 1984, Solar radiation and climatic data, in Proc Solar World Congress, Perth, Australia, S V Szokolay, (ed). Pergamon Press, Oxford, Vol 4, 2066-2082.

Page, J K, Thomson, J L & Simmie, J, 1984, A meteorological data base system for architectural and building engineering designers, Handbook, Vol II, Algorithms for Building Climatology Applications, Department of Building Science, University of Sheffield.

Page, J K, Ed. 1986a, Solar Energy R&D in the European Community, Series F, Vol 3, Solar Radiation Data, Prediction of solar radiation on inclined surfaces, D Reidel Publishing Co, Dordrecht, Holland.

Page, J K & Lebens, R, 1986b, Climate in the United Kingdom, A Handbook of solar radiation, temperature and other data for thirteen principal cities and towns, Dept of Energy, published by HMSO, London.

Page, J K, 1986c, Surveying the solar energy resource, Proc Of Internal Conference on research and Development of renewable energy technologies in Africa, Mauritius, C Y Wereko-Brobby, (ed), Commonwealth Science Council, London.

Page, J K, 1987, How to prepare local based design manuals for passive solar heated and naturally cooled buildings, In preparation for UNCHS, Nairobi.

Palz, W, 1984, European Solar Radiation Atlas, Vol II: Global and diffuse radiation on vertical and inclined surfaces, Compiled by J K Page, & R J Flynn, Dogniaux, R and Preuveneers, G, Verlag TUV, Rheinland.

Reidat, R, 1960, Wettendaten fur das Bauwesen - Hamburg Einzelveroffentlichungen des Seewetteramtes, Hamburg, No 23, Hamburg.

Sattler, M A, Sharples, S, & Page, J K, 1987, The geometry of the shading of buildings by various tree shapes, Solar Energy, Vol 38, 187-201.

Shellard, H C, 1959, Averages of accumulated temperature, Meteorological Office Professional Note No 125, HMSO, London.

Thom, H C S, 1954a, The rational relationship between heating degree days and weather, Monthly Weather Review, Vol 82 (1), 1-6.

Thom, H C S, 1954b, Normal degree days below any base, Monthly Weather Review, Vol 82, 111-115.

Thom, H C S, 1966, Normal degree days above any base by the universal truncation coefficient, Monthly Weather Review, Vol 94, 461-465.

UNCHS, 1984, Report of the expert group meeting on the use of solar energy and natural cooling in the design of buildings, Sept 1983, UNCHS, Nairobi.

Unsworth, M H and Monteith, J L, 1975, Long wave radiation at the ground (1), Angular distribution of incoming radiation, Q J Roy Met Soc, Vol 101, 13-24.

Unsworth, M H, 1975, Long wave radiation at the ground (II), Geometry of interception by slopes, Q J Roy Met Soc, Vol 101, 25-34.

WMO, 1981a, Meteorological aspects of the utilization of solar energy as an energy source, WMO Technical Note No 172, World Meteor. Organiz, Geneva.

WMO, 1981b, Meteorological aspects of the utilization of wind as an energy source. WMO Technical Note No 175, World Meteor. Organiz, Geneva.

WMO, 1986, Urban climatology and its applications with special regard to tropical areas, T R Oke (ed), WMO Technical Note No 654, World Meteor. Organiz, Geneva.

Additional WMO publications relating to energy not specifically cited in the text

WMO, 1975, Report of the expert group meeting on meteorology and energy production, Geneva, 1975, World Meteor. Organiz, Geneva.

WMO, 1981, Energy and special applications programme, Report No 1. Report of the meeting of experts to review WMO plan of action in the field of energy problems, Geneva, 1980, World Meteor. Organiz, Geneva.

WMO, 1981, Energy and special applications programme, Report No 2, papers presented at the WMO Technical Conference, Mexico City, Nov 1981, World Meteor. Organiz, Geneva.

20

Applications of meteorology to industrial activity

D. M. Houghton, Meteorological Office, Bracknell, UK

Most aspects of life on earth are sensitive to weather. Individuals and nations, economies, industries and agriculture all respond in varying degrees to the weather. Examples of obviously high sensitivity are found particularly in agriculture, aviation and shipping. Media interest is primarily in weather-related catastrophes and it seems their number is increasing. This increase may possibly be real, but it is more likely a quirk of improving communications. However, whether or not there is any change in the frequency of storms I am sure that we are experiencing a worldwide increase in sensitivity to weather in consequence of our activities being geared to increasingly tight margins. Even relatively minor variations in weather, let alone extremes, can now seriously interfere with leisure and business activities.

Mankind's sensitivity to weather, i.e. to variations and extremes of rainfall and temperature, strong winds, fog and electrical storms, is increasing by virtue of -

a. Increasing population leading to increasing pressure on existing resources of food, water, etc.

b. Increasing expectation of permanency of supplies in developed and developing countries.

c. Increasing complexity of infrastructure particularly in commerce and industry.

d. Increasing efficiency in agriculture so that the percentage impact of weather on production is higher in volume terms.

e. Increasing efficiency in the use of existing natural resources such as water so that the percentage impact of weather on supply is higher in volume terms.

f. Reducing margins in design and construction so that tolerances and allowances for extreme weather events are minimal.

Quantitative estimates of sensitivities have been derived for particular countries in relation to particular activities. W J Maunder is probably the best known source. Attempts have also been made to measure the total economic impact of particular weather events, for instance a severe winter. In this context we must remind ourselves that 'it is an ill-wind that blows nobody good', and there are pluses as well as minuses on the balance sheet. Energy industries for instance normally benefit from severe cold, as do producers of sledges, shovels, snow chains and salt.

Filtering out weather-related variations in demand and purchasing in particular market sectors is an important first step in the definition of an appropriate weather advice service. A study conducted last year into the retail sector in Britain revealed weather impacts on sales amounting to tens of millions of pounds in women's clothes, footwear, chemists (drug stores), and DIY (Do It Yourself) materials, and a staggering 80% weather dependency of the sales variance in the summer half year in the fresh fruit and vegetables sector, which has a total turnover of £3 billion.

As meteorologists we find all this very interesting, but we cannot respond directly to sensitivity to weather because we are not in the weather control business. Our business is weather information. Sensitivity to weather information is not the same as sensitivity to weather. The same basic considerations apply but with the proviso that for weather information to be useful there must be the opportunity to act on it. The opportunity facing meteorologists today has a number or complimentary facets:

a. With increasing sensitivity to weather there is a corresponding increase in sensitivity to weather information.

b. The quality of information and advice which meteorologists are capable of giving has improved out of all recognition in the past few years.

c. The capability of industry and commerce to make use of and respond to this better weather information is itself increasing because of improving communications and the introduction of automation.

The weather information required by commerce and industry comprises historical and forecast information. Traditionally we have tended to put the two types of information into completely separate compartments calling one climatology and the other forecast. More often than not however they are required together as a total package of weather advice.

The meteorologist has the data and the capability of interpretation to support design and planning activities for virtually any weather-sensitive enterprise the world over, be it roads, dams, oil rigs, boats, or whatever. Sadly many designers and planners still do not appreciate the scope or quality of information available, nor the integrity of judgement of which meteorologists are capable when it comes to the interpretation and application of that information.

Weather forecast information is often still thought of as a fairly detailed description of the weather expected today followed by an outlook for tomorrow. There is all too little appreciation of the accuracy of detail available for a few hours ahead, or the extent of the useful outlook.

The agricultural scene provides a good example of the changing opportunities for the application of weather information. Agriculture is sensitive to weather, and profits in both arable and livestock farming are highly weather dependent. The farmer is traditionally the doyen of weather forecasters. He spends much of his time out of doors and is the first to recognise signs in the sky of changes in the weather a few hours ahead. The extent to which he can react with profit to that information depends on the particular activity in which he is engaged and the time of year. If the signs are warnings of conditions likely to affect the safety of his animals he will take whatever actions are possible within the time available. The farmer's major sensitivities however, are on a longer time scale, far outside his own capability based on his observations of the sky around him, and also well outside the capabilities of the weather forecaster prior to the present decade, whose horizons were not much more than a couple of days at best. Neither was able to contribute greatly to the planning of sowing and harvesting, operations highly sensitive to weather and requiring at least several days information ahead. The meteorologist has recently crossed this hurdle and opened up completely new avenues for providing useful advice to farmers the world over.

When it comes to very short period forecasts the traditional warnings of rain, snow, gales, frost and the like are now highly accurate weather information over very short periods has made possible activities such as application of fertilizers and pesticides in marginal conditions which were out of the question only a few years ago.

Equally in regard to historical weather information new areas of opportunity have been opened in agriculture in recent years particularly in regard to the control of pests and diseases. The emergence of some ailments has been linked directly to particular sequences of weather over quite long periods, making it possible to plan and effect corrective action at the appropriate time.

Thus the outstanding problem faced by the meteorologist today is that few appreciate the quality of his product, its applicability to the majority of weather sensitive activities, and the benefits which can accrue from using it. In short, we have an excellent product, it is not properly used, and the benefits are not realised.

To the businessman faced with the challenge to correct this sort of situation the answer lies in marketing, since:

a. It will impose on the meteorologist the discipline of tailoring his products to the needs of the user.

b. It will make the meteorologist take adequate steps to ensure that prospective users know of the wide range of information which is available.

c. By putting a price on the product it will ensure that its value is realised so that it is used to maximum benefit.

The development of meteorological services to aviation is a good example of the development of a service based on the principles of marketing. It has been accomplished over many years without any mention of the word marketing, but with an overriding requirement of safety directing efforts to the same goal i.e.:

a. The best possible product designed to meet the needs of the user.

b. Adequate steps taken to make sure the user knows how to use it, and uses it to maximum advantage.

c. A continuing effort to ensure that the user's changing requirements are understood and satisfied.

I believe we must now repeat this process in developing our services to agriculture and to all weather-sensitive areas of commerce and industry so that our national economies derive maximum benefit from the skills we have acquired and the progress we are achieving. A rough estimate of the already realised benefit of weather information to the community, both nationally and globally is in the order of 10:1, and the potential benefit is in the order of 100:1. The application of marketing principles will help us to achieve this potential so that it will become abundantly clear to all that a meteorological service is not a luxury but a necessity.

REFERENCE

W J Maunder The Uncertainty Business, Methuen 1986.

21

Role of meteorology in connection with areas such as natural resources, human health, tourism and sports

G. Jendritzky, Deutscher Wetterdienst, Zentrale Medizin-Meteorologische Forschungsstelle, Freiburg, FRG

1. INTRODUCTION

Weather and climate influence the general health, well-being and physical performance of the human being in a multitude of ways. This is because the atmosphere is an essential component of the environmental conditions in which the human organism must react, in order to maintain an equilibrium (homoeostasis) of its vital functions. During its evolution, the human organism developed autonomic functioning regulating mechanisms, for example, thermal regulation. These autonomic regulatory mechanisms ensure an extraordinarily high adaptability even to large or rapidly changing meteorological conditions. Long-term adaptations result from a longer exposure to stimuli, for example, the acclimatization of the human organism to differing bioclimates.

The adaptability of the organism is significantly dependent upon age, sex, constitution, reaction type and degree of acclimatization. Due to this, large differences between individuals occur. In addition, sick people lose their acclimatization and their ability for adaptations. Physiological adaptation is assisted by behavioural adjustments, for example, clothing, living habits and climate modifications e.g. urban climate. These factors are essentially determined by moods and sensations (Weihe, 1986) as well as by experiences and expectations (Auliciems, 1981). Due to this unique ability of the human being to overcome adverse environmental conditions, the human race has been able to spread throughout the world. Further the human being is not only capable of surviving in all climates but is also capable of settling permanently. This means that no climate is completely unsuitable but on the other hand, no climate is suitable for any and all types of human activity nor completely suitable fore health and general well-being.

One can differentiate the predominating meteorological conditions according to load, unload and stimulus factors. Weak atmospheric stimuli produce a poor stimulus condition and this leads to a loss of adaptation and to an overprotected or even pampered organism. Mid-intensity stimuli serve to stimulate, activate, harden or even to heal, for example, the purposeful use of climate stimuli in climatotherapy. Strong stimuli can cause damage due to a load which is too great: distress, for example, skin cancer or erythema from UV-B-radiation, frostbite, heatstroke, air pollution etc. For a given population the boundaries between a healthy state and a state where clinical damages can be observed are fluid, i.e., interindividual variability.

In addition to the direct physical and psychological effects produced by the atmosphere, there are also indirect effects on the human being that are of importance:

- living habits in general (clothing, recreational behaviour, manner in which settlements are configured)
- food (an area belonging to agricultural meteorology)
- drinking water
- the initiation and spreading of diseases (parasitic, bacterial, viral diseases
- a particular problem in developing countries and for tourists travelling to distant countries such as these)
- socio-economic conditions, for example, transportation, agriculture, energy consumption etc., but also natural catastrophes such as hurricanes, drought, floods, avalanches.

2. THE BIOMETEOROLOGICAL CAUSE-EFFECT RELATIONS

The significance of meteorology becomes apparent, when one examines the cause-effect relations. The main points of impact by direct atmospheric influences - the receptor organs - are the skin surface, the visual system and the respiratory system (Hentschel, 1985a).

2.1 Heat Loss Conditions

As a homoisothermal being, the human being possesses the ability to maintain a constant body-core temperature within a very small range. This is largely independent of changing environmental conditions and of varying metabolism of the human body. Heat-loss and heat production are adjusted to one another and to environmental conditions by autonomic regulating mechanisms. In addition, manners of behaviour determined by discomfort-sensations (clothing, activity, heating or air conditioning, moving into the shade, wind protection etc.) support the thermo-regulating mechanism. The energy exchange of the organism with its environment takes place by means of convection of sensible and latent heat (air temperature, humidity, air movement), radiation (radiation balance, short and long-wave, sun elevation, length of optical path, turbidity, cloudiness, type of land use with albedo values, emission coefficients and surface temperatures) and respiration (air temperature and humidity) (Fig 1). The meteorological elements or physical characteristics of the surroundings (which appear in parentheses above) play a decisive role in the corresponding transport processes. In addition, the choice of clothing can change the conditions of heat loss considerably - behaviour adjustment.

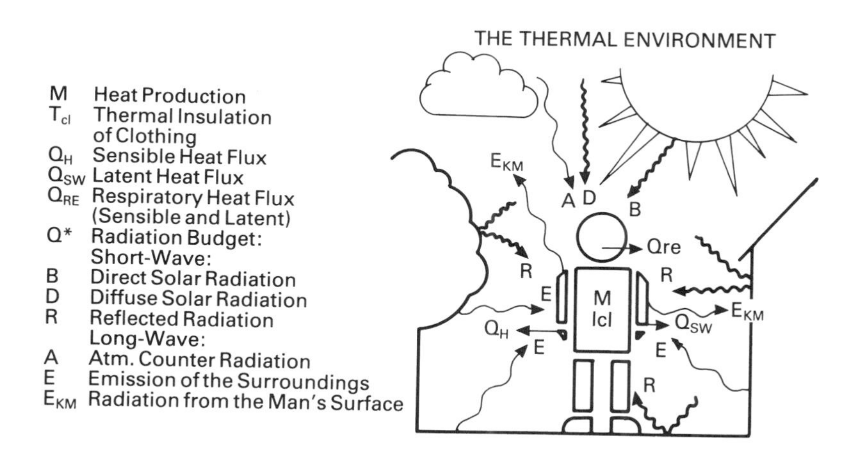

Figure 1 The heat balance of human beings

Under neutral and cold conditions, heat loss is essentially a factor of radiation and of the turbulent transport of sensible heat. Energy flux results, in general, in a cooling of the organism, however, radiation especially with the effects of direct sunlight can be an important energy source. The influence of direct sunlight must be considered with respect to the human surface exposed to the direct sunlight. This surface is dependent upon the posture, relation to the sun (azimuth angle) and especially upon the sun elevation (Fig 2).

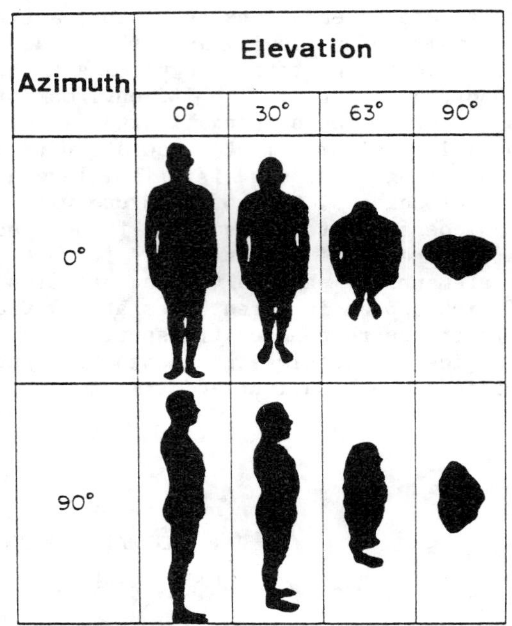

Figure 2 Affected surface of an erect male by direct-beam short-wave radiation at different values of solar azimuth and elevation (after Underwood and Ward, 1966, cit. after Oke, 1978)

An objective and physiologically relevant treatment of the problem of the meteorological conditions of heat loss requires a complete computation of the energy balance:

$$M + Q_{SHIV} + W + Q^* + Q_H + Q_L + Q_{SW} + Q_{RE} + N + S = 0$$

where
M = Metabolic rate of heat production
Q_{SHIV} = Increased metabolic rate due to shivering from cold stress
W = Mechanical work
Q^* = Radiation balance
Q_H = Flow of sensible heat
Q_L = Latent heat transfer due to water vapour diffusion
Q_{SW} = Latent heat transfer due to sweat evaporation
Q_{RE} = Respiratory heat transfer
N = Heat transfer due to food intake
S = Net body heat storage

Operationally usable equations exist. Fanger (1972) expresses the degree of thermal discomfort in his simplified equation by the PMV-value on a psycho-physical scale. This equation also forms the basis for the Klima-Michel-Model (Jendritzky et al, 1979), which using normal meteorological data, among other things parameterizes, the radiation fluxes for outdoor applications. Höppe (1984) calculates the skin surface temperature and the wetness of the skin as a measure of thermal load.

Previous attempts have been made to objectively represent the complex conditions of heat-loss of the human being using simple complex-variables, or thermal indices. Such procedures exist for cold conditions as a combination of air temperature and wind speed. These cooling indices or wind-chill factors are limited to the flow of sensible heat. For the warm conditions, combinations of air temperature and various measures of the water vapour content of the air are used. This is due to the significance of evaporation when the resultant heat-loss increasingly becomes a problem to the human being.

None of these variables which are too simply constructed consider all mechanisms which determine the heat balance of the human being. Frequently, however, in the practical application, the meteorological input data which are necessary for the use of complete heat balance models are missing. One is therefore forced to use a simple index in spite of its limited validity. An overview of the usefulness of such indices can be found in Landsberg (1972), Givoni (1976), Tromp (1980), or Hentschel (1985b).

2.2 The Actinic Conditions

This area encompasses all components of the biologically relevant solar radiation. Approximately half of the radiation intensity is in the range of visible light (380-780 nm). About 40% is in the (long-wave) infrared portion of the spectrum (IR-A .78-1.4 µm and IR-B 1.4-3.0 µm). The remainder is in the UV-area at the short-wave end of the spectrum of solar radiation.

The intensity of global radiation is dependent upon the turbidity of the atmosphere which is considerably higher in industrial areas as compared to other clearer air areas. The global radiation is also dependent upon distance through the atmosphere through which the radiation must travel. As a result, using the sun's elevation, there is a pronounced yearly and daily variation for radiation. There is also a clear dependency upon the altitude ASL. The influence of cloud cover is rather complicated. For a 100% cloud cover, however, a reduction of 80% of the short wave radiation fluxes can be assumed.

The role of radiation in the visible light range for vision requires no further explanation. Beyond this, however, there is also an effect on the hormone system. Bright light and strong contrasts produce an ergotropic condition, i.e., a mood conducive to activity and work.

UV-radiation will be divided according to biological criteria:

UV-A (315-500 nm): immediate pigmentation, psoriasis treatment.
UV-B (280-315 nm): erythema (sunburn), secondary tanning, anti-rachitis effect (synthesis of vitamin D), skin cancer, bactericidal effect.

In addition, there exist nonspecific effects of UV-radiation such as improvement of resistance to diseases, a greater feeling of well-being, an increase in physical performance, stabilisation of the vegetative-nervous system, to mention just a few. This occurs through an influence on metabolism, respiration, haemoglobin and internal secretions.

The dangerous radiation in the UV-C area does not reach the lower atmosphere due to the complete absorption by ozone. The UV-B radiation is also attenuated to a degree corresponding to the variable distribution of the ozone. Because short-wave radiation is strongly diffused by the atmosphere, the diffused portion of UV-radiation is generally greater than the direct portion so that tanning even in the shade is possible.

2.3 Air Hygiene

The atmosphere consists of natural and anthropogenic solid, liquid and gaseous components. In this connection, the effects of the components of the atmosphere and the purity of the air on the health of the human being is of interest to us.

Just as the combustion of fossil fuels requires oxygen, so does the oxidation of food in the organism. Due to the large supply of oxygen in the atmosphere and its production by vegetation during daylight, no oxygen deficit has been observed in industrial areas.

The partial pressure of oxygen decreases with altitude proportionally as does the total air pressure. Additionally the partial pressure of water vapour in the lungs must be considered. At altitudes between 1000 and 2000 meters ASL hyperventilation and an increase in heart rate occurs as a result of an adjustment reaction. This is barely noticeable and varies according to the individual.

Above about 4000 meters ASL acute altitude reactions much be expected. During longer stays at higher altitudes the circulatory and respiratory regulatory adjustments are of less significance due to the acclimatization through multiplication of the erythrocytes.

Air pollution is, according to the WHO, substances which are damaging to humans, animals, plants and property or that contribute to damage or disturb the general well-being or affect the possession of property to an unacceptable degree. In general a disturbance to the health exists when the functional and/or morphological changes of the organism exceed the natural range of variability.

Natural air pollution comes from sources which are not directly influenced by the human being. These pollutants occur more or less in all parts of the world. Illustrative examples are: inorganic substances from the stirring up of surface particles, volcanic particles, sea salt particles, organic mixtures predominately from vegetation (pollen, spores, bacteria, hydrocarbons from plants such as terpenes etc.). These represent the greatest share of the air pollutants.

Anthropogenic mixtures result from human activities. Their main characteristic is, in certain respects, their extreme localization and their extreme accumulation at certain times. The sources for the main anthropogenic components SO_2, NO_x, CO and aerosols are:

- Industry and power plants
- Home heating etc.
- Traffic

The effects of air pollutants are brought about by:

- Emissions (entry into the atmosphere)
- Transport (dispersion and transformation in the atmosphere)
- Emissions (affecting an active or passive receptor)

Major pollutants:

- CO_2 Main component of atmospheric trace gases (ca. 330 ppm). The human organism can tolerate concentrations of over 5000 ppm without damage. CO_2 absorbs radiation in the infrared region (green house effect, climate changes).
- CO Comes from the incomplete combustion in combustion engines. It has a strong affinity for haemoglobin (COHb) and because of this it reduces the ability of the blood to transport oxygen.

- SO_2 An irritant gas, very hygroscopic and therefore remains, even with high concentrations, in the upper areas of the respiratory system. Only heavier breathing can result in SO_2 deposits which can be found in the alveolar areas of the lungs. As a result we have increased respiratory problems. It is also transported by small dust particles and it is seen frequently in droplet form in the lungs. It is a significant component of London smog.
- NO_x (NO, NO_2) Only slightly hygroscopic and can extend as far as the alveolar area of the lungs. NO_2 has a strong physiological effect as an irritant on the eyes and the respiratory organs and presents the danger of lung oedema.
- O_3 In the vicinity of the earth's surface and with NO_x and CH as catalyzers it is produced through the effects of solar radiation. It is an irritant gas and very zytotoxic. The variable range for extreme levels is 0.3-05. mg/m^3 (photochemical smog, Los Angeles). It occurs naturally with concentrations from 0 to 0.1 mg/m^3.
- Aerosol: Solid (dust) and liquid mixtures in the air. Diameter from 0.01 to 100 µm. Chemically consisting of 2 essential components. Inorganic mineral, organic hydrocarbons and among other things soot particles and plant substances. Particles larger than 5 µm are deposited in the upper respiratory tract and eliminated within 24 hours. Smaller particles of up to 0.5 µm reach the bronchial and the alveolar areas. They may remain for months or years. Particles from 0.1-1 µm contain water soluble acids and organic aerosols in particular can have gases adsorbed onto them. The particle mixtures serve as vehicles for the transport of damaging substances into the lungs.

In many countries there are so-called smog alarm plans in existence. These plans are designed to limit the emissions from vehicles, industry and home heating when high emissions exist during inversion types of weather.

2.4 The Biotropy of Weather

Numerous studies have established stochastic relations between medical data and weather events in climatic areas which are characterised by frequent weather changes due to the location of the frontal zone. These medical data include respiratory diseases, heart and circulatory diseases, effects upon the blood clotting system, infectious diseases, neurological diseases, spastic disease forms, disturbances in general well-being and also traffic and industrial accidents (for more detail see e.g.. Tromp, 1980, Faust, 1977).

Apparently the weather changes play a decisive role here. Weather changes produce irritation for the human organism. This effect can be assimilated by a healthy human with no symptoms or at the most with very minor discomfort. This effect probably even has a positive side in the sense of a stimulation and an exercising of the vegetative function. If the vegetative adaptability is impaired stress may occur and its effects may extend from harmless subjective symptoms to a reduction of physical performance; or to an impairment of the body's natural resistance to diseases caused by external noxious substances; or to hindrance of the body's own healing mechanism; or even to a manifestation of previously latent pathological reactions. The biological effect of weather is expressed by term "biotropy".

MAIN CHARACTERISTICS OF BIOTROPY

- Biotropy acts as an additional stress, i.e. cybernetically as a disturbance variable, to which the organism must react to maintain homoeostasis

- Weather is not the cause but simply the initiating factor of acute meteorotropic reactions, i.e., meteorotropy is a fundamental property of the organism

- Biotropy maximum occurs in the areas of the strongest weather changes

- Cyclonic weather is biologically unfavourable, anticyclonic weather predominately favourable (exception: heat load, air pollution)

- "Hypotensive" and "spastic-hypertensive" reaction forms are associated with opposing atmospheric conditions

- Biotropy is accentuated by individual, pathogenic and climatic factors as well as time of year and day.

Statements about the biotropy and the meteorotropic reaction forms which are dependent upon it are only valid for larger collections of people and not necessarily for an individual (generality-individuality-dilemma). This is due to the nature of stochastic relations themselves. Present knowledge about the biotropy is, however, sufficient to allow its use in medical prognosis procedures.

3. CONSEQUENCES OF THE IMPACT OF THE ATMOSPHERE ON THE HUMAN BEING

3.1 Human Health

Since the human being is clearly influenced by atmospheric environmental conditions it must be the goal of human biometeorology to help to avoid stress situations or to reduce stress if avoidance is not possible. This can be accomplished by an assessment of conventional meteorological data with respect to human health and the determination of limiting factors.

The main stress factors are:

- Excessive heat or cold
- Too low oxygen partial pressure
- High emission concentrations from air pollutants
- Pronounced weather changes at a given location or pronounced changes due to a changing of location
- Unadjusted behaviour e.g., wrong clothing, exposure to solar radiation)

In the year 2000 approximately 50% of the world's population will live in cities and in the developing countries the percentage will be considerably higher. The urban climate represents the most significant anthropogenic climate change. Appropriate measures which are dependent upon the macro and meso-climate, must therefore be taken in the micro-scale field of urban planning (Jendritzky and Sievers, 1987) and building design to minimize stress from heat and air pollution (WMO, 1986). One must build in harmony with the climate and not fight against it. Locations for settlements, hospitals, sanatoriums, and senior citizens' homes should therefore be selected with consideration for the available bioclimatological knowledge.

In the area of prevention and rehabilitation the health services can make use of the knowledge of the importance of adaptive processes and of the corresponding possibilities to train the reactions of the organism to atmospheric stimuli. This includes applications in climatotherapy, physical therapy and even in both anaesthesiology and post-operative prophylaxis. Human biometeorology should play an important role in the general health education of the population since through proper use of this knowledge numerous erroneous developments can be avoided. Therefore all national weather services should offer medical-meteorological information for a wide range of applications, e.g. following the material summarized by Hentschel (1985a,b).

3.2 Tourism

Meteorology plays a role in tourism in three aspects which do not necessarily exclude one another:

- Recuperation (convalescence, relaxation)
- Sports activities
- Visiting a foreign country

There are at least two uses of a recreation meteorology (Reifsnyder, 1983): as a basis for facility planning which responds to the question as to where a recreation facility should be located for optimal recreational conditions and second, as a basis for individual trip planning in response to weather conditions most likely to be expected.

For a more passively oriented vacation on the beach relaxation is the main goal but for active vacations such as skiing, sailing, hiking, backpacking, adventure etc. athletic activities are the main interest. Therefore the relationship to the meteorological conditions are very specific depending upon the type of activity and in addition attention must be paid to the fact that there is a subjective and expectational quality of the climate. This complicates the determination of appropriate combinations of meteorological elements that determine outdoor physical comfort under the conditions of the various exercises. The major questions to be answered include the duration of the suitable season with information on the frequency distribution of the elements of interest such as lack of precipitation, conditions of heat exchange, demands on clothing, water temperature, sunshine, wind conditions, risks of thunderstorms or other hazards. What is the probability of deviations from the normal conditions?

With the help of meteorological data tourist areas can be bioclimatologically characterized. But whether meteorological factors provide unsuitable conditions or act as stimuli depends upon the reactions of the tourist. The tourist should, if possible, consult with his doctor to inform himself about his reactive condition. The correct choice of season and an appropriate behaviour (clothing, activities) are necessary, especially in extreme climates, to minimize the health risks of loads. The consideration of the season is also essential with respect to the climate conditions at home since every climate change causes stress on the organism.

Bioclimatic information can assist the tourist in preparing his expectations on the climate and in selecting equipment that is optimum for the area to be visited. But beyond that, such information can sometimes be vital to the safety and survival of the tourist (Reifsnyder, 1983). Furthermore current meteorological information has to be available to enable the tourist to adjust to the current weather in every respect.

3.3 Sport

When examining the significance of meteorology for sports one must differentiate the direct effects of meteorological factors on sports equipment, sports facilities and on the athlete by e.g. rain, wind, iciness, snow type, from the effects on the organism with the resulting indirect effects on the physical performance.

The dependency of the individual types of sport upon the weather varies considerably (TABLE 1) and the demands on meteorology are very specific. One only needs to think of the waxing problems for skiing; the aerodynamic conditions for ski jumping such as the resultant wind, turbulence; wind, thermals, visibility and precipitation for flying; the dynamics of air movement in sport stadiums which affect the trajectory for javelin and discuss throwing; the recognition of records where a maximum tail wind is allowed or the wind conditions for sailing.

TABLE 1

The degree of impact of selected meteorological elements on performance in sports (after Lobozewicz, 1981)

events	air pressure	temperature	wind	precipitation	fog
sailing	1	4	5	3	4
ice sailing	1	5	5	4	3
canoe racing	2	4	5	3	1
rowing	2	4	5	3	1
swimming	1	5	4	1	1
cross country skiing	2	5	4	5	2
ski jumping	1	4	5	5	5
down hill skiing	1	4	3	5	4
speed ice skating	2	5	3	4	1
figure skating and ice hockey	2	5	-	-	-
jumping	2	4	5	3	1
running	3	3	5	3	1
throwing	3	1	5	2	1
bicycle racing	3	3	5	4	1
bow and arrow shooting	-	3	5	3	4
shooting	1	1	5	3	4
soccer	1	2	4	5	1
flying	2	5	5	1	4
parachuting	2	3	5	4	4
bob sled racing	1	5	3	4	4
The point total showing the most important factor	33	75	86	65	47

Influence: 1 – very small, 2 – barely noticeable, 3 – periodically significant, 4 – significant, 5 – very large

For the sportsman himself the conditions of heat loss (as it affects optimal muscle performance), and water loss, are of great importance. The decrease of air density with altitude leads to a reduction of aerodynamic friction, i.e. it leads to a reduced exertion for all movements. The lower partial pressure of oxygen can, however, become a problem for sports which require a high continuous oxygen consumption. But during longer stays at these higher altitudes an acclimatization is attained which improves the oxygen intake capability of the organism at lower altitudes as well. A low air pollutant level is required in any case.

It is essential that sport facilities should be built with due consideration for meteorological conditions. Sporting events must also be scheduled with appropriate consideration for the conditions expected at that time of the year. If the athlete must travel long distances to a sporting event he must be given time for acclimatization.

4. FINAL REMARKS

The human being is directly and indirectly influenced by the atmospheric conditions in manifold ways. Numerous areas of everyday life are affected and consequently meteorology plays a significant role. The qualitative relationships are well known but beyond that a number of tools have already been developed to treat the man-environment problem quantitatively. For climate-related planning which deals with the demands of the population on land-use in connection with recreational activities, weekend holidays, tourism, and public health, a biometeorological analysis and evaluation of the space is indispensable. This applies to regional planning, urban planning, locations of hospitals, sanatoriums, homes, recreational centres, sports facilities, health resort climatology and environmental protection. Thus a problem-oriented adaptation of meteorological knowledge, experience and data supplies all who are involved with planning at different levels with valuable information and decision aids. This will help advantage to be taken of suitable conditions, avoidance of limiting conditions, and, in the broadest sense, the ensuring of the natural basis of life for human beings, regarding the atmosphere as a natural resource.

REFERENCES

Auliciems, A and de Freitas, C R 1976: Cold Stress in Canada. A Human Climatic Classification. International Journal of Biometeorology vol 20, 287-294.

Auliciems, A, 1981: Towards a Psycho-Physiological Model of Thermal Perception. International Journal of Biometeorology vol 25, 109-122

Burt, J E, O'Rourke, P A, and Terjung, W H, 1982: The Relative Influence of Urban Climates on Outdoor Human Energy Budgets and Skin Temperature. I: Modelling Considerations, II: Man in the Urban Environment. International Journal of Biometeorology vol 26, 3-35.

Fanger, P O, 1972: Thermal Comfort. Analysis and Applications in Environmental Engineering. McGraw Hill, New York, 244 pp.

Faust, V, 1977: Biometeorologie - Der Einfluß von Wetter und Klima auf Gesunde und Kranke. Hippokrates, Stuttgart.

Givoni, B, 1976: Man, Climate and Architecture. Applied Science Publications Ltd, London, 364 pp.

Hentschel, G, 1985 a: Curriculum for Training and Education of Meteorological Personnel in the Field of Human Biometeorology. Report of the Rapporteur on Human Biometeorology to CCl-IX, WMO, Geneva, 22 pp.

Hentschel, G, 1985 b: Material and Methods Available for Use in the Field of Human Biometeorology. Report of the Rapporteur on Human Biometeorology to CCl-IX, WMO, Geneva, 29 pp.

Höppe, P, 1984: Die Energiebilanz des Menschen. Wiss. Mitt. Meteorol. Inst. Univ. München, Nr 49, 171 pp.

Jendritzky, G, Sönning, W, Swantes, H-J, 1979: Ein objektives Bewertungsverfahren zur Beschreibung des thermischen Milieus in der Stadt und Landschaftsplanung ("Klima-Michel-Modell"). Beitr. d. Akad. f. Raumforschung u. Landesplanung Bd. 28, Schroedel, Hannover, 85 pp.

Jendritzky, G, 1987: The Human Biometeorological Forecast Procedures of the German Weather Service, Int. Symposium on Climate and Human Health, Leningrad, Sept 22-26, 1986, WMO, Geneva, in press.

Jendritzky, G, and Menz, G, 1987: Bioclimatic Maps of Heat Exchange of the Human Being in Different Scales, Int. Symposium on Climate and Human Health, Leningrad, Sept 22-26, 1986. WMO, Geneva, in press.

Jendritzky, G, and Sievers, U, 1987: Numerical Simulation of the Thermal Environment of the Human Being in Street Canyons. Int. Symposium on Climate and Human Health,

Leningrad, Sept 22-26, 1986. WMO, Geneva, in press.

Landsberg, H E, 1972: The Assessment of Human Bioclimate. WMO Techn. Note No 123, WMO-No 331, WMO Geneva.

Lobozewicz, T, 1981: Meteorologie im Sport. Meteorologische Probleme in der Sportpraxis. Sportverlag, Berlin, 1973 pp.

Oke, T R, 1978: Boundary Layer Climates. Methuen & Sons, London, 372 pp.

Reifsnyder, W E, 1983: A Climate Analysis for Backcountry Recreation. Proc. 9th Int. Biomet. Congress Int. f. Bio-met., Suppl. 2, vol 26, 87-99.

Taesler, R, 1987: Climate Characteristics and Human Health - the Problem of Climate Classification. WMO/WHO/UNEP-Symposium on Climate and Human Health, Leningrad, Sept. 22-26, 1986. WMO, in press.

Terjung, W H, 1968: World Pattern of the Distribution of the Monthly Comfort Index. International Journal of Biometeorology, vol 12 119-151.

Tromp, S W, 1980: Biometeorology, Heyden, London, 346 pp.

Weihe, W H, 1986: Klima, Wetter and Verhalten. In: V Faust (ed), Proceedings of Klima, Wetter und Gesundheit. Hippokrates, Stuttgart, 74-91.

WMO, 1984: Report of the Meeting of Experts on Climate and Human Health, Geneva, Dec 5-9, 1983. WCP-78 WMO, Geneva.

WMO, 1986: Urban Climatology and its Applications with Special Regard to Tropical Areas. Proc. Techn. Conf. Mexico City, Nov 26-30, 1984. WMO No 652, WMO, Geneva, 534 pp.

22

Japanese system of the meteorological information service to user communities including education and training

K. Nakai, Japan Meteorological Agency (JMA)

1. INTRODUCTION

The current trend of the users' needs of meteorological information is characterized in their specialization and diversification. The JMA has been making every effort to meet their wide-ranging requirements with improvements in forecasting techniques, the introduction of new telecommunication systems and the development of new techniques in information processing. Furthermore, the private meteorological services, established under the supervision of the JMA, are playing important roles in preparing and disseminating specific meteorological information based on the requirements of the customers.

The present system of meteorological information services of JMA and private firms authorized by the JMA is described in this paper.

2. THE PRESENT SYSTEM OF METEOROLOGICAL INFORMATION SERVICES IN JAPAN

2.1 The users' needs for meteorological information

The JMA has been directly disseminating meteorological information such as forecasts, advice and warnings to the organizations and companies concerned with aviation, railways, shipping, electric power and agriculture, which form the bases of social activity. As well as these areas, the users' requirements for meteorological information services are increasing since the optimal use of meteorological information can significantly take the risk out of activities affected by weather variations. Table 1 indicates the users' need for meteorological information in recent years.

In this connection, in 1986 the JMA established a forum - the "Users Panel on Meteorological Data Application" - for exploring the users' needs. The specific requirements from the panel have been kept under the constant review of the JMA and passed on to the operational meteorological services.

2.2 Data acquisition/dissemination system in the JMA

The meteorological information service of the JMA is made through the Automated Data Editing and Switching System (ADESS). This system has been in operation since 1980. The system is a nationwide high speed computerized meteorological telecommunications network consisting of the Central ADESS (C-ADESS) and the Local-ADESS (L-ADESS) as shown in Fig 1.

The L-ADESS is installed at 6 local forecasting centres throughout the country. They are linked with the C-ADESS located at the JMA HQ at the speed of 4800 bits/sec for domestic meteorological data exchange. The C-ADESS performs the functions of an integrated domestic data exchange and also a data exchange linked to foreign meteorological services by the Global Telecommunication System (GTS).

The JMA is now delivering meteorological information to the mass-media and a JMA-supervised private meteorological service through the ADESS.

2.3 "Meteorological Information Comprehensive On-line Service (MICOS)" of the JWA (Japan Weather Association)

To meet individual requirements of users the JMA has authorised the private meteorological services and has allowed them to serve the users with a specific area of information. Fifteen private meteorological information services are now operating under the supervision of the JMA.

The JWA, one of the leading private services, runs an information service in a user-friendly form to contracted customers in the various fields such as transportation, sightseeing, construction, natural resources/energy industries, agriculture and fishery through their system: "Meteorological Information Comprehensive On-line Service (MICOS)". They take advantage of getting more specific and precise information through the JWA. Fig 2 depicts configuration of the JWA's information service.

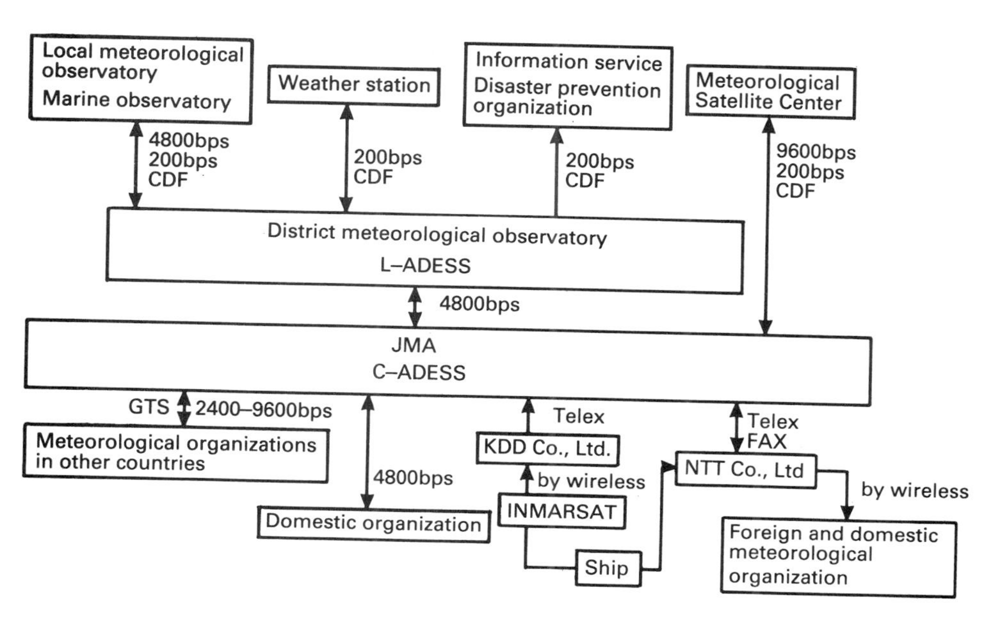

Fig 1 Schematic diagram representing the flow of the meteorological information through ADESS of the JMA

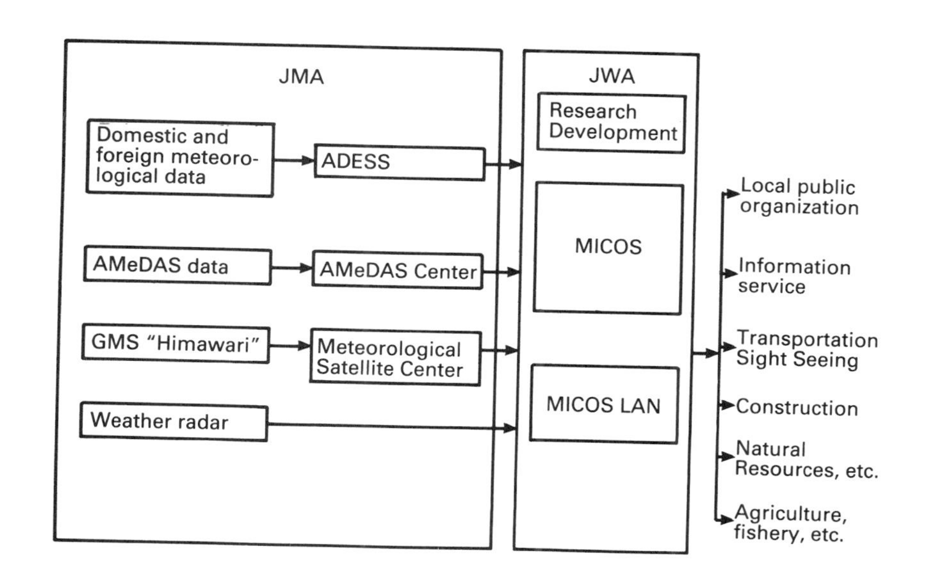

Fig 2 Schematic diagram of the JWA's meteorological information service

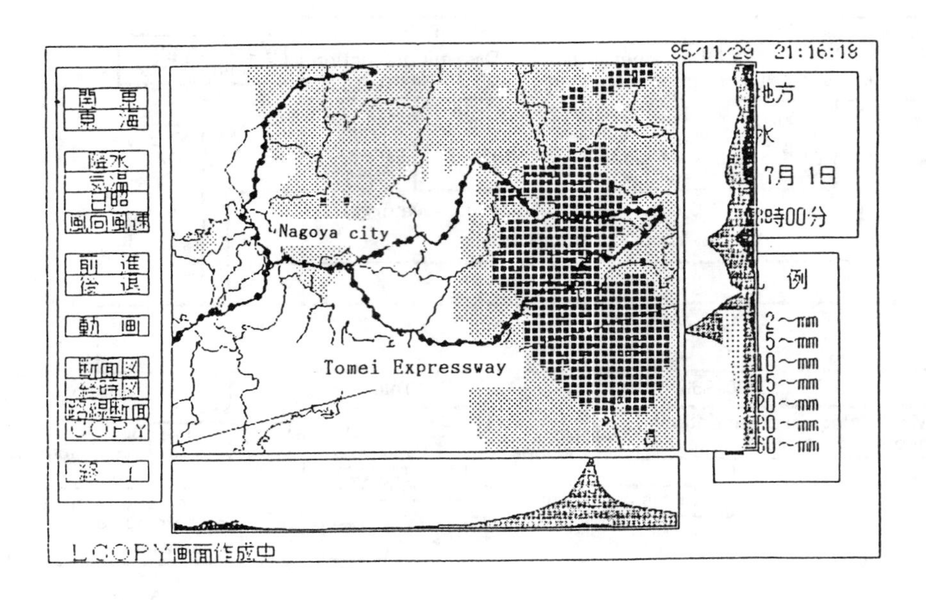

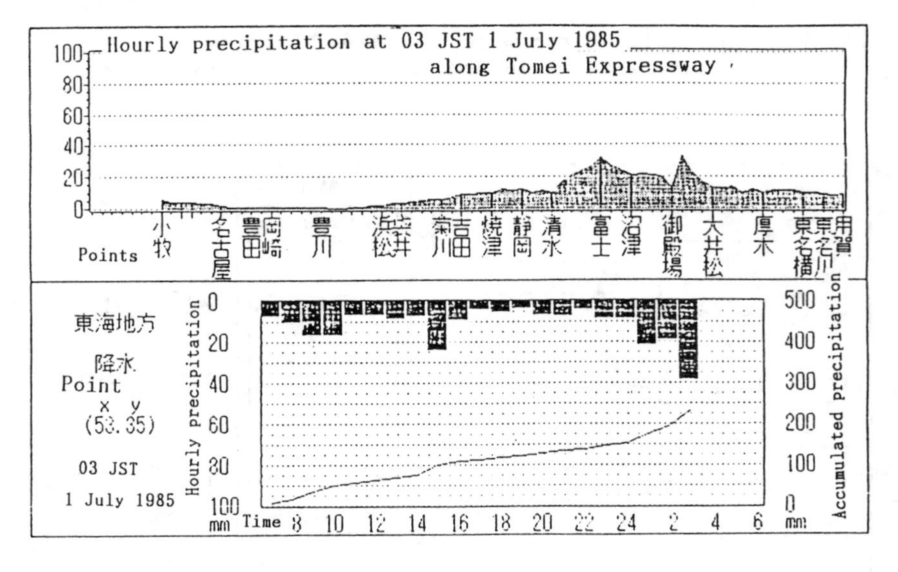

Fig 3 Fine distribution of hourly precipitation and time change of hourly and accumulated precipitation at a point along Tomei Expressway, shown in the upper and the lower parts respectively. These figures are the hard copies from a colour display (courtesy of the JWA).

Fig 3 is an example of meteorological information produced by the JWA. The figure shows fine distribution of precipitation on Tomei Expressway. This information is extremely useful for the management of road traffic.

3. NEW METEOROLOGICAL INFORMATION PRODUCTS

The JMA has introduced new meteorological information products in the form of the mesh climate charts, and a composite radar/AMeDAS precipitation map.

3.1 Mesh climate charts

It is well known that the climate of Japan is widely variable because of its complex geographical features. For a long time many efforts have been made to estimate the climate values for finding micro climate resources with fine horizontal resolution by the use of multivariate analysis. The principle of estimating the climate values on a point other than the meteorological stations is described in Appendix 1.

In order to obtain the climate values by means of this method we have to handle a lot of observational data and geographical factors. We have been able to calculate these since the appearance and spread of the electronic computer.

In Japan many kinds of climate value have recently been calculated and the result of our efforts are the completion of "the mesh climate data" or "the mesh climate charts" with 1 km resolution. Until now charts of precipitation, temperature, snow and wind energy have been made by the JMA and/or local governments.

The data of the national mesoscale meteorological observation network - AMeDAS (Automated Meteorological Data System, see Appendix 2) - are utilized to take the mesh climate data/charts as the basic data. These mesh climate data charts are provided to users through the JMA's information service. Fig 4 is an example of the mesh climate charts made by the JMA. This chart shows the normals of annual total precipitation amount.

On the other hand local governments have been active in producing and using the mesh climate data/charts under the JMA's advice. Fig 5 shows a schematic explanation of the use of the mesh climate data/charts in agriculture at local public organizations level.

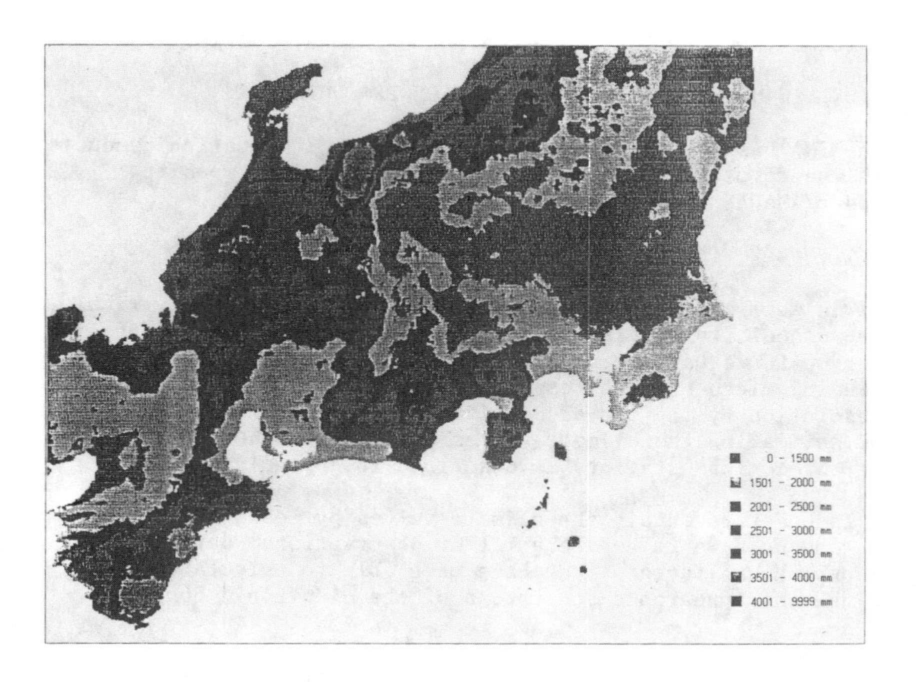

Fig 4 **Mesh climate chart made by the JMA: Normals of annual total precipitation amount within 1 km x 1 km resolution**

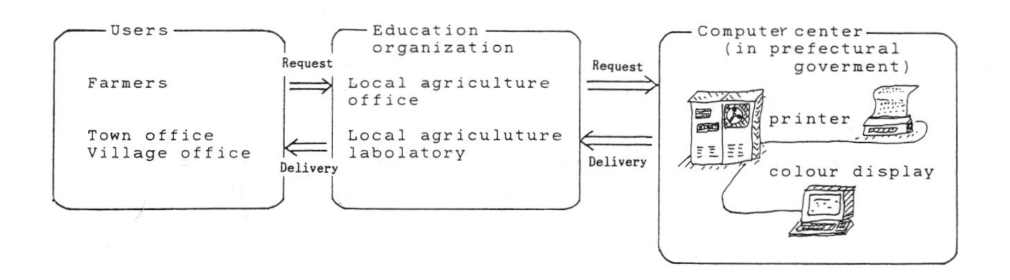

Fig 5 **Climate data/charts service of local organizations in the field of agriculture.**

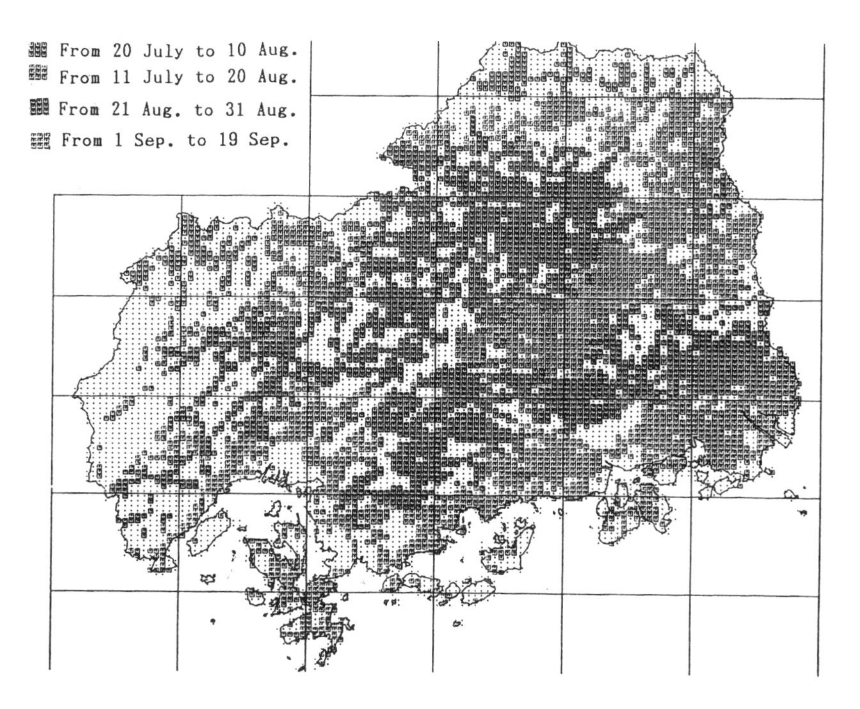

Fig 6 Application of the climate charts on air temperature to the management of agricultural cultivation. Distribution of the latest date of rice-earing in Hiroshima Prefecture are shown. Each symbol represents 1 km x 1 km (the leaflet The system of application of the mesh climate charts, published by Hiroshima Prefecture.

Fig 6 is an example of applying the mesh climate chart on air temperature to the management of agricultural cultivation. The growth of rice depends on air temperature during the 40-day period after coming into ear. A necessary condition for a good harvest is that the accumulated temperature during the 40-day period should amount to more than 880°C. If farmers can know in advance the latest date before which the accumulated temperature is assumed to be over 880°C, they can plan a good planting schedule and can avoid the risk of a poor harvest. This chart shows distribution of the latest date of rice-earing in Hiroshima Prefecture.

The above mentioned is an example of applying the mesh climate charts to agriculture. We can expect that the climate charts will be of use to various other areas besides agriculture.

3.2 Composite radar/AMeDAS precipitation map

Japan occasionally suffers from great damage caused by severe local rain. The AMeDAS is very effective in detecting heavy precipitation. It is recognized, however, that extremely localized rain may escape from this observation network. To supplement the AMeDAS network, weather radar information is readily available. The JMA has made up a plan for improved utilization of radar information by automatic observation and high speed data processing together with the raingauge data of AMeDAS.

The radar observation network of the JMA consists of 20 radar observatories and covers the whole country. The radar data are digitized into 5 km x 5 km resolutions and are transmitted through ADESS to the data processing center in the JMA HQ. Here they are processed into a composite map. To combine radar and raingauge data, precipitation data are converted into a logarithmic scale, and the radar-estimated precipitation field is calibrated by raingauge data.

RAAMAP (composite Radar/AMeDAS precipitation MAP) has been in operation in the JMA HQ since 1983. It has been demonstrated that RAAMAP is better than the individual information of AMeDAS or radar in detection of small scale precipitation.

An example of RAAMAP is shown in Fig 7. This composite map is disseminated to local forecasting centres, and is fully used in detecting heavy rain. This information is transferred to the TV station of the Japan Broadcasting Corporation for timely broadcasting as a colour-assigned map. Since the map covers all the Japanese archipelago showing accurate amounts of precipitation, the individual authorities such as disaster prevention activities, and the road management use it as central information for their activities. To meet their requirements the JMA is planning to deliver this information to them through the ADESS or the JMA information service from 1988.

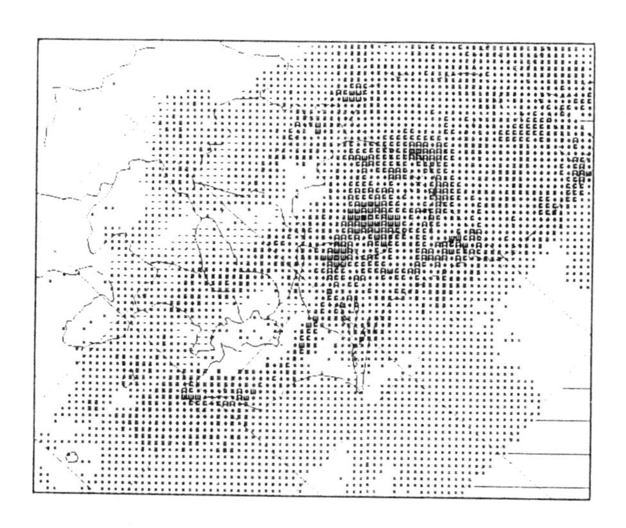

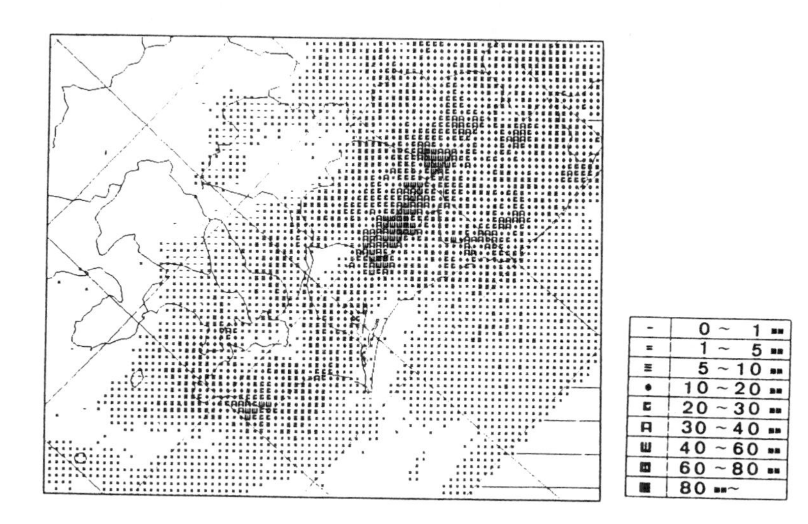

Fig 7 Composite radar/AMeDAS precipitation map at 1600 GMT and at 1600 GMT 4 August 1986. One hour precipitation amount is depicted by 9-level symbols. Each symbol represents 5 km x 5 km area.

Very short-range forecasts of precipitation are now in a developmental stage in the JMA with the use of RAAMAP as the initial data. An example of a very short-range forecasting experiment is shown in Fig 8. It is planned to commence this type of forecast in 1988.

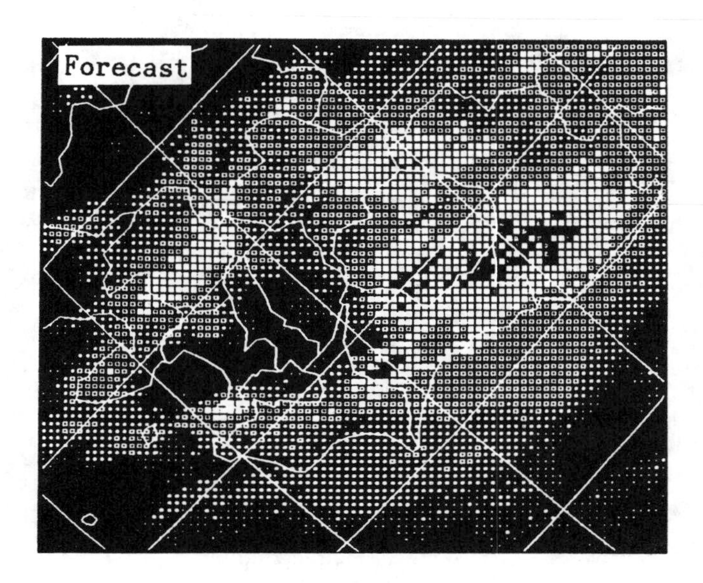

Fig 8 An example of very short-range forecast of precipitation valid for 1600 GMT on 4 August 1986, 3 hours after initial time (at 1400 GMT). Verification is in Fig 7.

4. EDUCATION AND TRAINING ACTIVITIES OF THE JMA

The JMA puts high priority on the education and training activities for strengthening the national infrastructure for better use of the meteorological information. School teachers and meteorological information users have an opportunity to bring their knowledge up to date in the "Summer Meteorological School" organized by the JMA and the Meteorological Society of Japan. The Summer Meteorological School is held in a few cities each year. Table 2 shows the themes of the "Summer Meteorological School" in 1986.

Users in local authorities also get up to date knowledge of the practical use of meteorological information through the meteorological workshops hosted by the JMA some 50 times a year. Furthermore, we have a unique TV programme titled "TV Observatory", in which the application of techniques of meteorological information and current development of meteorology are introduced. Weather/climate information from all over the world is also introduced. This is a weekly 30 minute programme technically supported by the JMA. Main themes picked up by "TV Observatory" are listed in Table 3.

5. CONCLUDING REMARKS

The meteorological information service, which is mainly aimed at disaster prevention activities, is now playing a vital role in industrial activities and making the individual's life more comfortable. In response to the situation, the JMA is going to establish an effective service scheme by strengthening the private meteorological services.

New networks for meteorological information services are now being introduced and extended by use of new information media, such as videotex, CATV (Cable TV) and personal computer. The JMA will provide a network of the basic meteorological information and allow the private services to deliver specific meteorological information for individual needs through the network.

APPENDIX 1

We define the criterion variable Y and predictors $X_i(i=1,2,...,m)$, where Y is the climate value at the observation station and $X_i(i=1,2...,m)$ are geographical factors at the observation station such as the elevation above the sea, its degree of relative relief and its degree of slope. For Y and $X_i(i=1,2,...,m)$, we assume the following relation (linear regression equation).

$$Y = \beta_0 + \beta_1 X_1 + \beta_2 X_2 + \dots\dots\dots\dots + \beta_m X_m \quad (1)$$

where $\beta_i(i=0,1,2,...,m)$ are partial regression coefficients and β_0 is residual. $\beta_i(i=0,1,2,...,m)$ are determined by using method of multivariate analysis. After determination of $i(i=0,1,2,...,m)$, we can obtain the climatic value at an optional point if its real values upon geographical factors (X_i) are given into the equation (1).

APPENDIX 2

The AMeDAS was designed to collect weather data observed by unmanned observation instruments (sensors) located at about 1,300 observation-points with 17 km interval throughout Japan and send them back to about 60 weather forecasting centres. 840 of the observation-points monitor precipitation, wind direction/speed, air temperature and sunshine duration. 460 of the observation-points monitor only precipitation.

The telephone switching network is used to connect the AMeDAS Center and the observation-points. The AMeDAS Center automatically calls all observation-points every hour and collects data from each point. Fig 9 shows the scheme of AMeDAS. All of AMeDAS data are stored at the JMA HQ where statistical analyses are produced for many purposes such as making the mesh climate charts.

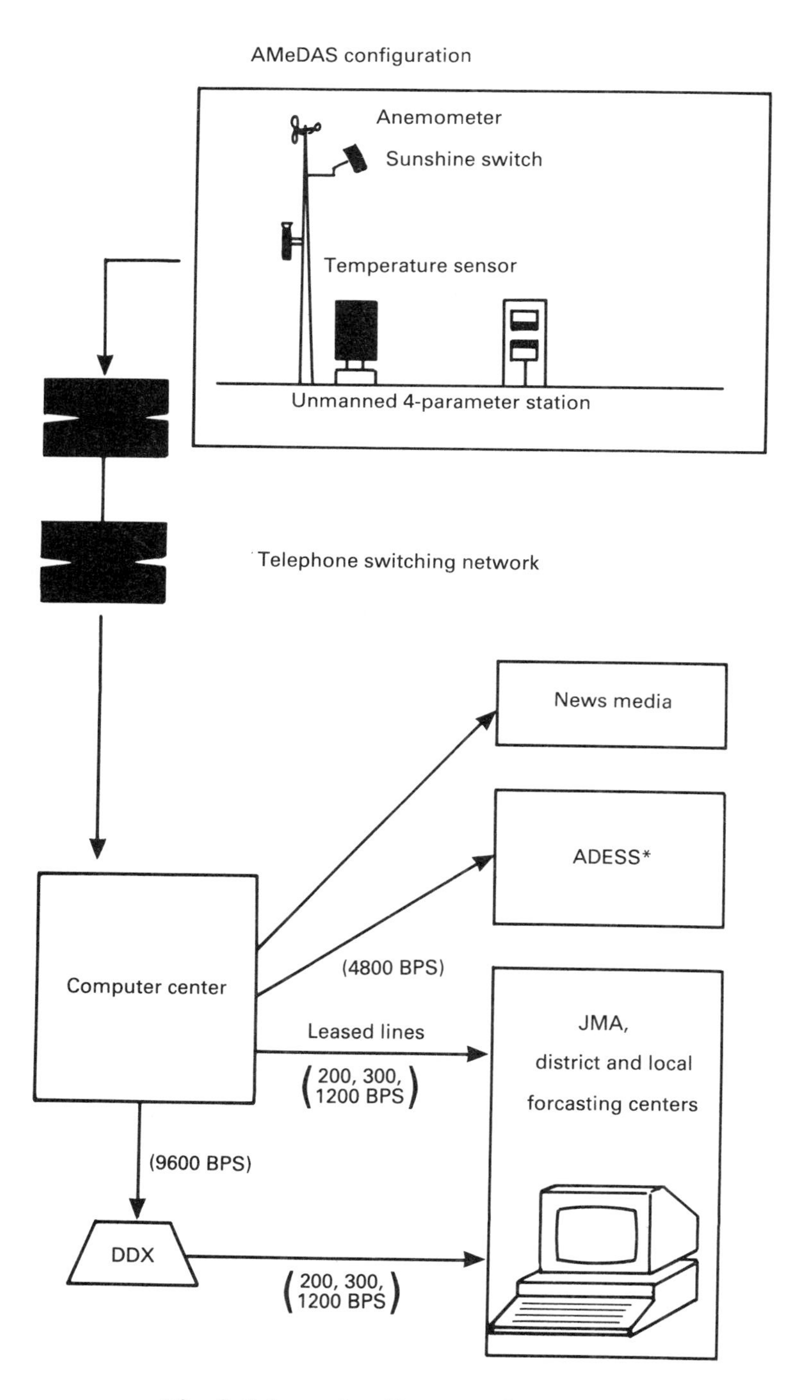

Fig 9 Schematic diagram of the AMeDAS

Table 1

Typical users' needs to meteorological information

Fields	User's needs
Transportation	Safe and economical transportation
Sightseeing	Meteorological information service to customers. Prediction of the number of visitors to skiing grounds.
Agriculture	Prevention of frost damage to oranges. Prevention of insect infestation. Management of agricultural cultivation.
Fishery	Grasp of good fishing areas. Safe fishing work.
Electric power	Prediction of demand for electric industry power supply. Prevention of lightning damage to facilities. Management of dams for power generation.
Manufacturing industry	Prediction of demand for clothes in summer and winter. Prediction of demand for electrical products. Prediction of demand for beer and fruit juice.
Pastimes	Prediction of rainfall on a baseball stadium.

Table 2

Themes of the "Summer Meteorological School" organized by the JMA and the Meteorological Society of Japan in 1986

In Tokyo	In Osaka
1. Cyclone and front as the precipitation system	1. Recent climate variation and forecast of the future
2. Nowcasting of precipitation	2. Abnormal climate judging from the weather map
3. Rainfall and Flood damage	3. Simulation of palaeoclimate
4. Application of NOAA data data in meteorological and oceanographic observation	4. Why does climate change?
5. Training of the upper air chart	
6. Films	
(1) Meteorological service of Japan	
(2) Landslide disaster	

Table 3

Recent themes of "TV Observatory"

1.	The experiment on nowcasting on precipitation
2.	This is "local severe rain"
3.	Fog in Kusiro city
4.	Launching of the meteorological sounding rocket
5.	All about artificial snow
6.	The riddle of acid rain
7.	The mechanism of snow storm on the ground
8.	Heavy snow in Sapporo city
9.	Prevention of landslide disaster
10.	The observation station on the top of Mt Ibuki

23

Forecasting the peak loads for electric power systems

G. Haijtink, Royal Netherlands Meteorological Institute (KNMI)

Public electricity supply companies are obliged to respond instantly and unfailingly to the ever changing demands of their customers.

It has been shown that the prevailing weather conditions have a substantial bearing on these demands. In the daily consumption two maxima appear.

It is evident that companies are very interested in the expected demand. They want to anticipate the number of production units needed, particularly at peak hours.

Using a semi-objective method it is possible to forecast maximum loads.

The first question which might arise is whether there is much fluctuation in the demand for electricity from a supply system. There is no doubt that the answer is yes. The fluctuations in demand depends on the consumers. The graphs below show this. Load as a function of time can be presented in the load curve.

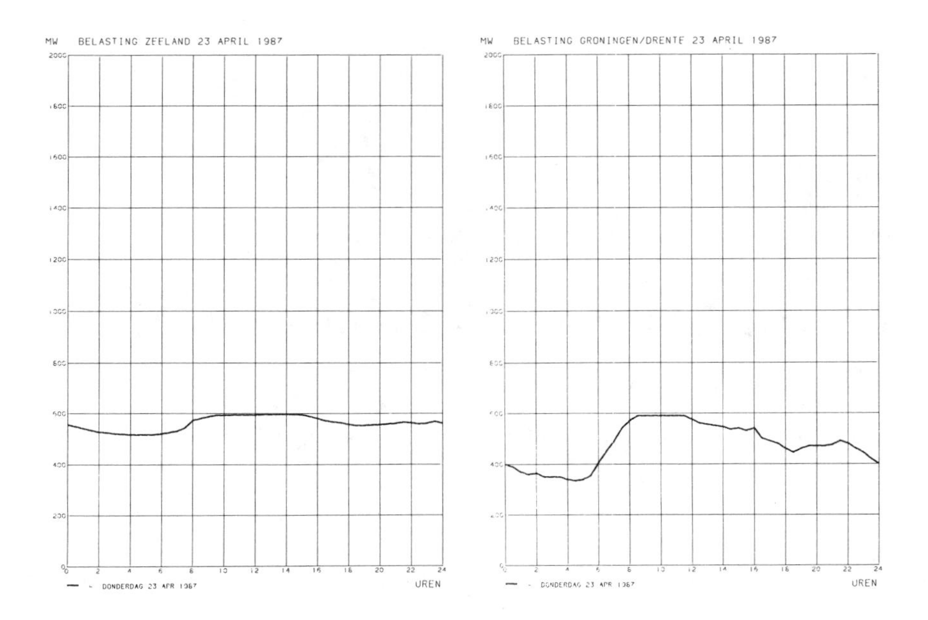

Graph 1 Graph 2

The first graph is that of a small company in the southwest of the country. Their load shows only a little fluctuation. There is an increase in daytime but that is about 25% of the daily average. The most important consumer is industrial. The second graph is of a company from the northeast of the country, an area with little industry and more agriculture. There are only two cities with more than 100,000 inhabitants. It shows more differences during the day than the first graph.

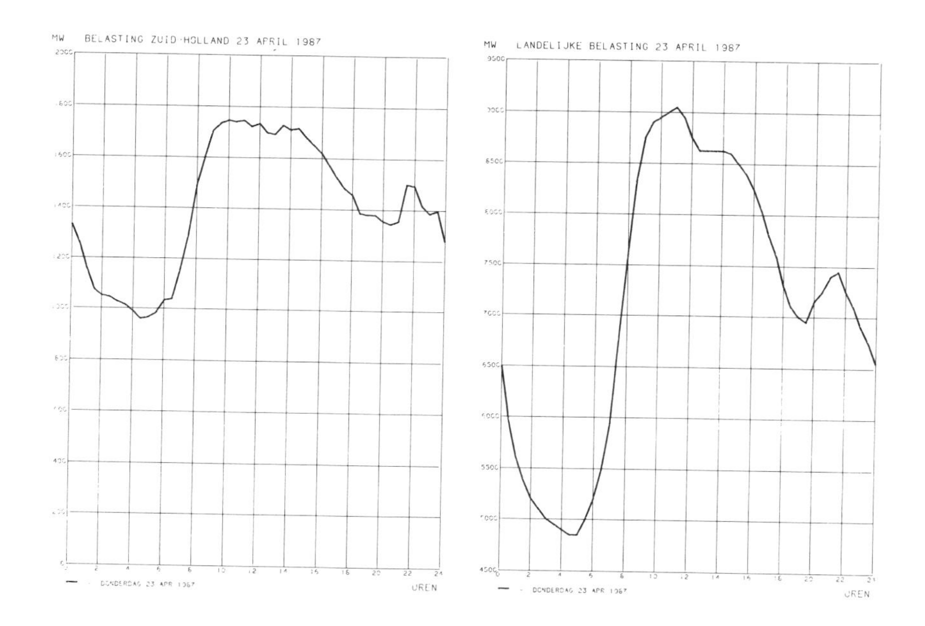

Graph 3 Graph 4

The third graph is that of the biggest company in the Netherlands, which serves the west part of the country with the cities of the Hague, the capital, and Rotterdam, the biggest port in Europe.

The fourth graph shows the total demand of the country that day, April 23 1987 a day of average weather for the time of year.

These graphs make clear that it depends on the time of the day and the kind of consumers to account for great differences in the demand.

Looking at the graphs as a meteorologist you could form the idea that the third and fourth might be the most sensitive as regards the weather.

To see if weather contributes to the load you have to compare several load curves. This is done in the next graph using two successive days, both working days.

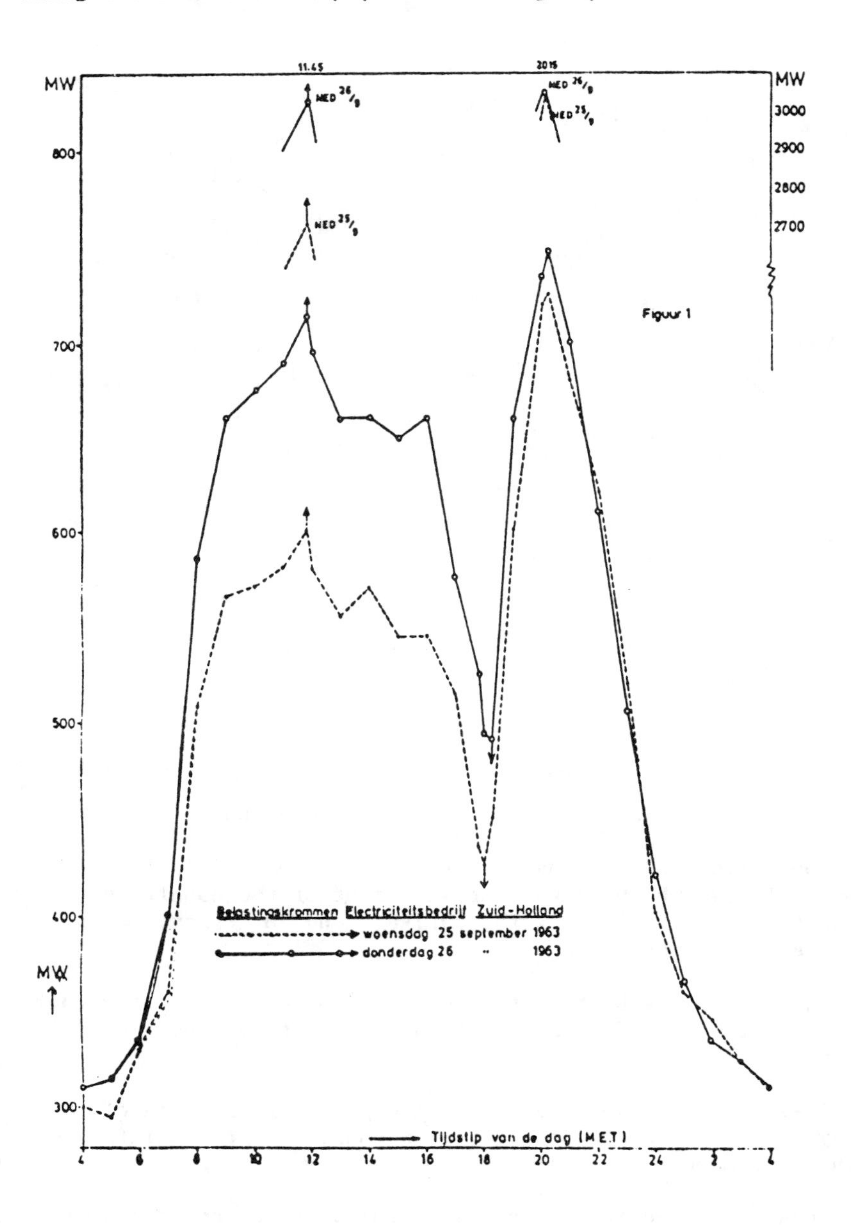

Graph 5

The curves show two maxima each, one in the morning and one in the evening. The morning maximum of the second day is about 20% higher than the day before. This difference was an effect of the weather. The first day, that shown by the dotted line, was dry and clear. The second was dark, rainy, stormy weather. It is clear the very difference in weather conditions might be considerable and therefore critical. If the company should have produced the same amount as the day before, there would have been a shortage of more than 15%. measures to overcome this shortage are to start up more generating units or to buy power from other companies. The sooner these measures are taken the more economical it is. Not only from the economic point of view should the company know what the demand will be. The company is obliged to respond instantly and unfailingly to the demand with constant voltage and frequency. Gas can be stored, electricity can not. In this case the shortage would have caused damage to equipment and interrupted the supply of electricity to the consumers if there had not been a good weather forecast and lack of understanding of the effect of weather on demand. In 1949 Dr Yan provided the thesis: inaccuracies in the daily load forecast are usually the result of failure to calculate accurately the effect of weather on the system load.

As soon as the significance of the day by day variations in the weather is established the question arises: how should these problems have been solved.?

First, make good weather forecasts. Second, the daily load forecast due to the weather can be done by the company itself. It also can be done by the Meteorological Service. In the Netherlands, KNMI, the National Meteorological Institute, is doing both. The Institute, which coordinates the load supply of all the companies provides a weather forecast twice a day for the following 12 to 24 hours and a list with the temperatures and visibility at the time of the maximum load for several places. The morning maximum in the Netherlands is about 09.00 UTC, the evening maximum about an hour after sunset.

For the biggest company in our country, which serves about 30% of the country, the KNMI forecasts the maximum load. These estimates are given by telephone, so that the loading engineer at the system control center is in touch with the KNMI. He is not only told the maximum X but also at which time the weather will change. So he gets an idea as to what the slope of the curve will be to the maximum. This is just as important as the maximum for the value of the weather lies principally in connection with the consumers demand. If the loading engineer has a correct forecast he can anticipate in advance what the consumers demand will be. Economically the information is very interesting for the company. They can make a choice which of the alternatives is the best, for instance either to start another generator and which one, or to buy power.

To study the influence of the weather conditions on electric power production and consumption it may be useful not to start with the weather. For example, the lightning load is a function of the daylight illumination and from "abnormal" weather conditions it varies throughout the year in a predictable manner. As was shown with the load curves earlier it might be useful to examine the different purpose for which electricity is used within the network and the manner in which these uses of electricity vary from time to time.

Associated with this is the importance of distinguishing the effect of weather and non-weather on power at different times during the day.

As KNMI is forecasting maximum loads, these were taken to determine the influence of the weather. These maxima should be mutually comparable. The available data should be made homogeneously.

The factors, which influence the load maxima can be divided into two categories:

A: Non-meteorological factors

1. long term changes from year to year
2. seasonal variations
3. character of day: working day, Saturday, Sunday and public holiday
4. week by week change

B: Meteorological factors

1. visibility
2. temperature
3. precipitation
4. windspeed in combination with temperature

At first, corrections were made for non-meteorological factors. This is necessary in order to be able to quantify the influence of the weather. Information as to the influence of the weather is valid for a limited number of years. The influence of the weather to the load curve is accomplished by the reaction of people to the weather and this is determined by the available possibilities and the social pattern of the people. In the Netherlands, for example, a great change took place when natural gas was found. In spring and autumn especially, the heating component of the total load became smaller. And in 1973, the oil crisis disturbed the continuously growing set of data.

When the data is compiled homogeneously, correction tables can be made which give the quantitative influence of the weather. These corrections are seasonally dependent and the data should be split up into months.

If you have data for 10 years it appears that these data are hardly sufficient to examine them for two weather factors. The visibility factor is divided into 5 classes, from clear to dark with rain or snow. Temperature is in 11 classes. Each month there are 22 working days. So in 10 years there are 220 data per month. As there are (5 times 11) 55 combinations the average in each combination is 4. Panofsky said in 1958: "the number of available data in a parameter section decreases rapidly with an increasing number of parameters and classes". Only by application of linear regression with temperature as an independent variable in spite of visibility was it possible to determine regression constants per month. These constants are determined for each month separately and then averaged over the period of years. Regression after splitting up visibility appeared thus not possible and is more over undesirable for the correlation between temperature and visibility itself (low temperatures and visibility are related positive).

The influence of visibility is examined as well each month independently of the temperature. On the score of examination of visibility and temperature one is able to compose monthly correction tables, which give the quantitative influence to the peak loads. If one constructs the annual course (determined out of the available years) one gets the so-called basis data. To this one can add by means of the monthly tables, the corrections for the influence of the expected visibility, the temperature and possible precipitation and so make estimates for the morning and evening maximum loads. The basic data give the load with the normal temperature and clear weather. In this manner we have done so since 1952, for 35 years.

24

The problem of defining exceptionally severe weather for social security payments

B. D. Giles, University of Birmingham, UK

There has been a long-standing tradition in the United Kingdom to make payments to the old and needy either as supplementary benefit or as single payments when specific needs arose. Over twenty years ago discretionary payments were made under the National Assistance scheme during the severe winter of 1962/63. This kind of payment continued until the Supplementary Benefit Act of 1976 was amended by the Social Security Act of 1980 just after the fourth coldest winter this century in Birmingham, England. Perhaps I should add that all the subsequent examples will use the Birmingham University data base since we have continuous records since 1885. The 1980 Act was rather badly drafted in places, as were the Regulations that accompanied it. It is in Regulation 26 that the evocative phrase "exceptionally severe weather" appears but with no attempt at definition. The nearest approach to explanation is in a later phrase "consumption greater than normal," referring to fuel. Both of these phrases had to be interpreted by the Chief Supplementary Benefits Officer when this part of the Act was invoked.

The winters of 1979/80 and 1980/81 were not particularly cold (Fig 1) but thereafter on any criterion parts of the United Kingdom experienced a series of cold winters with 1981/82, 1984/85 and 1985/86 ranking as ninth and tenth equal in severity during this century. In the 1981/82 winter a single countryside criterion was used, namely if a claimant could show (by means of a fuel bill) that more fuel had been consumed than in 1980/81 then the difference would be paid. In other words, the payment was made on fuel consumption rather than on climatological grounds. Naturally problems arose because there was no guidance as to how much of the extra fuel had been used in any undefined period of exceptionally severe weather. It was clear, in retrospect, that there had been additional fuel consumption but it was not clear whether when it, or the causal severe weather, had occurred. The mean monthly temperature data indicated that December had the lowest mean temperature this century with a value of 0.5°C In fact it was the second coldest since records began in 1985, so was 'exceptional' by any criterion.

Because of these problems consultations were held with the Meteorological Office during 1982 to devise more suitable criteria for operating the regulations. The advice given and accepted was to use degree days to the base 15.5°C at seventeen stations in the United Kingdom. The Meteorological Office agreed to provide the data and to fix 'trigger' levels for 'exceptional' based on an analysis of the coldest periods over the last twenty years. 'Exceptional' was taken as a level that would be reached one year in five, i.e. the 20% probability level. This system did not have to be used until the winter of 1984/85. Then when the number of degree days in a week exceeded the trigger value the fact was advertised. Unfortunately there were a number of appeals against the consequences and in October 1985 a Tribunal of Social Security Commissioners strongly criticised the definitions of 'exceptionally severe' on a number of grounds, particularly because

(i) of the arbitrary decision that 'exceptional' was at the 20% level;

(ii) too much variation could be expected in areas far from the seventeen stations;

(iii) 'what is abnormal or exceptional at one time of year may be quite normal at another';

(iv) the system was too complicated and the concept of degree days was not clearly understood by the public.

Consequently in November 1985 the Department of Health and Social Security devolved responsibility to local offices who made a decision on exceptionally severe weather on the basis of local (unspecified) information. This ad hoc arrangement was tested in the 1985/86 winter (Fig 1) and again was found wanting, since February 1986 was the second coldest of this century and contained seventeen seven-day periods when mean temperatures were below -1.5°C On 1 March 1986 an editorial in The Guardian noted that in spite of this only 360 areas out of 451 had approved special fuel payments.

Once again advice from the Meteorological Office was obtained and in the summer of 1986 the Minister announced that in future the criterion would be simply in terms of temperature and the trigger value would be a weekly mean of -1.5°C This is close to the actual value of -1.64°C which is the equivalent of 120 degree days over a seven-day period. Thus the Government again accepted the 20% level for 'exceptional' and the temperature equivalent of degree days that had been criticised by the Tribunal. As I have shown elsewhere (Giles, 1986) there were 96 seven-day periods with mean temperatures below -1.5°C in the twenty-seven winters between 1959/60 and 1985/86. This is about 4% or a chance of occurrence of 1 in 25; slightly worse than the Government's 1 in 5. But when a day of the week, e.g. Monday, is specified then there were only 13 occurrences in the 27 winters - about 0.5% or 1 in 180. If last winter is added there will have been 106 seven-day periods at or below -1.5°C which is about 2.6%. This was due to the exceptionally cold January, the sixth coldest since 1900. Consequently after only a few cold days public pressure persuaded the Government to authorised payments both before the actual 'trigger' had been reached and in the subsequent week which turned out to be comparatively warmer. Yet when cold spells occurred in February and March (Fig 3) at the 20 percentile level no one commented!

Clearly during the 1980s the various criteria used to delimit exceptionally severe weather were unsuccessful from the public point of view and abandoned. They were either too difficult to understand or too stringent, and all were based on temperature alone.

How can climatologists help? If reasonably long records are available various alternative scenarios can be provided for the customer. From daily values sequences of any length can be compiled so that a variety of periods, either finite or overlapping, can be calculated. The climatologist can advise on the suitability of different lengths of period from both climatological and economic or administrative viewpoints. In the context of additional fuel payments the administrative inconvenience of periods not beginning on a Monday need to be weighed against the recipients need for immediate reimbursement. Should they have to wait for seven days or spend money in the hope they may get it back? They are faced with the gamble that after five exceptionally severe days the last two may be sufficiently warm to prevent any trigger being activated.

While the period may not be a problem, the words

'exceptionally severe weather' are the major concern of the climatologist. Is the element used during the 1980s in the United Kingdom - temperature - the best? When used in the context of severity should the minimum, maximum or mean be used? If payments are to be made for extra fuel used are low maximum temperatures more important than low minimum temperatures? Meteorologically speaking the areal differences will be much greater for minimum temperatures because of frost hollows and valley sinks, whereas low maximum temperatures are more commonly the result of widespread anticyclonic conditions over Europe. Much greater care needs to be taken in choosing representative stations if minimum temperatures are used. The politician might prefer to avoid stations in cold sinks, yet many large British population centres are to be found in valleys. The climatologist can once again offer alternatives by illustrating how these two temperatures differ across the country. From the public point of view, and to smooth out local idiosyncrasies, the mean temperature for each day would seem to be the best solution since at many stations it is calculated from both the maximum and minimum temperatures.

When a temperature has been agreed, the next step is to provide a frequency distribution for an agreed period of the various temperature levels. The period has to be agreed because of the fluctuations in temperature that have occurred over the last 150 years. If a thirty year period is used, which thirty years? The most recent, even if this is out of step, or an earlier? Fig 2 shows the differences in the various 30 year periods in Birmingham since 1891. The alternative is to use the total data set and continuously update it - much more feasible with computers than in the era of mechanical computation. Similarly the season needs to be defined specifically for the purpose in hand. The traditional winter from December to February may not be long enough to include all periods of exceptionally severe weather. There is a case for extending it by two months, November and March, to ensure all likely cold spells are noted (Figs 3 and 4). Only after such an analysis is it possible to decide the length of period.

At Birmingham mean daily temperatures from 1 November 1959 to 31 March 1987 have been used as a data base for an analysis of recent winters. The temperature range was from -7.9°C (13 January 1987) to 15.5°C (29 March 1986) with a slightly skewed distribution with a longer tail at the lower temperatures. From this data base seven-day running averages were abstracted starting each year on 1 November. In twenty-eight winters there are 4060 seven-day periods. On ranking these various percentile limits are obtained (Table 1). The climatologist can then offer trigger values at any percentile level or at specified values of temperature calculated over a variety of time sequences. The Table includes, for illustrative purposes, percentile values of 1-day, 3-day and 7-day running averages. It becomes a political decision, taken in the light of public opinion, whether to opt for a 1 in 5 occurrence based on x years of data or whether to specify an absolute temperature level irrespective of its chances of occurrence. The two sets of data are shown in Figs 1 and 3. In the former sequences of seven-day periods with mean temperatures of 0°C or colder are plotted and those colder than -1.5°C are shaded. This is clearly a stringent test of severity as the lack of occurrences in the twenty-eight winters clearly shows. The diagram could equally be used to demonstrate that raising the trigger from -1.5°C to 0°C would not produce a great many extra seven-day periods, and the latter is a temperature to which the public can relate. But if one compares it with Fig 3, which shows the percentile level occurrences, a markedly different picture appears. The 20 percentile trigger was reached at some time in most winters and a political case could easily be made for the 10 or even 5 percentile value being used as a trigger for exceptional severity.

I won't dwell for very long on the alternative temperature criterion that has been used, namely degree days. These units are not understood by the general public and are laborious to calculate for maximum and minimum temperatures - three different equations have to be used depending on the shape of the daily temperature curve (Energy Efficiency Office, 1985). However for illustrative purposes they have been calculated and plotted for twenty-seven winters in Fig 4. There is very little difference between Figs 3 and 4 so clearly the extra computational effort is not cost effective in any sense of the word.

Finally I would like to mention an alternative measure of severity which I have put forward on two occasions recently (Giles 1986; Giles & Kings 1987), namely wind chill. This raises other problems. Which equation should be used? The Siple-Passel equation which is well-known in North America, or the more elegant and theoretically correct Steadman equation? The arguments have been analysed by Dixon (1986, 1987) and need not detain us here. The Siple-Passel equation, for all its basic faults, is easy of compute, uses only temperature and wind, and can be understood by the public. Some levels of severity have been recognised but I have made a case, in the context of Birmingham at least, that a value of 900 might be adequate to define 'severe'. An extended analysis of daily windchill values from 1 January 1940 to 31 December 1986 showed that the 900 level was exceeded on 645 occasions - about 4% of the year or 9% of the winter. Five of these values occurred in April and two in October so there is little need to extend the winter analysis beyond November to March. The value of 1000 was reached on 125 days or about 2% of the long winter and on 115 days in the short winter - about 3%.

Clearly over the last forty-seven years a value of 1000 (qualitatively called 'very cold') is too stringent and the value of 900 has more to commend it. 900 is of course half way through the 'cold' band which describes wind chills between 800 and 1000.

There are other strategies for defining exceptionally severe and some of these are explored in Giles & Kings (1987). They include using the mean date values which show cold spells at the end of January and the middle of February, with March being generally colder than November. Mean date wind chill values have a similar pattern but a greater spread with high (i.e. severe) values at the end of December, the middle of January and February. Threshold values based on percentiles can be obtained from such a data set but it must be remembered that a mean date value hides a considerable spread of individual data points. Another strategy is to treat each month individually and have different criteria based on that month's climatology. One might even divide each month into ten day periods and consider threshold values for these (Table 2); while this is scientifically more sensible it may not be acceptable to the public.

I have tried to illustrate, within the context of a very specific example, the various ways that climatological data can be analysed and manipulated to provide definitions of what our parliamentary draftsmen called exceptionally severe weather. The same methods can be used for any similar purposes where specific trigger values need to be defined or how often absolute values are exceeded. From the description I have given you I think two lessons can be learnt. If there are several ways of delimiting critical values it is usually better to choose the simplest, even if it is not too accurate or scientifically elegant, so that it can be explained to the non-specialist. Secondly climatologists need to enter the economic and social world more confidently and show how they can assist in those many problems that are weather-related.

ACKNOWLEDGEMENT

I wish to thank my colleague John Kings for his expertise in setting up the Birmingham University meteorological data base.

REFERENCES

Dixon, J C and Prior, M J 1986 'Wind chill indices - their history, calculation and applications in the construction industry'. Meteorological Office, UK. Unpublished.

Dixon, J C and Prior, M J 1987 'Wind chill indices - a review', Meteorological Magazine, 116, January 1987, 1-17.

Energy Efficiency Office, 1985 Degree Days, Fuel Efficiency Booklet No 7 HMSO.

Giles, B D 1986 'Exceptionally severe weather: a problem for the DHSS. The situation in February 1986'. Weather, 41(7), 226-229.

Giles, B D 1987 'Winter fuel payments - the proposals for winter 1986/87', Weather, 42(1), 27.

Giles, B D and Kings, J 1987 'Good King Wenceslas: alternative strategies for the UK. Department of Health and Social Security (DHSS)', Journal of Climatology, 7, 129-143.

TABLE 1

Temperature ranges (°C) for various percentile levels at Birmingham, UK, November 1959 to March 1987

Percentile level	7-day running average	3-day running average	Daily Temperature
20	2.3 to 0.8	2.1 to 0.5	1.8 to 0.2
10	0.7 to -0.3	0.4 to -0.6	0.1 to -1.1
5	-0.4 to -1.7	0.7 to -2.2	-1.2 to -2.5
2	-1.8 to -2.5	-2.3 to -3.0	-2.6 to -3.6
1	>-2.5	>-3.0	>-3.6

TABLE 2

Threshold temperature used to determine exceptionally severe weather during winter 1985-1986

(60-year averages based on observations made at Edgbaston Meteorological Observatory 1920-1979).

Ten-day period	Temperature Deg C
1-10 October	9.7
11-20 October	8.2
21-31 October	6.7
1-10 November	5.3
11-20 November	4.0
21-30 November	2.9
1-10 December	2.0
11-20 December	1.2
21-31 December	0.7
1-10 January	0.4
11-20 January	0.3
21-21 January	0.6
1-10 February	0.9
11-20 February	1.5
21-31 February	2.1
1-10 March	2.8
11-20 March	3.7
21-31 March	4.8

Note:
Threshold temperature = -4
(10-day average maximum temperature)

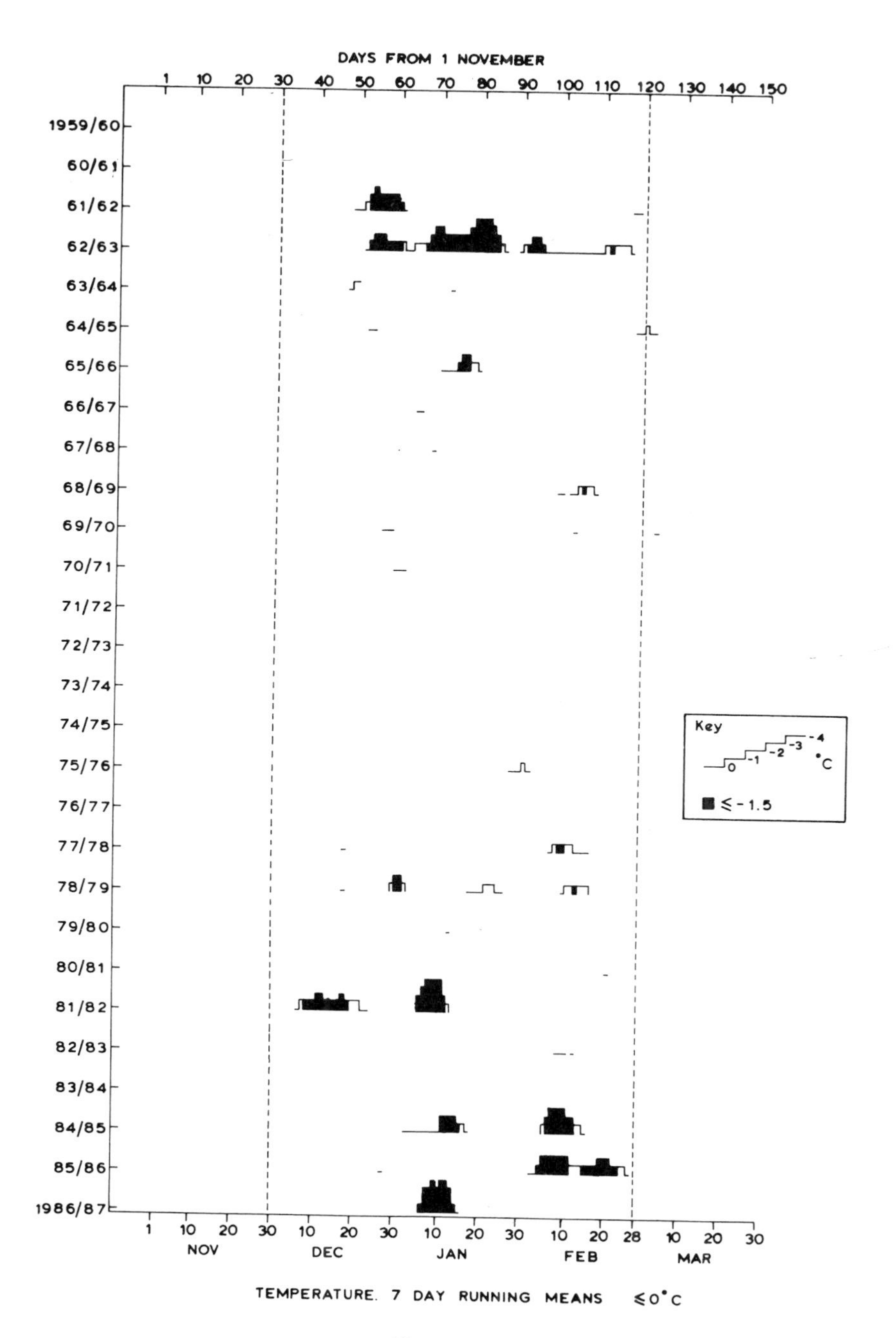
DAYS FROM 1 NOVEMBER
1 10 20 30 40 50 60 70 80 90 100 110 120 130 140 150
1959/60
60/61
61/62
62/63
63/64
64/65
65/66
66/67
67/68
68/69
69/70
70/71
71/72
72/73
73/74
74/75
75/76
76/77
77/78
78/79
79/80
80/81
81/82
82/83
83/84
84/85
85/86
1986/87
Key
0 -1 -2 -3 -4 °C
≤ -1.5
1 10 20 30 10 20 30 10 20 30 10 20 28 10 20 30
NOV DEC JAN FEB MAR
TEMPERATURE. 7 DAY RUNNING MEANS ≤ 0 °C

Figure 1

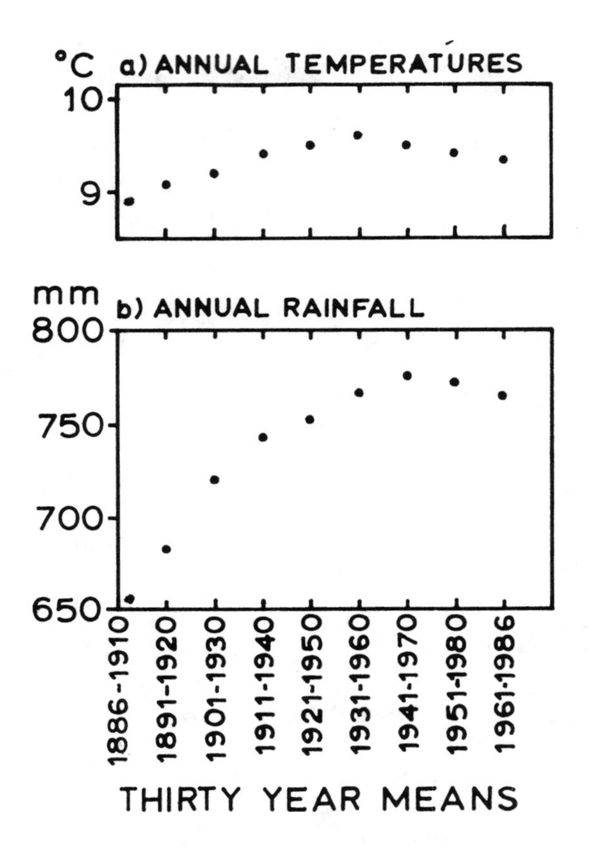
°C a) ANNUAL TEMPERATURES
10
9
mm b) ANNUAL RAINFALL
800
750
700
650
1886-1910
1891-1920
1901-1930
1911-1940
1921-1950
1931-1960
1941-1970
1951-1980
1961-1986
THIRTY YEAR MEANS

Figure 2

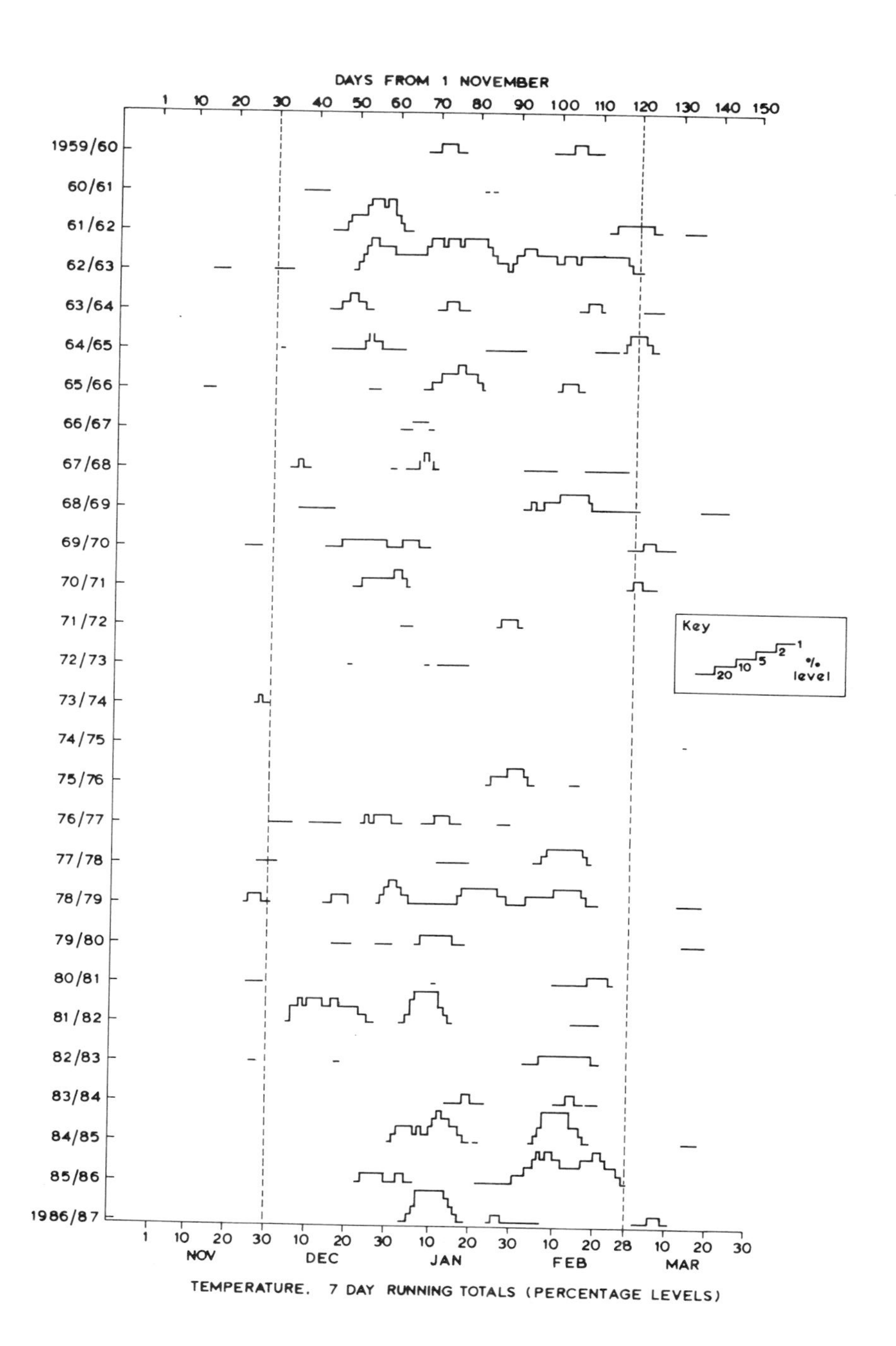

TEMPERATURE. 7 DAY RUNNING TOTALS (PERCENTAGE LEVELS)

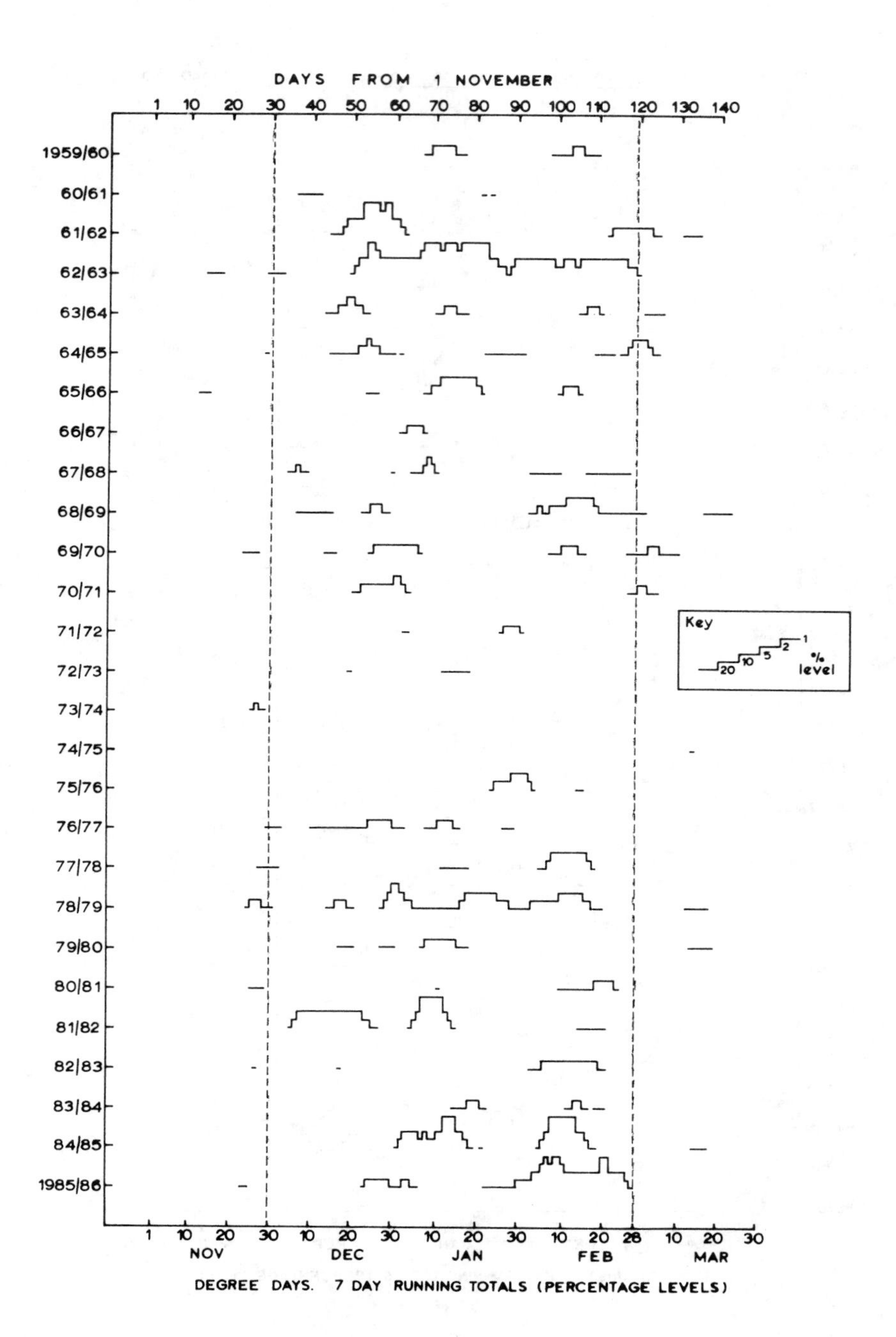
DAYS FROM 1 NOVEMBER
1 10 20 30 40 50 60 70 80 90 100 110 120 130 140
1959/60
60/61
61/62
62/63
63/64
64/65
65/66
66/67
67/68
68/69
69/70
70/71
71/72
72/73
73/74
74/75
75/76
76/77
77/78
78/79
79/80
80/81
81/82
82/83
83/84
84/85
1985/86
Key
20 10 5 2 1
% level
1 10 20 30 10 20 30 10 20 30 10 20 28 10 20 30
NOV DEC JAN FEB MAR
DEGREE DAYS. 7 DAY RUNNING TOTALS (PERCENTAGE LEVELS)

25

The relevancy of meteorology to the development and utilization of aquatic resources

E. A. Mukolwe, Meteorological Department, Nairobi, Kenya

1. INTRODUCTION

The importance of aquatic resources in the world of today has been recognized and, in this connection, there is an urgent need to put into practical use meteorological knowledge and information in the development of these resources. In parallel, intensive efforts should continue to be made with a view to refining the efficiency of the application of meteorology to the field of aquatic resources.

The aquatic resources that need meteorological input for their development and utilization can be classified as follows:

(a)	Aquatic life	- that can be developed and utilized for food and clothing.
		- Tourism
(b)	Aquatic deposits	- in terms of mineral resources including fossil fuels that can be utilized as a source of energy.
(c)	Ocean water	- as a source of fresh water and energy.

The different activities and processes that are required in the development and utilization of these aquatic resources determine the meteorological considerations that come into play. In this respect, the meteorological input in the development and utilization of the various aquatic resources, each of which is considered separately, are considered below. Also considered is the meteorological input in the prevention of the spread of pollutants over water surfaces.

2. AQUATIC RESOURCES THAT NEED METEOROLOGICAL INPUT

2.1 Aquatic life

These include fish, crocodiles, birds etc. Each species of life has its own suitable weather conditions for survival at its various stages of development. For instance, the species whose life begins with the laying of eggs requires calm waters where there are no strong ocean currents and where there are suitable temperatures for incubation. By the time the eggs hatch, there must be sufficient food nearby, possibly in the form of plant life (algae), to sustain the creatures who are incapable of hunting further afield. The growth of such plants is determined by weather parameters such as temperature, light and salinity of sea water.

It is important, however, to note that particular weather conditions may be favourable for more than one species of life. This is necessary for the species which prey on others for survival.

Meteorological input in the form of studies of surface wind systems and water temperatures is definitely vital for the development and utilization of these various species. Through such knowledge, it is possible to map out breeding areas of any given species.

It is normal for some of these creatures to migrate from one area to another in search of food, hibernation and safety. In most cases, the time for migration is chosen to coincide with relatively strong prevailing surface winds and/or ocean currents for ease of movement towards the required localities. If the weather conditions of the eventual settlement areas and those en route are known, a qualitative estimate of the population of those species can be made. It should be noted that, in addition to mapping out the actual weather conditions, prediction (forecast) of weather conditions such as cyclones, cold or warm air masses and frontal systems which may pass through breeding and settlement areas greatly helps in estimating the population yield of aquatic life. It is then possible to estimate the harvests for food and/or clothing and to plan and establish tourist resorts in advance. In the case of species which do not migrate from their breeding areas, there are certain weather conditions during which reproduction does not take place. Consequently, the harvest of such species should be planned carefully.

Having determined fishing areas, fishermen are kept informed of the existing weather conditions by the issuing of regular weather bulletins by meteorological centres. These bulletins include such weather phenomena as the prevailing wind direction and speed, sea surface temperatures, thunderstorms, sea fog and visibility, and the general state of the sea. Accordingly, fishermen are able to plan their routes including the amount of fuel required and the weight they can carry. Warnings of approaching storms and any adverse weather conditions are also issued on a regular basis to avoid disasters.

It might also be necessary to set up processing substations near fishing grounds. To establish such substations, the weather conditions of the area have to be known, including likely weather hazards (storms, rough seas, etc.). In this respect, accumulated climatological data presently available within the Meteorological Department can be used.

2.2 Aquatic Deposits

These include minerals (both suspended and dissolved in sea water, and also at the bottom of oceans) and oil.

Minerals are likely to be deposited and accumulated in regions where the seas are normally calm with a large deceleration of water-flow. After predetermining sea areas where deposits are likely to be accumulated, surveys are often conducted to determine how much of the deposits are available and whether exploitation would be profitable. The surveys are normally carried out through sampling by ships, photography by aircraft or by satellite. If the survey is by ship, the shipping forecast may be issued to the crew. In the case of surveys by aircraft which must fly low, special weather forecasts (as opposed to the usual forecasts for normal aviation purposes) must be issued. Areas with sea fog, low clouds, frontal systems (associated with adverse weather conditions), low level jetstreams (associated with marked turbulence), etc. are predicted in advance by the meteorologists and should be avoided by the crew. Photography by satellite is also possible where there is not much cloud cover.

Offshore drilling for oil depends very much on ocean wave systems. Studies have to be carried out to relate wind strength to the ocean waves which can also be harnessed to generate energy. Any installation that goes with the offshore drilling should be strong enough to withstand adverse conditions of wind, temperature, water acidity and storms. Meteorological information on these conditions has to be available for any proper planning and construction. The transportation of any product to the shore has to be planned for when the weather conditions are favourable.

2.3 Ocean Water as a Source of Fresh Water and Energy

The fresh water processed from ocean water is used mainly for irrigation and domestic purposes.

The required meteorological information is similar to that required for installation of an offshore drilling site, as mentioned above.

3. POLLUTION

In all these aquatic activities, water pollution as a result of factors such as oil spills, can be a major hazard particularly to aquatic life and to the use of ocean water as a source of fresh water. The study of surface wind systems helps to predict the movement of pollutants over water surfaces and, hence, precautionary measures can be taken to prevent their spread over large areas.

Man-made emissions into the atmosphere significantly affect the chemical composition of the air and precipitation and, consequently the ocean, lake and river waters. This has made it important to study the distribution of these emissions and to establish their natural concentration levels. To do this, it is vital to carry out a study of the atmospheric circulation of the area of interest.

It is important to note that, at present, the Kenya Meteorological Department is running a precipitation chemistry project and is playing a leading role in the setting up of a baseline pollution monitoring station on Mount Kenya.

4. CONCLUSION

It is clear that an understanding of Marine Meteorology is a necessary requirement in the development and utilization of aquatic resources. As a result of inadequate meteorological data over ocean areas, this branch of meteorology has lagged behind other branches of the science in Kenya.

However, with the limited data now available through satellite, buoys and ships, some of the difficulties are being overcome and a larger number of meteorologists are becoming increasingly more interested in Marine Meteorology. It is consoling to note that the Kenya Meteorological Department has already taken steps to train personnel in the field of Marine Meteorology. This will definitely contribute to the development and utilization of aquatic resources in this country. This is a welcome move, due to the fact that all human and animal activities are influenced, to a great extent, by meteorological factors.

Acknowledgement

The author is grateful to Mr K N Mutaku for his helpful comments and suggestions. The paper is published with permission of the Director, Kenya Meteorological Department.

26

The routine use of wind and solar data in the sizing of small wind turbine and photovoltaic power plants

W. Grainger and J. Rapson, Northumbrian Energy Workshop Ltd, Hexham, UK

1. INTRODUCTION

Renewable energy, power from atmospheric energy flows, is being increasingly exploited all over the world to produce power for individuals, communities, agriculture and industry. The economics of such systems depend on the availability of the 'fuel' - wind, sun or rain - both for the size of the system and also for the energy store since by their nature they are not available all of the time. Obviously meteorological data is required in the design of these systems. This paper outlines what data set is required, what is available and how it is processed. We are concerned with the routine analysis of meteorological data for commercial purposes not for the design of these energy converters. This constraint limits the amount of money available to collect or to purchase data and also the time taken to process data.

2. WIND ENERGY CONVERSION SYSTEMS

2.1 Power output from Wind Turbines

On a small scale, say for wind turbines rated up to 10 kW, energy is stored in a battery for periods of calm or low wind. To appreciate the data and analysis required, consider Figure 1, the output characteristic for a 10 kW wind turbine. The salient features are:

a. Power is only generated for winds greater than 3 m/s

b. Power climbs rapidly at windspeeds above 3 m/s but below 12 m/s

c. Power is constant for most of the range 12 to 15 m/s

d. Power falls for windspeeds in the range 15 to 16 m/s

e. Power output rises again for windspeeds above 16 m/s

For details of why the characteristic has this shape see the wind turbine design literature (Lysen, 1983).

The implication of this is the effect of a change in windspeed. It is well known that the power in the wind varies as the cube of the windspeed, but the above shows that power output variation is more complicated still. The magnitude of the variation in power output for a given change in wind speed not only depends on the magnitude of the change in windspeed but also on the windspeed itself. To predict the energy produced at a site it can be seen that the mean windspeed at the site is not enough information. The distribution of windspeeds about the mean is also required, usually in the form of the percentage of the time the wind blows at each windspeed.

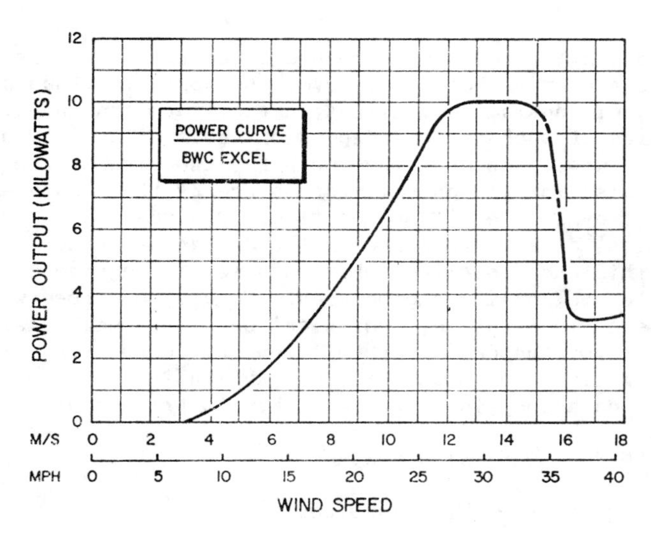

Figure 1

Using a windspeed distribution is simpler than using raw data and our standard practice is to reduce raw data to a distribution. For some windspeed data processing an analytic approximation for this distribution is convenient and a 2 parameter statistical function, usually a Weibull distribution is used (Bowden, 1983). In many cases a good fit to the observed data can be achieved using a Weibull distribution, but sometimes it can be spectacularly bad and thus should never be used automatically.

What averaging period is used to produce such distribution? Wind turbines respond to changes in windspeed of less than a second in duration. However, the windspeed characteristic shown in Figure 1 is typically based on a ten minute average windspeed. This removes inertial effects from the rotor and corresponds to a frequency in the wind speed spectrum with no turbulence or meteorological significance. On a practical level for wind energy conversion system design, the most commonly available wind data is that based on hourly means. However in many cases not even data of this interval is available.

2.2 Meteorological station data

Lucky is the user who finds data for the site under investigation. Usually there is no anemometer at the site and the potential wind energy user can try to use data from a neighbouring meteorological station, always making appropriate corrections (Jensen, 1984). The effects of topography and altitude have to be included in these corrections. This can be quite complicated in mixed terrain such as the northern UK, but is not so much of a problem in a flat country like Denmark.

In the UK summary data in the form of windspeed distribution based on hourly means is available from a large number of meteorological stations at reasonable cost. However, a word of caution should be stated in connection with this data. The experience of NEW engineers has shown that a detailed knowledge of the exposure of the meteorological station is important. Changes in the siting of anemometers from roof tops to open sites and a change of measuring instrument can dramatically effect the data. When historic meteorological data is used for quantitative analysis care has to be exercised in its interpretation. It must be remembered that meteorological data has been collected for a long time and during most of this time the purpose of collection was not energy prediction.

2.3 Wind Atlas

Another approach to obtaining wind data is to use a wind atlas. These are available for Denmark (Petersen, 1981) and other countries and are produced by extrapolation from high altitude wind data down to wind turbine heights (15 to 25 m typically). As with raw meteorological data care must be taken in using these atlases, as the effects of rapidly varying terrain are not easily taken into account in their production and only a small number of stations are used to produce the data. Nevertheless, these maps can provide a very useful source of inexpensive wind data for those countries. The success of using this data is well documented in the Danish case.

2.4 On-site Measurements

Often no data is available which can be used with confidence. Recourse must be made to measurements in the field. What do you record? Time and money are limited when doing system design for small systems, so specially designed dataloggers are often used, such as the NEW Windlogger. These wind dataloggers are designed only to record the data that is actually required, in most cases the windspeed distribution. Using a specialist datalogger saves analysis time by only recording a few calculated numbers rather than large quantities of raw data. In fact in some cases as few as thirty numbers are all that are required. Readings are usually based on one minute or ten minute averages.

Wind direction can be recorded, but the direction distribution is often known or can be deduced from the site, and thus there is little point in the added expense of measuring it. Turbulence data may be useful, but again topographic features likely to induce turbulence can be identified when the site is selected and may be able to be avoided, e.g. buildings. Turbulence is of more interest to designers of wind turbines than installers, and is usually not measured. Site measurements should ideally be made for a year, and then an attempt be made to correlate the results with long-term weather records for the region.

2.5 Height correction to windspeed data

Windspeed measurements are usually made at 10 m height. Small wind turbines are usually mounted at a height between 15 and 25 m. To estimate the increase in windspeed due to this change in height in the boundary layer some model has to be assumed; NEW uses a formula which modifies the Weibull parameters. This is a great simplification of what happens and takes no account of many features of the boundary layer. Our experience of using this simple method has been good. To be certain, we always process the unmodified data and compare this with the predictions using modified data as a check.

2.6 Prediction of wind turbine output

When the windspeed data is in a useful form, the windspeed

distribution and the characteristics for the wind turbine under consideration can be combined to predict the energy output from the machines at a particular location. The monthly or weekly pattern of energy production throughout the year and the calm distribution are used to size the wind turbine and the battery. However, to specify the system fully the predicted load on the system must be taken into account. This can vary considerably from application to application. At telecommunication repeater stations for example the load is constant, day and night throughout the year, but in domestic or light agricultural loads variations with seasons are considerable.

3. SOLAR ENERGY CONVERSION SYSTEMS

3.1 The exploitation of solar energy

Solar energy is exploited in two ways. First in the direct generation of electricity using photovoltaic panels, and secondly in the form of heat using thermal collectors where a heat transfer medium is warmed by the sun. NEW concentrates on supplying photovoltaic systems usually for loads ranging from a few watts to 1 kilowatt. As with wind energy conversion systems the need to size the PV array and battery is very important and can determine whether the system works or is uneconomic by over-design. The significant differences between solar and battery is very important and can determine whether the system works or is uneconomic by over-design. The significant differences between solar and wind systems are as follows:

a. There is obviously a much stronger daily pattern in solar data

b. The photovoltaic's output characteristic is linear with respect to solar energy input. If the solar irradiance doubles then the energy output doubles.

3.2 Meteorological station data

Solar data is not so widely available as wind data. Local variations are not so great as with wind, although they are not insignificant and are the subject of current research. Often all that is recorded is the daily number of sunshine hours and not a proper energy measure. As with wind data solar atlases are available for some countries (Lof, 1966; Page, 1986; Palz, 1984). However, these atlases are often based on sunshine hours records and not on energy measurements.

Global solar irradiation measurements are usually made in the local horizontal plane. Photovoltaic panels are very rarely mounted in a horizontal plane, and what is required is the solar radiation on the plane on which the panels are being installed. Photovoltaic panels are usually tilted at approximately the angle of latitude for the site. Caution must be observed when using data derived from sunshine hours and also where corrections have been applied to calculate the solar irradiation on an inclined plane.

The situation is further confused because of the angular dependence of the sensors used to measure solar irradiation are different from the photovoltaic panels. This can lead to overestimation of the output from a photovoltaic panel in the early morning and late evening, and around the solstice when the sun is low in the sky.

As with wind data, solar data has been recorded for many years, but it was not originally envisaged that the records would be used in a quantitative manner.

3.3 On-site measurements

As with potential wind energy sites quite often there is no solar data which can be used reliably and measurements have to be made at the site. Again these are usually made for a year and some attempt is made to correlate these readings with regional historic weather data. The recording is much simpler with solar data. All that is required is the energy incident per unit area on the surface of interest as measured by the NEW Sunspot datalogger for example. The orientation of the PV array can easily be set before the measurements are made. The array is usually tilted at the angle of latitude, although the annual load demand pattern may alter this angle slightly. Where possible, solar panels face South, although they will often be mounted on a roof and then the angle or orientation is set by the building (Buresch, 1983).

3.4 Prediction of PV Panel Output

The energy output from a solar system can easily be predicted by using the manufacturer's data which gives the efficiency of conversion of solar energy into electrical energy per unit area. Typical figures are in the range of 6 - 12% on an area basis.

Again the batteries are sized to cope with periods of low sun (typically in winter in Northern latitudes and summer in Southern latitudes) but also for the ever present daily variation in the solar intensity.

3.5 Hybrid power systems

In the last few years there has been increasing use of hybrid energy conversion systems which use both photovoltaic and wind turbines for the power input. This has the advantage of diversity of sources which increases reliability, but also it can mean that smaller batteries are used since the power sources are to some extent complimentary, although this is not always as simple as it may appear.

4. WATER ENERGY CONVERSION SYSTEMS

Hydro power has long been exploited by man, and the technology at a mechanical level has not changed much in the last 80 years. However advances in electronics have meant that modern control systems are much simpler and systems can now be installed more cheaply than was the case a few years ago. Hydro power is characterised by a very consistent output in comparison with wind or solar, the variations being over a much longer time scale. However the calms can also be much longer, and dry seasons in many parts of the world exclude the use of hydro if the load happens to occur during these periods or is constant throughout the year.

To predict the output of a hydro system the flow in the watercourse must be determined. In some instances, detailed flow records are available. Alternatively, historic rainfall and evaporation data can be used, together with a detailed assessment of the catchment area feeding the river or stream to be exploited. Methods for processing the data are given in several places (Gibson, 1921; Inversin, 1986). River flow or rainfall data from the neighbourhood are of course very valuable, but as with other renewable energy flows often no data is available and recourse must be made to measurement at the site. Water flow data are commonly measured using a notched weir and measuring the height of the water in the pond behind the weir. Modern advances in transducer technology mean that this is performed using pressure transducers. The distribution of height gives the energy prediction for the site.

5. CONCLUSION

This paper has outlined the meteorological data which is required for use in predicting the output of renewable energy power sources. The characteristics of the various methods of extracting power from the environment are different and require different techniques for measuring and processing the raw data. The emphasis here has been on the routine use of meteorological data by a company designing and installing energy systems using wind, solar and hydro power. Extensive meteorological analysis on a long time scale is not economically feasible in most cases but by using the available data with care, accurate estimates can be made.

6. REFERENCES

Bowden, G J, Barker, P R, Shestopal, V O, and Twidell, J W The Weibull distribution function and wind power statistics. Wind Power Engineering, 7, 85-98.

Buresch, M, 1983 Photovoltaic Energy Systems. McGraw-Hill, New York.

Gibson, 1921 Hydroelectirc Engineering. Blackie and Sons.

Inverin, A R, 1986 Microhydropower Sourcebook. NRECA International Foundation, Washington DC

Jensen, N O, Petersen, E L and Troen, I, 1984 Extrapolation of mean wind statistics with special regard to wind energy applications. WMO - World Climate Programme.

Lof, G O G, Duffie, J A, Duffie, Smith C O, 1966 World Distribution of Solar Radiation. Report 21. Solar Energy Laboratory, The University of Wisconsin.

Lysen, E H, 1983 Introduction to Wind Energy. 2nd edition. Consultancy Services Wind Energy Developing Countries, Amersfoort.

Page, J and R Lebens (eds), 1986 Climate in the United Kingdom. HMSO London

Palz, W (ed), 1984 European Solar Radiation Atlas. Vol I and Vol II, CEC Verlag TUV, Rheinland.

Petersen, E L, Troen, I, Frandsen, S, Hedegaard K, 1981 Windatlas for Denmark. Riso National Laboratory, Roskilde.

27

The Caribbean Institute for Meteorology and Hydrology and some aspects of its work in the applications of meteorology

C. A. Depradine, Caribbean Institute for Meteorology and Hydrology, Bridgetown, Barbados

INTRODUCTION

The Caribbean Institute for Meteorology and Hydrology (CIMH) was created in 1986. It represents an amalgamation of the Caribbean Meteorological Institute (CMI) and the Caribbean Operational Hydrology Institute (COHI). The CMI was established in 1967 through a project funded by UNDP/WMO and fourteen regional governments in the Caribbean. The project had, as its primary objectives, the training of personnel at all four WMO classified levels for the meteorological services in the region, and investigations and applied research in meteorology and its allied branches. The Institute is affiliated to the University of the West Indies which awards the BSc degree with meteorology as one of its major subjects. Apart from its training and research activities, the Institute also has two additional roles that are of considerable importance to the meteorological services. It provides a maintenance, repair and calibration service for the meteorological equipment in the region, including six 10 cm weather radars, and maintains a stock of equipment and spare parts for loan and/or sale to the meteorological services. It also acts as a data depository for several islands. In more recent times the Institute has been involved with the coordination of activities and the maintenance of the improved meteorological telecommunications system - ANMET.

The COHI was established in 1982 using the infrastructural

facilities of the CMI. This had the advantage of utilising existing teaching programmes and data processing facilities. The Institute was established through a project funded jointly by the Netherlands Government and the UNDP/WMO. Like the CMI its primary objectives are to train personnel for the hydrological services in the region and to undertake investigation of various hydrological problems. Training is offered at two levels - a General Technician course of four months duration and a Higher Technician course of eighteen months duration. These courses provide a level of technician training that is available in very few English-speaking countries and we have had enquiries from countries as far away as the Sudan and India.

In 1985 the regional governments decided to amalgamate the two Institutes to form the CIMH to serve the needs of both hydrology and meteorology. To date over eight hundred students have been trained in meteorology and sixty-five in hydrology. Various courses are offered every year and most of the available places are taken up. Some of the other activities of the CIMH are described below.

ALTERNATIVE ENERGY

The spiralling cost of oil in the late seventies and early eighties triggered a renewed interest in the development of alternative sources of energy. This increasing cost was particularly severe on the developing countries of the Third World and put considerable stress on the limited hard currency foreign exchange that was available in these countries.

The Caribbean countries, like many others, turned their attention to wind, and to a lesser extent, solar power. It was generally recognised that most of these countries were in the relatively constant trade wind regime and could thus make use of wind power. The use of wind power was not unknown in the region since, prior to the development of electrical services, many sugar plantations used windmills for pumping water from wells both for irrigation and drinking purposes. Even today this practice still continues with the water being used primarily for irrigation purposes.

However, the actual wind resource which varies across the islands, was never measured and documented to a degree where it could be determined whether it would be economically feasible to install wind turbines for the generation of electricity. Similarly, the distribution of solar radiation was unknown.

These problems were recognised by the Caribbean Development Bank which established a fund for studies to be carried out and contracted the CMI to undertake a programme aimed at producing the necessary information. A unit was established at the Institute for this purpose and the work was implemented and completed over a three-year period.

The method used consisted essentially of the installation of ten metre towers at various points on the island for periods varying from several weeks to a year. The wind data obtained was used in conjunction with the long-term data available from the meteorological station at the Airport to generate the required statistics. Thirty metre towers were also used at the most promising sites.

This has resulted in the identification of a number of sites on six islands for the location of wind turbines. To date a vertical axis Darieus machine has been installed on the island of Antigua and a horizontal axis 250 kW Hawden machine on Barbados. These installations are intended as pilot projects with the long-term objective of establishing wind farms for the large-scale production of electricity. The Institute will also be involved in the monitoring phase of the Barbados turbine and has erected a 100-metre tower to provide measurements of wind speeds and other parameters at various heights for this purpose. In the area of solar radiation several maps have been produced showing distribution across the islands. However, we are not aware of any project in which these have been used to any great extent.

In addition to this project, the Institute has carried out on-site measurements for various industrial concerns to determine whether it would be economically feasible to install wind-power turbines. At least one company has carried out such an installation to date and appears to be satisfied with its output.

WATER RESOURCES

In several islands of the region, there are many rivers but little data on the available water-resource which could be used in the production of hydro-power. In several of these islands the electricity grid does not extend to cover the entire island and consequently, several rural villages and areas are without any form of electricity supply. Dominica is one such island in which there are reputedly 365 rivers. In an effort to solve this problem the Government, with the assistance of the Caribbean Development Bank, devised a feasibility project to determine the mini hydro-electrical potential of six rivers in the eastern part of the island. The Institute was contracted to carry out the stream-gauging and other hydrological measurements necessary to obtain data for the project, which had as its primary aim, the determination of the availability of water primarily during the low-flow season.

Following this initial project, the CDB, after discussions with the Governments concerned, decided that the project should be extended to set up hydrological stations on three large rivers on the islands of Grenada and Dominica and on five large rivers on Belize. These installations were carried out either by the Institute or by personnel working in conjunction with the Institute's staff. The projects have recently been completed. The project output included actual and model-generated data. Based on the reports submitted decisions will be taken by the relevant authorities on the economic feasibility of installing small scale hydroelectric turbines. The data will also be available for use in other areas such as irrigation. The institute has also participated in various ground water studies in several islands, particularly in resistivity measurements, aimed at determining feasible sites for locating water-wells.

HUMAN HEALTH

The Institute has been involved in two areas or research in the area of human health. The incidence of Leptospirosis, a disease caused by rodents, is relatively high in Barbados and research has shown that disease is transmitted mainly during the rainy season. A research project was prepared by a local medical team and the Institute participated in its work primarily through the provision of analysed rainfall data during the periods of interest. The Project has recently been extended and discussions have taken place with a view to further participation.

In another study, the Institute's staff joined with medical staff at the local hospital to study the relationship between weather and bronchial asthma in Barbados. Since asthma is generally believed to be caused by airborne allergens, a part of this study was to determine the possible influence of Sahara dust, which reaches the Caribbean in large quantities, on the incidence of asthmatic attacks. Other areas of the effects of weather on human health are under active consideration.

ENVIRONMENTAL EFFECTS

The Institute has participated in at least one study which may have prevented possible damage to the environment. On the east coast of Barbados there is a sand dune which is generally used in the construction industry. Because of the high silica content a private company devised a project to mine the sand for export to a glass producing company. It was decided, however, that it would be necessary for them to show that this removal of the sand would cause no environmental damage. The Institute was engaged to carry out this study and our results indicated that the sand 'hill' formed an effective barrier to the surface-level salt-laden winds and its removal would probably have created problems further inland. As a consequence the project was abandoned.

In recent times beach erosion has become a major problem for Barbados. Experts in that area have been employed by the Government to provide a solution to the problem. The Institute has been engaged in discussions with project personnel with a view to making measurements of stream discharges and water quality to the sea, since it is believed that polluted water or water containing herbicides may destroy the coral producing organisms and eventually affect the production of sand.

In the northeastern part of Barbados (the Scotland District) considerable land erosion occurs, and the Government has been involved in various projects to try to solve the problem. The Institute has participated in one such Government-sponsored erosion control project by installing a river flow and sediment transport measurement station on one of the rivers in the area, to provide raw and analysed data to the engineers concerned.

Also in some areas of Barbados, sewage and some industrial wastes are disposed of in wells. Because the island's water supply is obtained through pumping water from underground streams, there is concern that the water could become contaminated. The Institute expects to become involved in a project with PAHO to carry out a systematic monitoring and testing of water from several wells to determine the water quality on an on-going basis. These are primarily chemical tests since the Institute is not equipped for biological testing at the present time.

ENGINEERING

The Institute has been providing both raw and analysed data to engineering firms involved in the construction of roads and buildings for many years. In the most recent example we were contracted to carry out an extreme wind analysis study which was one of the major inputs into an engineering project designed to produce a building code for the region. It is expected that this code will be of great value to builders in an area which is exposed annually to destruction by tropical storms.

Other areas in which the Institute has made indirect contributions are tourism and sport. Tourism forms a major industry in the Caribbean region primarily because of the year-round sunshine and warmth. The Institute has provided several statistical values of various meteorological parameters and these are available to several interests in tourism. Similarly weather forecasts are provided on a daily basis in most of the islands. These forecasts are used both by tourists and sports officials and are produced by forecasters, many of whom would have been trained at the Institute. There is therefore an indirect contribution in these areas.

SUMMARY

In summary, the CIMH has had successful training programmes both in meteorology and hydrology. These programmes continue to be in demand by countries in the region and some further afield. Areas of investigation and research at the Institute are usually of an applied nature, and several useful contributions have been made in the applications of meteorology.

28

Complex meteorological information for the national economy especially for agriculture, transport and energy

L. Špaček, Czech Hydrometeorological Institute, Prague, Czechoslovakia

INTRODUCTION

The way in which meteorological information is provided, its contents, and its form are important factors that determine its effective utilization. Human activity has been and always will be affected by atmospheric conditions. We can considerably enhance the success of some of our activities by acquaintance with atmospheric conditions. Described briefly here is the way in which meteorological information is provided in Czechoslovakia since the introduction of a new system and some of our first experiences with this new system.

SOCIOECONOMIC ASPECTS OF METEOROLOGICAL INFORMATION

The social and economic organization of any society is reflected in all spheres of life. The same is true therefore for a Meteorological Service. Specific kinds of information are provided in developed countries special emphasis is laid on climatic information which makes possible the optimum exploitation of the natural resources which hitherto have not been used. Developed countries, having experience over a much longer period are so well-acquainted with basic climatic characteristics that they lay emphasis on operational information. Climatic information is only important insofar as it concerns a new constituent of the environment or when it contains information on changing climatic conditions, especially as a result of the influence of human activity on the earth's atmosphere.

The quality of information depends on the level of the national telecommunications. Lesser-developed countries which have not sufficient equipment for high speed communications are not, in general, able to disseminate the information that contains very short-range forecasts quickly enough before this information loses its topical relevance. In developed countries we are now able to disseminate these forecasts and the methods of nowcasting are widely used when necessary.

It is possible to state that the influence of social factors on meteorological information is the more pronounced the more that special information is disseminated. This influence is of two kinds: on the one hand specialists issuing the information endeavour to provide as useful information as possible; on the other hand users demand such information that would meet only their needs. This pressure depends on the degree of organization of each specific human activity. For example, let us mention the important activity of agriculture which is the most affected by weather conditions.

In Czechoslovakia agricultural production is organised in the form of Standard Farming Cooperatives and State Farms. The break-up of our agricultural production into thousands of small farms has been halted by the introduction of units of production. This varies from several thousands of small farms has been halted by the introduction of units of production. This varies from several thousands to several tens of thousands of hectares of arable land which from the administrative point of view may represent an area of up to 100 km^2, depending on the intensity of the agricultural production in the region. In comparison with earlier times this means that the Meteorological Service does not have tens of thousands of small farmers as a partner but, in fact, several tens of hundreds of qualified agronomists who manage the activities of these huge areas. It is clear that the requirements of the Meteorological Service by these persons in authority carry greater weight with the Meteorological Service than a lot of small farmers.

In the last few decades industrial production has reached such proportions in some regions that its aftermath affects the quality of the environment to the point that basic conditions of like are violated. The forecaster of the meteorological conditions that are responsible for the deteriorated dispersion of pollutants in some agglomerations, has become so important that it is used as the basic criterion in the decision-taking process, as to whether the industrial pollutants of the region should be regulated or not. This represents a general restraint upon industry, transport, energy, etc.

METEOROLOGICAL INFORMATION IN CZECHOSLOVAKIA

Meteorological information is provided for the national economy at several levels in Czechoslovakia, all being organised within the framework of the Meteorological Service. In addition to the general information which is available for the mass media, users from selected sectors of the national economy are provided with special meteorological information. The most important sectors are agriculture, transport, energy, water management and quality control of the environment.

Interest in meteorological information has been growing in recent years, especially in the sectors of the national economy where it can be effectively used in decreasing:

1. the high degree of vulnerability of those sectors with respect to atmospheric conditions/agriculture, civil engineering, transport, energy,

2. the adverse impact of human activities on the environment and consequently on public health/local authorities.

The supply of the meteorological information has gradually become the most important activity within our Meteorological Service. This evolution led necessarily to the state where the original organizational structure of the Meteorological Service ceased to match up to the requirements.

It was necessary to change the organisation and structure. We choose inputs and outputs, as well as the processing methods of the services provided, as the basic criterion. Hence we left the traditional division in accordance with specialization and we adopted a new way of organization on the basis of the type of the information provided. In this way we made a step towards the users. We modelled the basic sections of the technological line: acquisition and collection of data, their processing in operational and climatic routines. We have followed our concept even more closely by creating Regional Centres. Their task is to give precision to the information sent from the Meteorological Service Headquarters and to interpret it for the users in the area of jurisdiction of each Regional Centre.

We have also looked at our data acquisition system. We have unified our internal report INTER which serves us today not only for the transmission of the information on predominant characteristics of the weather, which has existed at the station during the last 24 hours, but with the help of this report we are able to obtain hydrological data, agrometeorological data, air pollution characteristics, and even information on skiing grounds, on the temperature of basins, etc. This report prevents duplication of information which occurred before, e.g. in measurement of precipitation at meteorological and hydrological stations. At the same time it warranted an effective use of this data as a whole which did not exist before in meteorology or hydrology.

The system of the data handling is fully automated. It includes an automated control procedure which informs either observer or operators of errors and asks automatically for a correction from the Regional Centre responsible or directly from the synoptic stations.

This control system is administered by the operational information division which represents the first processing unit of our technological line. During processing it is not possible to insist on the absolute completeness and accuracy of data as the main objective is to provide operational information in a complex manner, with the least possible delay. Consequently, emphasis is laid on full automation. We have worked out a statistical-analogical method of interpretation which we apply to our NWP products from the RMCs in Offenbach and Bracknell, from the WMC in Washington, from the ECMWF, and from our own limited area model products. As our Meteorological Service is the seat of the RTH Prague, we have no problems with access to the data from the GTS and we make full use of this.

What cannot be automated is the work of weather forecaster. The role of meteorologists consists of considering all the materials, describing the expected outcome and in formulating weather forecasts, both general and special. This is done in consultation with the specialists in aviation meteorology, hydrology, atmospheric chemistry and agrometeorology who work in the different use of this.

What cannot be automated is the work of weather forecaster. The role of meteorologists consists of considering all the materials, describing the expected outcome and in formulating weather forecasts, both general and special. This is done in consultation with the specialists in aviation meteorology, hydrology, atmospheric chemistry and agrometeorology who work in the different sections of the division.

A fully-automated system of processing observed data has been developed. Any user can have a review of weather characteristics prevailing over the last 24 hours, the last week or month in the form of tables. He does this by means of telex to the central computer. The transmission of facsimile charts is fully automated as well and users can have hourly weather reports for our republic, supplemented with the weather in a major section of Europe at all the synoptic hours.

This stream of data is interpreted by our Regional Centres and modified by local specialists for regional purposes. However, any user can also obtain all the data directly.

The data undergoes climatic processing after the operational processing. For this purpose the climatic division has been created. One should realize that the sense of the word climatic must be perceived in a broader sense as this division has almost the same structure of sections as the previous one with some exceptions. Its main task is to create and keep a data bank and to process the data stored, both for needs of the Meteorological Service and for needs of users. The completeness and accuracy of data enjoy priority over the time factor. The dissemination of products does not require as a rule, special telecommunication devices and is carried out by correspondence.

As has already been mentioned, the centre of interest has been shifting more and more towards operational information though the activity of both sides of the division is carried out with the same professional skill. The demand on the operation component of meteorological information is greater, and this is why the number of people employed by the operational information division is approximately one third greater. The importance of individual needs of special sectors: aeronautics, water management, agriculture, surface transport and energy show the importance of the individual needs of special sectors. Recently we have seen an ever growing requirement on the quantity and quality of services supplied to the needs of the ecological control.

CONCLUSION

Our Meteorological Service is unified with the Hydrological Service. The main reason was that precipitation is almost the unique source of moisture for our region. Our region, though not the highest, is one of the roof-tops of Europe. Practically no water flows into it. Consequently, we have to manage the moisture as economically as possible.

A system of river dams has been built up in Czechoslovakia. It supplies water to agriculture, industry, dwellings, etc. The necessity to manage this system under such a severe climatological constraint led to the constitution of the complex Hydrophobia Service. It is clear that hydrological interests have affected the structure of the Service, especially as the technical equipment and the area of responsibility of each of the Regional Centres had to be determined by the borders of catchment areas.

The rationale of our Hydrometeorological Service was based on geographical features so there is little room for error. Using our experience we have recently unified a range of disciplines allied to meteorology. Here the geographical features did not play a major role and we operated partly social reasons where the degree of uncertainty is somewhat higher. However, our experiences confirmed that we have made steps in the right direction. We have organized a complex consultative process between users and the Meteorological Service in an economic manner which suits the requirements of all users.

29

Contribution of the "Teletel" system towards a better use of meteorological information

M. Roques, Western Interregional Meteorological Service (France)

The potential value of a piece of meteorological information is naturally related to the quality of its content, and also, of course, to the way in which it is made available to the user. Verbal contact between the user and a meteorologist either directly or by telephone, or by the user's listening to the radio at a set time, dialling recorded information or subscribing to a postal information service, have been for many years the principal methods used in France to transmit meteorological information to particular social and professional groups (shipping and aviation) or to a wide audience. The Teletel system is a new means of communicating information to the public at large, and has been made available for some two years not by the national telecommunication service. Any telephone subscriber can have a small terminal with a screen, the Minitel, on which he can call up pages of information put out by the "datacommunication distributors", computers which transmit information using the "videotex" transmission standard. At the end of 1986, 2.3 million minitels had been distributed in France, some 10% of the total number of telephone subscribers.

The system makes it possible to offer very cheaply, to a very large number of users, benefits which had until then been more costly and restricted to a few subscribers. These benefits are:

- Real-time availability of information;

- Information storage (printer, microcomputer);

- Interactive capability in some magazines.

The Meteorological Office in France has six regional datacommunication distributors, each of which houses one or more magazines aimed at various categories of user:

The public at large

Under these headings are offered forecast bulletins and recently observed data. For forecast bulletins, the Teletel systems has a number of advantages:

For the user:

- the information is in written form, is therefore easier to digest and can also be stored;

- there are several bulletins in the same magazine which cover different areas, which makes it possible, simply by changing the page, to obtain meteorological forecasts for areas which are quite extensive or remote;

- the updating rate is the same as for telephone recorded information services (two or three times a day) and the information is immediately available.

For the meteorologist:

- The system provides facilities for recording information and makes it possible to make more frequent partial updates to a bulletin. For example, forecasting bulletins cover a five-day period. The part covering the last four days is updated once a day, while the first 24 hours are updated two or three times a day;

- However, the Teletel system is of greatest interest in that it can make available very recent observational data, as data acquisition can be carried out completely automatically using meteorological databanks. Some magazines can thus provide data on temperatures, winds and relative humidity, updated every three hours, against the background of a geographical map.

Shipping

Sailors, whether professional fishermen or amateurs out in yachts, are well aware of how dependent they are on meteorological conditions. In their magazine we find the same types of information as above, coastal forecasts (safety and ordinary) and recent observational data from all the signal stations in the range of the author of the magazine. These data are updated every three hours and are available 15 minutes after they have been observed. The advantages here are potentially greater than for the public at large, particularly in terms of safety, before the boats weigh anchor for the fishing grounds or set off on a cruise. Moreover, there have been tests which have shown positively that information put out through Teletel would also be accessible on board a ship at sea; the storage or printing capability for the bulletins would solve the problems surrounding listening to information broadcast at set times by radio stations.

Farming

In most magazines, the benefits of the transmission system make themselves felt in the speed at which processed data, such as potential evapotranspiration, water balances or temperature totals relative to various thresholds, are made available.

Some of the more sophisticated magazines have interactive applications which make it possible for farmers to use the meteorological databank for calculations concerning their own crops; an example of this is the IRRITEL application discussed in the paper by S Dervaux which is also in this volume.

Aviation

A magazine for this specific group has recently been set up; although, in our opinion, it should not be a substitute for flight sheets or direct contact with a meteorologist, this magazine seems very popular with VFR pilots, as transmission over Teletel makes it possible.

- To provide information over a preset geographical overlay;

- To store the information as hard copy;

- To have immediate access to the latest airfield METAR or TAF

The Teletel system is a system of computerized data telecommunication which has the original feature of opening up to the public at large the advantages of computer data processing. The Meteorological Office, which is largely organized on the basis of using computerized data transmission and processing, has found in Teletel the means to put its information potential to good use. It was also, incidentally, one of the first bodies to offer services using this form of communication. The general satisfaction on the part of users, who consult these magazines more and more often and regularly expressing the wish to have new and more and more specific magazines, seems to give proof of the positive contribution of the system.

30

Climate data applications in the ASEAN region

Mr. A. David and Dr. L. J. Tick, ASEAN Coordinators of the ASEAN Users' Manual Project (Malaysian Meteorological Service)

1. INTRODUCTION

The compilation of the ASEAN (Association of South East Asian Nations) climate data of the five ASEAN countries - Indonesia, Malaysia, Philippines, Singapore and Thailand - was accomplished under a joint cooperative project during the period of years 1978-1982. The project which was funded by both the ASEAN nations and UNDP (United Nations Development Programme) resulted in the publication of the _ASEAN Climatic Atlas and Compendium of Climatic Statistics_ (1982a and 1982b). A unique feature of the publication is that it contains climate data sets and basic statistical analyses of a standard period of years, 1951-1975, for all the five countries. In addition to this, wherever possible, climate records of some stations dating back to the late eighteen hundreds have been included especially to assist researchers involved in studying climatic trends as well as effects of urbanization and rapid industrialisation taking place in the region.

In recognition of the need to encourage the use of the wealth of data contained in the ASEAN Atlas and Compendium, the ASEAN Sub-Committee on Climatology and the World Meteorological Organization (WMO) formulated, as a follow-up project, the preparation of a guidebook, known as the _ASEAN Users' Manual for the ASEAN Climatic Atlas and Compendium of Climatic Statistics_. This project, which was started in late 1986, is expected to be completed in late 1988. While most of the funding support for the project comes from UNDP, the ASEAN nations and WMO provide the technical and administrative support so essential in any regional project.

It has been recognized that the preparation of this Manual is one of the first undertakings by the ASEAN Meteorological Services to assist the users of climate data and information in the statistical techniques employed by climatologists, and more importantly, to provide a future forum of interaction between the users and climatologists so as to ensure that economically relevant information will be made available or formulated in such a way to best suit the users' requirements.

2. GENERAL ASPECTS

Geographically speaking, the ASEAN region has a rather complex land-sea distribution comprising a peninsula and thousands of islands of varying shapes and sizes separated by seas/straits/oceans of varying distances. The various scales of atmospheric motions that ensue from this land-sea distribution as well as its interaction with the monsoon and global-scale motions bring about many sub-climates as described by the Dutch meteorologist C Break (Ramage 1971).

Over the years though the ASEAN climate has been given a fairly thorough treatment by geographers and climatologists, there is still much more research to be done in order to understand the subject in terms of the day-to-day weather and its associated phenomena. A primary reason for the lack of interest in the subject is the poor network of climate/meteorological stations especially in the interior parts of most of the large islands in the region - the Borneo and Celebes Islands can be quoted as typical examples. Combined with this shortfall is the high variability, both temporal and spatial, rainfall regime found in the region. Such a variability imposes a stringent demand of a dense network of climate monitoring observational programme in order to better understand the climate system of the region.

In addition there has been little interest, until of late, in conducting studies on applying climate statistics to weather sensitive activities except those connected with the agricultural sector. Consequently this has led to a lack of reference materials for guidance purposes, and most importantly, the failure of users to appreciate the intricacies of local climate and weather. Many instances can be quoted in which not only foreign but also local consultants have, inadvertently or otherwise, resorted to using imported techniques and models to derive meteorological inputs to the projects. Such techniques and models, though they have been proven to be effective in mid-latitude countries, have been found to be inadequately treated by researchers specifically targeted to answer the questions asked by planners.

3. REVIEW OF APPLICATIONS

3.1 Wind Loadings

In early 1985, a draft standard code of practice on wind loading was prepared under the authority of the Building and Civil Engineering Industry Standards Committee of Malaysia. This draft standard, which was derived from the British Standard Code of Practice, proposed that the assessment of wind load be based on the following equation:

$$V_s = V\, S_1 S_2 S_3 \quad \ldots\ldots \qquad (1)$$

where V_s is the design wind speed;

V is the basic wind speed which is taken as the 3-second gust estimated to be exceeded on the average once in 50 years;

S_1 is the topography factor;

S_2 is the ground roughness, building size and height above ground factor; and

S_3 is the statistical factor which takes account of the degree of security required and the period of time in years during which there will be exposure to wind.

Without dwelling on the details of estimating V, S_1, S_2 and S_3 adopted by the British, it is obvious that appropriate modifications have to be made to them based on local climatology. In Malaysia, estimates for V and S_3 can be obtained from data source available in the Malaysian Meteorological Service (MMS). For S_1, the British procedure can be employed without reservations. The question that comes to mind is for estimating S_2 - whether its values for variation of wind speed with height above ground are governed by the power law in the equatorial region. As one considers a large number of high-rise buildings being erected in the large cities, such as in Kuala Lumpur in Malaysia, the importance of making tolerably precise estimates of S_2 cannot be overly emphasized.

At the outset, it would be prudent to ask whether or not we should have considered the nature of wind gusts in Malaysia and some other ASEAN nations for which the local wind-gusts (maximum 3-second gusts) are associated with thunderstorm activity which possesses both spatial and temporal variabilities. To take into account this feature, it may even be prudent to introduce an additional factor S_4 in Equation (1) such that:

$$V_s = V\ S_1 S_2 S_2 S_3 S_4$$

where S_4 is to be taken as a function of the frequency of thunderstorms at a place where the wind load is to be assessed (f_p) and the frequency of thunderstorms at the station where the basic wind speed V is obtained (f_s). This function may well be a simple ratio of f_p to f_s.

The need for S_4 can be exemplified further. The assessment of wind load in a large city, such as Kuala Lumpur, has to be generally based on wind measurements at a distant airport station in a relatively less developed and less hilly environment, for example Sunbang Airport Station which is located 15 km west of Kuala Lumpur. Thus the incidence of thunderstorms and, by inference, the maximum gusts at the city would be expected to be significantly higher than those observed at the distant station. To estimate S_4, however, the frequency of occurrence of thunderstorms in the city has to be known and this may be estimated from observations made at another station in an urbanized environment (for example Ipoh Station in Malaysia) or through the use of results based on related observational studies.

It is clear by now that the power law relationship between wind speed and height does not hold in the presence of thunderstorms. Unfortunately, there is no easy way to overcome this problem except through special wind profile measurements using a tower. Such measurements have yet to be carried out in the region. However, a joint cooperative project between MMS and the University Science Malaysia, Penang has been initiated to make tower measurements at Perai Station (located opposite Penang Island on the mainland of Peninsular Malaysia) and collection of data for wind, temperature and humidity elements at three levels up to 45 meters will begin sometime in later part of 1987. Results obtained from this joint project will give an insight into the variations of wind speed with height during thunderstorm activities in the Malaysian region and the findings could be used for the other ASEAN sites.

3.2 Applications in Architecture

It would be natural to expect the styles adopted for the design of buildings and for planning of cities be responsive to the local climate (Oke, 1984). Often this practice is not strictly adhered to specially in developing countries. For example, a feature article entitled "In Search of Architecture" published in a Malaysian newspaper, the Malay Mail, on 1st January 1985, reported that,

"In Kuala Lumpur, some new skyscrapers are a very good example of architecture for temperate climate being transplanted to a tropical environment. Their designers did not take into account the intensity of tropical sunlight, the monsoonal rain and stifling heat".

The article went on to recount how the use of heavily tinted glass could produce energy inefficient buildings which require huge electricity bills for their lighting and cooling purposes. Neither do the flat-roof designs conform to the local weather conditions. Such designs are said to be ill-equipped to cope with the high intensity and short duration rainfall associated with thunderstorms which, as a consequence, cause major leaks in relatively new buildings.

Throughout the ASEAN region, it is not an uncommon sight to see the older buildings and houses which were, in general, built prior to the Second World War, looking distinctly different to the present ones. What immediately catches the attention of an observer are the large slopes of the roof tops and spacious verandahs. While these large slopes and verandahs may be a costly investment these days, it is regrettable that the requirement for them in terms of the climatic needs of tropical areas is not being considered by present architects and planners. The inadequate slope of roof tops for the newly-built houses has led to roof leaks that are generally beyond repair and this has led to a stage where, what has been saved in planning time and initial capital building costs, may be worthless and uneconomical. In the case of verandahs, it has been well established that they are the main contributing factors in reducing the solar insolation falling directly onto the walls and in providing natural ventilation which causes a cooling effect on the building because of its low specific heat capacity compared to concrete surfaces. Once again the savings effected on the capital cost on investment by not providing verandahs will certainly be offset by the increased operational costs needed in air conditioning for the building. In these days of escalating costs of energy production there is therefore a need to address these problems by taking into account the local climatic conditions.

The tropical climate may in some ways pose more severe problem in the design of buildings than that in the mid-latitudes. Indoor climate control in modern tropical buildings, largely of mid-latitude design, requires extensive use of energy for cooling all year round. An important aspect that is overlooked is that cooling as approximately four times as costly as heating, in terms of energy use per degree of temperature change. High temperatures and high humidities in combination with strong solar radiation and large emissions of air pollutants, especially from traffic in the urban tropical areas, can create a potentially dangerous situation with a high potential for formation of photochemical smog. This should be avoided at all costs due to its devastating effects on materials as well as on human health.

It is hardly necessary here to emphasize the importance of initiating studies for determining the slope of roof tops for efficient discharge of rainfall based on rainfall intensities in the tropics. A separate study is also deemed essential for deriving the spatial extent of verandahs for various types of buildings based on local wind distribution and solar insolation. Such studies can lead to legislation that need to be incorporated into the building bylaws.

It is just unthinkable to believe that architects, especially the local ones, are not aware of the need to adopt designs which are functionally ideal for the local climate. Most likely the case would be that the demands and wishes of their clients far outweigh any other considerations. Perhaps if there were ample literature, particularly the promotional brochures (WMO, undated; Joyce, 1982), it could make it easier for architects to convince their clients of the fallacy of imitating totally designs employed in temperate countries and the desirability of adapting western designs and technology to suit local environment. It is critical that designers appreciate the nuances of the climate in which they build, since the severity of each climate liability and the availability of each climate asset, will be different according to the project location (Loftness V, 1982).

3.3 Urban and Building Climatology

A national project 'Special Climate Applications Programme (SCAP)' has been proposed in Malaysia to initiate some studies particularly relating to the urban and building climatology. The original plan is currently being reviewed and reformulated by a WMO expert who completed a short-term mission to Malaysia in late June 1987.

3.4 Application in Water Resources Management and Planning

Linsley (1982) has pointed out that there are social and political (or economic) aspects of drought and water resources, as well as its management. Landsberg (1982) has added yet another important aspect - the climatic effect which is often neglected by many. These three aspects are felt in Malaysia during most of the period of rainfall deficiency as well as immediately thereafter. Landsberg has further attributed the problems of water resources and management to lack of coordination in the planning of cities, towns, agricultural and industrial sites. For an effective planning and implementation, it is essential to involve the services of a wide range of specialists who may directly or indirectly be involved in water management and resources. It is hardly necessary to emphasize the role of a climatologist as one of the specialists in this area. His contribution in the form of analyses of rainfall from the context of climate system will constitute towards an efficient utilization of water resources and its management for the well-being of future mankind.

While the overall capacity of supply is above the demand for water in Peninsular Malaysia, water shortages are frequently experienced in urbanizing regions such as the Klang Valley in the Federal Territory of Kuala Lumpur and the State of Selangor and Seremban township in the State of Negeri Sembilan. Singapore does not have adequate water resources to meet its requirements and is dependent for more than half its fresh water supply on the State of Johore whose southern region is regarded the second-fastest urbanizing area in Malaysia.

Most of the readily available water resources close to major demand regions in Peninsular Malaysia have already been developed and it is realized that future demands could only be met from more inaccessible sources. Hence to plan for a comfortable supply and demand balance, strategies adopted for the water resources development include the construction of more storage dams for the purpose of retaining water in wet seasons for use in dry seasons, and facilities for diverting water from rivers located in either the same (intra-basin) or another state (inter-basin). As was reported by the Far Eastern Economic Review (3 October 1985), the joint national water resources study undertaken by the Japan International Cooperation Agency (JICA) and the Malaysian Government has proposed the construction of 43 storage dams and 15 diversion facilities up to the year 2000, estimated to cost around US $4.2 billion.

Major water resources projects undertaken in the past were mostly farmed out to foreign companies which invariably gave insufficient or inappropriate consideration to the meteorological/climatological inputs. It is time that climate applications be taken seriously as a worthwhile undertaking especially for projects which are being planned on a scale as large as the one proposed for the water resources development. Indeed only through systematic use of specialized information, such as those mentioned below, one would expect optimum returns on such a huge investment made on the project.

a) Characteristics of local rainstorms;

b) Possible climate modification during a change in environment at the site of storage-reservoir;

c) Analysis of dry spells (Stern and Dale, undated) including impact of El Nino; and

d) Climatic trend analyses.

Among the items listed above, item (b) would require the use of computer models. Through computer simulations it is possible to derive invaluable information on changes in the cumulus-scale convection or for that matter, the rain pattern over the site of the storage-reservoir and its neighbouring areas.

3.5 Other Applications

It was reported in a Malaysian local newspaper (December 1984) that during the half-year ending September 1984, the operating profit of one of the largest local soft-drinks manufacturers took a 17 percent tumble to about Malaysian Ringgit (MR) $26 million (US $10.5 million) from MR $32 million (US $12.8 million) in the previous corresponding period. This relatively poor performance was attributed by the same manufacturer to the inclement weather experienced in the first half year. The higher sales however, over the previous corresponding period, were linked to the exceptionally hot weather.

This statement in the press associating weather with sales naturally attracted the attention of the Malaysian Meteorological Service (MMS) which submitted an enquiry to the manufacturer for more information on any correlation studies undertaken by its researchers before arriving at the conclusion adopted by the Directors of the Company Group. It was explicitly pointed out to the Group that the purpose of the MMS enquiry was to evaluate the particular studies of the Group and in this way to provide the necessary technical climatological support and advice, through close interaction with users of climatic data, for mutual benefits of both parties. A critical review of the analytical technique employed by the Group revealed that its conclusion was barely tenable because of lack of proper understanding of the variability of rainfall pattern of the tropics. It was also clear that the climate data and analysis were used in a manner to conceal the truth behind the failure of the economic activity.

The necessity of representative reliable weather data and analyses for agricultural planning can be seen from another case. Before the initiation of a large sugar cane plantation in the late 1970s in an area of Peninsular Malaysia, a foreign consultant group was asked to do a feasibility study on the climate of a pre-selected cultivation areas. Having no weather data in that area, the consultants used the data from the nearest meteorological station which was about 100 km away. The report of the consultants implied that the area was suitable for the cultivation for sugar cane. The project was launched as a joint cooperative one between the Government and private sector. Some years later, the plantation encountered several problems. Production was always low. A second group of consultants was assigned to study the viability of the whole project once again. This group of consultants, in collaboration with MMS, set up a few site observational stations to collect the necessary data for the study. Their study revealed that the area was not suitable for sugar cane cultivation. Soon after that the trouble-ridden plantation was wound-up, incurring large financial losses to the Government and causing much misery and suffering to the plantation workers.

A third remarkable case which needs a particular reference is linked to strategies which were considered for the cultivation of crops other than rubber and oil palm - the mainstay of economy in Malaysia. A decrease in the annual rainfall in most parts of Peninsular Malaysia in the late 1970's promoted a large plantation group to investigate whether the observed rainfall deficiency was a 'climatic change' resulting from the extensive deforestation due to timber logging prevalent during the preceding years. By performing a simple statistical analysis on the rainfall data for a few stations covering the years in 1960's and 1970's, the plantation group concluded that the rainfall deficiency of the late 1970s was something 'real' and that alternate strategies for development of other crops might have to be considered in the future. The Government, being concerned about the situation, submitted the study made by the plantation group to MMS for its evaluation. After an in-depth analysis of the reliable rainfall records, varying over the past 48 to 95 years, of stations maintained by MMS, the researchers Todorov and David (1982) concluded that.

> "Annual amounts of rainfall have a high degree of variability. Large year-to-year deviations above or below mean rainfall amounts should be considered as normal phenomena. The rainfall records indicate that several declines in the annual amounts of rainfall-similar to that during the last decade (prior to the year 1980), have occurred also in the past."

This conclusion by MMS came as a 'relief' to the plantation group to continue with its on-going activities of cultivation of rubber and palm oil crops. The collaboration established between MMS and the plantation group paved the way to initiate and set up a network of climatological stations in several estates of the plantation group for effective planning of their future economic activities.

4. INTERACTIONS WITH USERS

One way to determine the awareness on the part of users in the utilization of climate data and information for their operational climate-sensitive activities is through the statistics available in the files of enquiries kept by the Climatological Division. In Malaysia, while the climate data and daily weather forecasts are supplied on a complimentary basis to the press, television/radio, airlines, Government and Statutory bodies (including the universities and institutions of higher learning), the private sectors are generally required to pay nominal charges for the supply of any climate data and information. Statistics maintained by the Climatological Division of MMS reveal an increase of about 25% per year in the number of requests over the period of years from 1973 (total no. 848) to 1985 (total no. 3658), and the revenue generated from the sale of data and services for the same period shows an increase of about 29% per year from MR $5,734.00 (US $2,389.00) to MR $27,652.00 (US $11,522.00). The actual figures quoted here may not be dramatic, but they go to show the increasing trend on the importance attached to climate resources in a developing country. It should also be emphasized that one of the factors that contributed towards an increase in the revenue collection in MMS is due to the initiative taken by a few meteorologists who were able to interact directly with the users and provide tailored climate services rather than raw climate data compared to the previous years. The introduction of a new computer in MMS for data processing and development of software tailored to meet specific users' requirements made the task of providing such climate services not only more fashionable but also desirable in order to enhance the contributions to the collaborating agencies.

The introduction of the CLICOM system which aims at achieving a transfer of technology in climate data management and user services through the provision of comprehensive microcomputer systems in the developing countries can extend a wide range of tasks beyond the normal bounds of climatological services and provide the necessary operational products tailored to applied purposes of the user. This approach will be able to motivate further the use of climate data and information to optimize resource management activities which can lead to visible socio-economic benefits.

5. CLIMATE PROJECTS (ON-GOING AND PLANNED) IN THE ASEAN REGION

5.1 ASEAN Users' Manual - (on-going Project)

The contents of the ASEAN Users' Manual are formulated in a manner to fully exploit the climate data source in the best possible way to benefit the activities of the users in the ASEAN Region and the most important are those activities related to the current socio-economic developments taking place rapidly in the Region. The Manual will appear in two parts - Part I and II. Part I will be devoted to guidance on available methods and statistical techniques used to derive useful climatological information from data in the Compendium, with explanations and examples of such methods and techniques as applied to the Compendium. Part II will concentrate on providing guidance material on ways and approaches to using the Compendium data as they relate to specific climate applications with relevance to the ASEAN region. Again examples will be used as illustration.

In identifying the key climate application areas, a sub-project activity was formulated through the services of focal persons in each ASEAN country. This was accomplished in two stages - firstly by surveying the files of climatological requests received by the meteorological services during a period of at least 1 to 5 years from 1981 to 1985 with the objective of identifying both the sectors/activities and the type of meteorological elements which were of interest to the user groups. The results of the survey for the ten types of selected activities are shown in Table I. Unfortunately, intercomparisons of the various activities from one country to another cannot be made because:

i) the period of years used for sampling the users is not common to all countries;

ii) the population of users (somewhat related to size of a country) in the countries varies greatly;

iii) the priority given to the different activities varies according to the natural and manpower resources of the countries.

Overall, the results depict that for most countries the industry/commerce activity receives a high priority, and this primarily arises due to the increasing applications of climatological data to the industrial, engineering and construction sectors. According to the strategy of the Thailand Economic and Social Development Plan for the period 1986-1991, the stress of increasing the efficiency of country development will be through the process of utilizing the results of 'technology and science' together in order to encourage the raw material production and marketing, thus improving the national resources and readjusting the environment.

Agriculture is the next important activity for at least three of the countries (Indonesia, Malaysia and Thailand). The structural/urban sector is to a large extent the prime activity in the Philippines. Other activities that require to be dealt with in the Manual are water resources management, land use, energy, transportation, environment/pollution and tourism.

An important aspect of the survey is that the investigation was carried out on past uses and it was unable to recognize areas of activities where significant applications of climate data can occur in the future. The contents of the Manual must not therefore be determined solely by the results of the survey of files. Application areas in the sectors/activities such as environment/pollution, energy, building climatology, air quality indices, will become increasingly important in the near future and an awareness of the usefulness of climate data and analysis remain to be developed.

The second stage of the user survey was aimed at obtaining information on the specific climate application areas as well as the type of analyses which the users require to be included in the Manual. For this purpose a special questionnaire was developed and distributed to selected users in each ASEAN country. A total number of approximately 435 users responded. Analysis of the completed questionnaires revealed that some of the following climate application areas and analyses have to be included in the Manual:

- Rainfall intensity
- duration
- frequency curve
- Probable maximum precipitation for thunderstorm and monsoon rains
- Atmospheric stability determination
- Wind speed
- duration
- frequency curve
- Return periods for rainfall and wind

- Computation of drought indices
- Comfort indices
- Correlation between solar radiation and sunshine
- Computation of diffuse radiation
- Water availability for particular tropical crops
- Crop responses as affected by climate
- Solar radiation on inclined surfaces
- Wind loading on buildings
- Computation of potential evaporation
- Water resources management
- Land use
- Urban planning design
- Pollution control
- Energy conservation in building
- Architectural design of buildings and homes
- Air quality indices

5.2 Preparation of a Climate Supplement of Brunei Darussalam (Planned Project)

The project for the ASEAN Climatic Atlas and Compendium of Climatic Statistics was completed in 1982. At that time, Brunei Darussalam was not a member of ASEAN. After taking cognizance of the potential benefits to be derived by economic planners and industrialists, the First Planning Conference (FPC) for the ASEAN users' Manual Project (1-3 October 1986), Petaling Jaya, Malaysia) felt that there was a need for the preparation of a separate Climate Supplement of Brunei Darussalam. A project proposal was accordingly formulated by one of the ASEAN Coordinators to the Users' Manual Project in consultation with the Permanent Representative of Brunei Darussalam with WMO. In responding favourably to the proposal, WMO decided that the funding part of the Supplement should be incorporated as an additional budget proposal into the overall current budget requirements of the Users' Manual Project.

5.3 Assessment of Solar Energy and Daylight Resource for the ASEAN Region (Planned Project)

The envisaged successful completion of the ASEAN Users' Manual Project will mark the second cooperative effort of the ASEAN Sub-Committee on Climatology (SCC) to fulfil the climate data user requirements in the ASEAN region. A stage has now been set for the SCC to initiate action for the next regional project - Assessment of Solar Energy and Daylight Resources for the ASEAN Region. This project aims to provide specific quantitative solar energy, daylighting and other derived information for practical and research applications in those human and national activities that will benefit the population, national development and national economics of ASEAN. The realization of these benefits is through the appropriate applications of the information in solar dependent or solar sensitive activities. Key application areas include agriculture, water resources management, town planning, building design and renewable energy resources development. Solar radiation and daylighting conditions are also important factors in considering the environmental impacts of both urbanisation and industrial development.

The basic data collection involves the six ASEAN Meteorological Services which will carry out special measurements of global solar radiation, its diffuse and direct components and sky luminance measurements at their centres for about two years. In some countries, observations of direct and diffuse components and sky luminance measurements will be carried out for the first time.

In addition to the special measurements mentioned above, a mesoscale monitoring network for global solar radiation with about 11 sets of equipment will be carried out at a selected ASEAN capital city. This network will have an approximate 5-10 km grid configuration and be sited partially within and partially adjacent to a large urban area. This network may be moved to a second ASEAN capital city after collecting sufficient amount of data for about a year at the first site.

The institutional framework for the implementation of the two-and-a-half project has been finalized at the Ninth SCC Meeting in 1986. The project proposal has been submitted to UNDP for its consideration of funding. WMO has been named once again to be the Executing Agency to the Project.

6. CONCLUSION

This paper illustrates a few cases which show that there is a need to support adequately climatological input in a wide variety of socio-economic activities in the ASEAN region.

Although the daily weather patterns of the countries may be largely simple and perhaps monotonous, the lack of precise knowledge of the governing mesoscale convective systems and the different temporal and spatial scales of air-sea interaction as well as the roles of orographic and local effects and the influence due to surrounding oceans, pose many difficulties in providing climate products effectively to the user communities. Inadvertent transportation of imported statistical techniques from mid-latitudes to the equatorial areas in several climate application areas have either caused large economic losses to specific past projects or created certain situations in which users tend to attach less importance to the local climate scenario.

The project ASEAN users' Manual currently undertaken by the six ASEAN nations aims at providing a guide book to serve a variety of users who will derive greater benefits through the proper applications of the techniques of climate-sensitive activities in the region.

ACKNOWLEDGEMENT

The authors wish to gratefully acknowledge the support and encouragement given by Mr. P Markandan, Acting Director-General of the Malaysian Meteorological Service (MMS) in the preparation of this paper and Mr. Ooi See Hai of MMS for his critical review of the paper. They also wish to thank Ms. Lim Mae Ai of MMS for her assistance in typing and checking the original manuscript.

Table 1 - Summary of Climate Data User Enquiries of ASEAN Nations Country

Activity	BRUNEI 1981-Oct86 No	%	INDONESIA 1985 No	%	MALAYSIA 1981-1985 No	%	PHILIPPINES 1981-1985 No	%
Agriculture	0	0	158	17	543	24	282	5
Land Use	0	0	34	3	31	1	263	5
Water Resources Management	2	4	70	7	59	2	526	9
Transport	4	7	162	17	17	-	75	1
Tourism	0	0	43	4	26	1	50	-
Industry/ Commerce	32	60	54	5	555	25	1625	30
Structural/ Urban	11	20	45	4	198	8	2630	46
Energy	4	7	73	7	32	1	163	3
Environment/ Pollution	0	0	34	3	50	2	60	1
Others*	1	2	329	33	814	36	0	0
Total	54	100	1002	100	2325	100	5674	100

Activity	SINGAPORE 1981-1985 No	%	THAILAND 1982-1986 No	%	ASEAN No	%
Agriculture	11	1	280	22	1202	11
Land Use	0	0	54	6	382	2
Water Resources Management	11	1	120	13	788	7
Transport	1	-	36	4	295	3
Tourism	3	-	17	2	139	1
Industry/ Commerce	1017	93	134	14	3417	32
Structural/ Urban	3	-	62	7	2949	62
Energy	2	-	68	7	342	3
Environment/ Pollution	2	-	95	10	241	2
Others*	56	5	146	15	1346	12
Total	1106	100	940	100	11101	100

Remarks:

1. - Means less than 1%
2. * Requests from the general public for weather forecasts and university researchers for unspecified activities.

REFERENCES

ASEAN Sub-Committee on Climatology, 1982a The ASEAN Climatic Atlas. ASEAN Secretariat, Jakarta.

ASEAN Sub-Committee on Climatology, 1982b The ASEAN Compendium of Climatic Statistics. ASEAN Secretariat, Jakarta.

Joyce, J, 1982 The economic impact of climate on cities. Vol X. Oklahoma Climatological Survey.

Landsberg, H E 1982 Climatic aspects of droughts. Bull Amer Met Soc, 63, 593-596.

Linsley, R K, 1982 Social and political aspects of drought. Bull Amer Met Soc, 63, 658-591

Loftness, V, 1982 Climate Energy Graphics, Climate Data Applications in Architecture. WCP-30, WMO Geneva.

Oke, T R, 1986 Urban Climatology and the Tropical City. Proceedings of the WMO/WHO Technical Conference on Urban Climatology and its Applications with special regard to Tropical Areas, Mexico D E, WMO Technical Note 652.

Ramage, C S, 1971 Monsoon Meteorology. Academic Press, London and New York, 296 pp.

Stern, R D and Dale, I C, undated. Statistical methods for tropical drought analysis based on rainfall data. WMO Programme on Research in Tropical Meteorology, WMO, Geneva.

Todorov, A V and David, A, 1982 Recent rainfall trend in Malaysia. MARDI Report No 79, MARDI, Kuala Lumpur.

WMO, undated Climate, urbanizations and man. World Climate Programme, WMO, Geneva.

31

Use of meteorological data and information in hydrological forecasting

E. A. Hassan, WMO Secretariat, Geneva, Switzerland

INTRODUCTION

A hydrological forecast is the prior estimate of a future state of hydrological phenomena. Hydrological forecasts and warnings are issued for many purposes, varying from those for short-term events like flash floods to seasonal outlooks of the potential water supply for irrigation, power production, or inland navigation. Techniques for forecasting range from the use of simple empirical formulae or correlations to the use of complex mathematical models representing all phases of the water balance of a river basin.

The basic water regime elements whose forecasting is of practical interest are as follows:

(a) volume of runoff in various periods of time (e.g. period of high and low flows, month, season, year);

(b) discharge or stage hydrograph;

(c) peak flood stage or discharge, and the time of its occurrence;

(d) maximum water level in lakes and the date this level will be reached;

(e) average and minimum water levels in navigable rivers

and lakes during various calendar periods;

(f) height of waves created by wind on lakes and large reservoirs;

(g) wind setup in lakes, coastal waters and estuaries;

(h) water quality parameters such as temperature and turbidity.

INFLUENCE OF METEOROLOGICAL FACTORS

The development of hydrological processes are influenced by meteorological factors, but the changes they bring about in the regime do not occur instantaneously or immediately. For example, the duration of the runoff caused by precipitation is often many times longer than that of the rainfall itself, and a time lag intervenes between the causative temperature rise, the melting of snow, and the consequential rise in river level. The relatively slow rate at which hydrological processes develop, and the fact that they lag behind the more rapid meteorological processes, make it possible to forecast elements of the hydrological regime.

Meteorological and climatic data and information are also extensively used in hydrological analyses such as in the preparation of monthly, seasonal and annual water balance studies including river forecasting, irrigation operations and the derivation of maps of mean annual rainfall, runoff and evapotranspiration. A convenient generalization and frequently a first step in studying the water resources of a region, is the relation between annual runoff and the meteorological and climatic components of which it is the residual. Ideally, runoff would be balanced in long-term averages by precipitation minus evapotranspiration, taking into account changes in soil-moisture and ground water storage.

All the basic factors governing runoff and other hydrological processes can be divided into the following groups:

(a) initial factors, which govern conditions existing at the time the forecast is made, and which can be calculated or estimated on the basis of current hydrological and meteorological observations or measurements;

(b) future factors, which influence the hydrological processes after the forecast has been issued. These factors, which include future weather conditions, can be taken into account explicitly only if a weather forecast is available. However, no reliable methods of quantitative forecasting of weather elements far in advance have yet been developed.

For this reason, the practical possibilities of extending the period of the forecasts is limited by the degree to which the future factors affect the development of the eventual forecast. Subject to this time limitation the primary factors which influence the accuracy and timeliness of hydrological forecasts are the accuracy, speed, and reliability of assembly of meteorological and hydrological initial factors, the adequacy of the forecasting model, and the size of river basins.

Forecasts can be issued beyond the period of reliable weather forecasts only if they are made contingent on possible or specified weather conditions. Such contingent or probability forecasts are quite common in seasonal weather supply forecasts.

METEOROLOGICAL DATA FOR FORECASTS

The meteorological data used in hydrological forecasts can be divided into two groups: the first includes the material required for developing the forecasting method; and the second group includes the information needed to operate the forecast.

The first group includes the conventional time-varying meteorological information necessary for testing and evaluating the trial forecast models, as well as other hydrological data. The second group includes the meteorological data specified by the forecasting scheme to characterize the state of the catchment immediately before the issue of the forecast. It may also include a measurement of the element itself which may be used to monitor the forecast performance or update the forecast model.

The meteorological data and information normally used in the development of hydrological forecasts are listed in the Table for this paper. In addition to precipitation, which is the most commonly required meteorological element used by general-purpose forecasting centres, there are a further 11 meteorological elements listed in the Table including temperature, wind, and other meteorological data related to evapotranspiration computations. The precision of observation of meteorological elements for hydrological purposes and the reporting interval for hydrological forecasting purposes are indicated in the Table. The standards are recommended as guidelines to follow in planning of equipment development and data acquisition programmes for hydrological forecasting.

TABLE

DESIRABLE PRECISION OF OBSERVATION OF METEOROLOGICAL ELEMENTS FOR HYDROLOGICAL PURPOSES AND THE REPORTING INTERVAL FOR HYDROLOGICAL FORECASTING PURPOSES

Elements	Precision	Reporting interval for hydrological forecasting purposes
(a) Precipitation - amount and form[1]	± 2 mm below 40mm ± 5% above 40mm	6 hours[2]
(b) Snow depth	± 2cm below 20cm ± 10% above 20cm	Daily
(c) Water equivalent of snow cover	± 2mm below 20mm ± 10% above 20cm	Daily
(d) Air temperature	± 0.1°C	6 hours
(e) Wet-bulb temperature	± 0.1°C	6 hours
(f) Net radiation	± 0.4 MJ m^{-2} d^{-1} below 8 MJ m^{-2} ± 5% above 8 MJ $m^{-2}d^{-1}$	Daily
(g) Pan evaporation	± 0.5 mm	Daily
(h) Surface temperatures - snow	± 1°C	Daily
(i) Temperature profiles - snow	± 1°C	Daily
(j) Wind speed and direction	± 10%	6 hours
(k) Sunshine duration	± 0.1 hour	Daily
(l) Relative humidity	± 1%	6 hours

Notes:

(1) In some locations it will be necessary to distinguish the form of precipitation (liquid or solid);

(2) The reporting interval in flash flood basins is often required to be two hours or less; in other locations, daily values may suffice.

However, it should be noted that meteorological data and information are not always required for the preparation and computation of hydrological forecasts. On large and seasonal rivers, hydrological forecasts are normally computed without using any meteorological data. For example, the real-time hydrological forecasting system on the River Niger in West Africa uses, for the time being, only models of flow in the river based on routing and correlation of upstream and downstream discharges.

Many successful forecasts depend on very simple linkages that have been empirically established between an observed variable, e.g. upstream stage, and forecast variable of interest, e.g. downstream stage at some later time. The mathematical technique by which unsteady (time varying) flows are predicted at points downstream in rivers, reservoirs, and estuaries, is known as 'flow routing'. A variety of techniques to route flows has developed over the years. None of these techniques use meteorological information or data and are based exclusively on hydrological data and information.

An extremely common requirement is for a forecast of crest stage, and on moderate-sized rivers a practical technique is to effect a simple graphical correlation with an upstream crest stage, thus providing a forecast with a lead time equal to the travel time of the flood wave. Thus no meteorological data are needed in crest stage forecasting. Recession forecasting relating to low flow forecasts and flow forecasts based upon storage volume depend exclusively on hydrological parameters and no meteorological data are used in the development of these forecasts.

Meteorological data for developing forecast procedures

An adequate data observation network is a prerequisite for the development of hydrological forecasts. At the development stage, alternative forecast methods may be under investigation and analysis of hydrological events is retrospective, so more observations are made than when the forecast becomes operational. It is nevertheless necessary to have in mind the eventual purpose so that siting and reading frequency should be realistically set to achieve consistency.

At the development stage, the type of hydrological analysis that is undertaken involves the determination of relationships between input and output variables, e.g. snowmelt as a function of degree days or unit hydrograph derivation from areal rainfall and catchment runoff. The network and instrumentation requirements are those quoted in the relevant sections of Reference (1).

Meteorological data required for a forecast operation

Once the forecast procedure has been determined, the data network used in development will probably be reduced to just those elements required to operate the procedure. There are numerous alternative requirements for data, depending on the interrelationship between forecast methods and basin types. The parameters and constants of forecast equations which were established during the development stage to reflect the climate, topography and morphology of the catchment must be available for forecast operation.

USE OF METEOROLOGICAL FORECASTS

The expected development of certain meteorological elements from a weather forecast is frequently used as a basis for short-term forecasts of snowmelt runoff, rainfall-produced floods and ice formation and break-up. In particular, the announcement of an initial alert is frequently based upon expected weather conditions. As a consequence, forecasts will be concerned with:

(a) quantitative precipitation forecasts (QPF) for periods of up to 72 hours;

(b) air temperature, humidity, dewpoint, wind and sky conditions for up to 5 days;

(c) height of the freezing level in mountainous regions;

(d) wind speed and direction for 24 hours or more

Reliability of meteorological forecasts decreases rapidly with the period of the forecast. Temperature is usually more reliably forecast than precipitation or wind. The amount of precipitation is forecast less reliably than the chance of precipitation. These are important considerations at the development stage when decisions are made on the level of meteorological forecast information to be incorporated into the hydrological forecast (5).

QUANTITATIVE PRECIPITATION FORECASTING (QPF)

Quantitative Precipitation Forecasting plays an important role in hydrological forecasting in particular in flash flood forecasting. QPF shows both the areal distribution and the amount of the rainfall expected to occur over a given time period. The most important feature of QPFs is that they alert the forecaster to the potential for excessive rainfall. Three basic approaches have been tried to provide QPF values - empirical, statistical and physical methods.

The empirical methods are based on the climatology of precipitation or an extrapolation of existing precipitation areas and their movements (6.7). Statistical methods essentially make use of the same parameters employed in the empirical methods but utilize stepwise multiple regression techniques to develop the prediction equations (8). The physical methods are based on the dynamic equations of motion, the first law of thermodynamics and the laws of conservation of mass and water in its different phases (9).

The above references state that there is currently no operational capability for numerically generating QPFs. They give the same reasons for problems with time and space resolution and lack of a suitable convective parameterization technique.

HYDROLOGICAL FORECASTING SERVICES

The organization of a hydrological forecasting service is an internal matter in each country. The organizational pattern with respect to such service varies widely from country to country. The main conditions for efficient operations are:

(a) a well-developed network of hydrological and meteorological stations;

(b) facilities for rapid and reliable communication for collecting and distributing hydrological and meteorological information;

(c) well-documented meteorological and hydrological records with facilities for data processing, storage, and rapid retrieval;

(d) a sufficient number of specialists.

It is desirable for hydrological forecasters to work in close contact with meteorologists in order to have immediate access to their observation data, forecasts and advice. This may be achieved either by combining services or, where the services are separate, by establishing appropriate administrative and operational links.

CONCLUSIONS

The development of hydrological forecasting is influenced by meteorological factors. Prior knowledge of the meteorological conditions increase the scope and efficacy of hydrological forecasting, lengthens the validity of such forecasts and increases their accuracy and reliability. Meteorological data and information are required both for developing forecast procedures and for forecast operation. An adequate meteorological data and information network is a prerequisite for the development of hydrological forecasts. Quantitative precipitation forecasts and other meteorological forecasts (temperature, wind, snow and sky conditions) constitute an important and essential input to the sophisticated present and future procedures and methods of hydrological forecasting. Considering the present trend in making use of rainfall-runoff models to simulate the catchment response to precipitation inputs, any improvements in timeliness of hydrological forecasting for any but the very large rivers will hinge chiefly on progress made in rainfall forecasting. However, in spite of the compelling need, no reliable method of quantitative forecasting of rainfall and other weather elements far in advance have yet been developed. More recently new tools such as the laser and Doppler radar and satellite infrared sensors led to the development of numerical weather prediction based on a large-scale multi-level representation of the atmosphere. Thus one of the most urgent problems in the field of hydrological forecasting is the need for more research in the development of reliable methods of quantitative forecasting of rainfall and other weather elements far in advance to show their effective use as inputs to hydrological forecasting techniques.

Although the organization of a hydrological forecasting service is an internal matter of each country, maximum effectiveness, efficiency and economy in the field of hydrological forecasting is obtained only through the establishment of national institutional arrangements for effective co-operation and co-ordination between national hydrological and Meteorological Services where such services are separated.

REFERENCES

1. WMO, 1983: Guide to Hydrological Practices, WMO-No. 168.

2. WMO, 1981: Flash Flood Forecasting, WMO-No. 577.

3. Kalinin, G P, Kuvilova, Yu V, and Kolosov, P L, 1977: Komiceskie metody v gidrologii (Space methods in hydrology). Gidrometeoizodat, Leningrad, 182 pp.

4. WMO, Proceedings, 1979: Workshop on remote sensing of snow and soil moisture by nuclear techniques, Voss, Norway, 23-27 April.

5. Bobinski, E, Piwecki, T and Zelanzinski, J, 1975: A mathematical model for forecasting on flow in the Solar River. Bulletin IAHS XX. 1, pp. 51-60.

6. Maddox, R A, Chappell, C F and Hoxit, L R, 1979: Synoptic and mesoscale aspects of flash flood events. Bulletin AMS, Vol. 60, No 2.

7. Muller B M and Maddox, R A 1979: A climatological comparison of heavy precipitation and flash flooding. Preprints, 11th Conference on Severe Local Storms (Kansas City), AMS, Boston, 8 pp.

8. Belville, J, Johnson, G A and Ward, J D, 1978: A flash flood aid - the limited area QPF. Preprints, Conference on Flash Floods: Hydrometeorological Aspects (Los Angeles), AMS Boston, pp. 21-28.

9. Fritsch, T M and Chappell, C F, 1978: Numerical prediction of heavy convective rainfall from mesoscale systems. Preprints, Conference on Flash Floods: Hydrometeorological Aspects (Los Angeles), AMS, Boston, pp. 34-43.

32

Application of climatic data to estimation, exploitation and management of water resources

N. P. Smirnov and B. N. Malinin, USSR

Climatic information may be used to the full in studying large-scale hydrological processes for the description of which it is worth while employing the system of equations of water balance of the atmosphere (1) and of the land surface (2), i.e.

$$\delta W/\delta t + \mathrm{div}\vec{F} = E - P \qquad (1)$$

$$\delta S/\delta t + \mathrm{div}\vec{Q} = P - E \qquad (2)$$

where W is the atmospheric moisture content (precipitable water), $\vec{F}$ is vertically integrated moisture flow, E is total evaporation, P is precipitation, S is total surface and groundwater storage, Q surface and groundwater flow (stream runoff).

The left side of the equation (1) can be immediately calculated from the aerological data. For large regions ($A \geq 10^6$ km^2) when the atmospheric water balance in the form of (1) is fulfilled most identically [Malinin 1977, Rasmusson 1977], using aerological data in calculations makes it possible to get reliable estimation of the total moistening (P-E), total evaporation and total surface and groundwater storage which are not determined exactly by the traditional hydrological methods, the errors often not being amenable to quantitative assessment.

Table 1 may serve as an illustration of the aforesaid, presenting the annual variation of the water balance terms of the drainage basin of the Volga down to Kuibyshev (A = 1.2 x 10^6 km^2) which are calculated for the eight-year period (1957-1965).

Month	Q	div $\vec{F}$	δW/δt	P	E	δS/δt
January	12	-42	0	45	3	30
February	12	-38	0	42	44	26
March	13	-35	2	43	10	20
April	21	-15	4	42	31	-10
May	49	18	5	56	79	-72
June	19	26	5	69	100	-50
July	13	14	1	76	101	-38
August	11	3	-4	71	70	-10
September	11	-13	-5	61	43	7
October	11	-34	-4	58	20	-27
November	10	-34	-3	50	13	27
December	12	-34	-1	60	5	43
Total for a year	194	-194	0	673	479	0

Table I

Terms of water balance of the Volga basin
for the period of 1957-1964, in mm

Instead of (1) for the areas of 4 x 10^4 < A < 10^6 km^2 it is worth while employing the equation showing the relation of the water balance of the atmosphere and of the land surface [Malinin, 1984]

$$(E - P)/P = \psi(Eo/P) \qquad (3)$$

where Eo is evaporativity, ψ is a function depending on the geobotanical zone.

The given expression has three characteristic properties:

1) it reflects the inter-relationship of all the parameters influencing moistening;

2) the difference E-P simultaneously enters into the balance equation of the atmospheric and land branches of the

hydrologic cycle, thus connecting the processes of moisture transfer in the free atmosphere and those in the soil;

3) it is fulfilled only at sufficiently large space-time averaging ($A >= 4 \times 10^4$ km^2, period of averaging $\tau >=$ one month), when the evaporation is determined first of all by the climatic factors, the importance of the local (landscape) faction being negligible.

For the normal annual period the relationship equation (3) may be presented as

$$\frac{E-P}{P} = \frac{-Q}{P} = \begin{cases} -1 + C_1(E_0/P) & , E_0/P <= d \\ -1 + C_2(P/E_0)^{C_3} th(E_0/P), & E_0/P >= d \end{cases} \quad (4)$$

where C_1, C_2, C_3 are constant ($C_1 = 0.9$; $C_2 = 1.06$; $C_3 = 0.06$); th is the hyperbolic tangent; d is the parameter limiting the conditions of the excessive and sufficient moistening.

As seen in (4), at $E_0/P < d$ the annual evaporation linearly depends on the evaporativity. Under other conditions of moistening ($E_0/P > d$) evaporation shows a nonlinear dependence on the moistening coefficient. Note that equation (4) indirectly reflects the action of the soil moisture on the evaporation process because the soil moisture may be expressed nonlinearly as a function P/EO [Volobuev, 1983].

Applying the relationship equation to normal monthly conditions of moistening we get

$$\frac{E-P}{P} \begin{cases} -1 + a_1 E_0/P & , \quad 0 <= E_0/P <= \zeta_1 \\ a_2 + a_3 l E_0/P & , \quad \zeta_1 <= E_0/P <= \zeta_2 \\ a_4 + a_5 \exp(E_0/P) & , \quad \zeta_3 <= E_0/P <= \zeta_2 \\ -1 + a_6 E_0/P & , \quad 0 <= E_0/P <= \zeta_3 \end{cases} \quad (5)$$

where $a_1, \ldots\ldots, a_6$ are constants, ζ_1, ζ_2, ζ_3 are the moistening parameters with a clear physical meaning (Figure 1).

The numerical value of the parameter ζ_1 corresponds to the soil moisture equal to the minimum moisture capacity ($\zeta_1 = 1.22$), that is why in the interval $(0, \zeta_1)$ one observes excessive and sufficient moistening which results in the evaporation rate being completely determined by the meteorological conditions (moistening stage I).

Parameter ζ_2 corresponds to the conditions of the soil moisture reaching the moisture content of the capillary burst ($\zeta_2 = 3.76$). In consequence of this, interval (ζ_1, ζ_2) represents insufficient moistening which is characterized by the evaporation rate being determined here in the main by the ascent of water to the soil surface (moistening stage II).

The numerical value of the parameter ζ_3 corresponds to the soil moisture promoting restoration of the capillary links when suspended water acquires the ability to move to the vaporization zone ($\zeta_3 = 2.17$). Interval (ζ_3, ζ_2) represents the arid conditions of moistening in which the evaporation rate is determined by the speed of the water vapour molecular diffusion through the dry upper soil layer (moistening stage III).

In the interval (0, ζ_3) corresponding to moisture storage in the soil it is again the meteorological conditions that evaporation depends on (moistening stage IV).

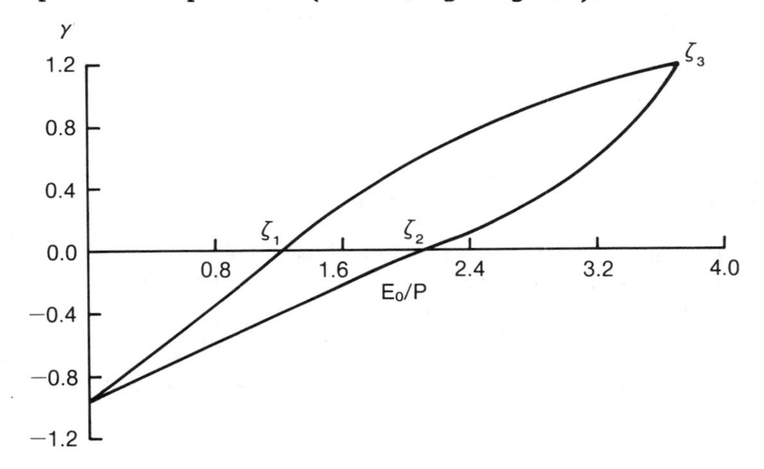

Figure 1 Dependence of relative moistening on evaporivity coefficient E_0/P for the steppe zone

Thus in the general case the annual cycle of the land surface moistening is a peculiar hysteresis loop with four points. Here the first two equations in (5) characterize the ascending branch of the hysteresis curve and the two others the descending one.

It is to be noted that seasonal variations of the land surface moistening in the forest zone and in the forest-steppe are completely determined by the ascending branch of the hysteresis curve. For coniferous and mixed forests $a_1 = 0.90$; $a_2 = 0.12$; $a_3 = 1.10$ and for deciduous forests and forest-steppe $a_1 = 0.85$; $a_2 = 0.16$; $a_3 = 1.05$. The difference in the numeric values of the above parameters is caused by the ground water level.

The phenomenon of hysteresis is characteristic only of the

steppe zone, and the transition from the ascending branch to the descending one is accomplished after the termination of the active transpiration of vegetative cover which makes up a considerable part of the total evaporation. Later on until snow cover is formed, the total evaporation may be considered in the main as evaporation of bare soil. The condition

$\sum^{m} (E_o/P) \geq 8$ may serve as a criterion of the transition to m, the arid conditions of moistening, where m is the number of months of the warm period of a year, April being the first month. The parameters of the descending branch of the hysteresis curve have the following values: $a_4 = -0.265$; $a_5 = 0.035$; $a_6 = 0.48$.

In (5) while passing from normal averaging to the real monthly period of time it is not worth while limiting the hysteresis loop by the parameter ζ_2.

This means that after crossing at point ζ_2 the branches of the hysteresis curve start to diverge, resulting in the values E_o/P becoming larger than ζ_2.

If one makes use, for instance, of M J Budyko's widely known complex method [Budyko, 1971] to evaluate monthly evaporation rates then it is not difficult to determine monthly and yearly evaporation rates and yearly runoff based on the climatological data (air temperature and humidity, cloudiness, precipitation) by the relationship equation. For example, the range of the spatial evaporation variations in the European part of the USSR is about 350 mm. The lowest evaporation values are registered in Nenets Autonomous Area (to the south from the Peckora Sea) - 265 mm per year, the highest ones in Krasnodar territory (to the east from the Black Sea) - 620 mm per year.

This agrees well with the analogous evaporation data obtained as a finite term of the water balance equation $E = P - Q$, where Q is determined from hydrometric fluctuations. The random error in calculating magnitude of the annual evaporation according to the relationship equation is 5-10%.

As for the "climatological" runoff, against the background of the generally zonal character of the runoff distribution one observes the non-zonal regions associated mostly with the influence of large-scale orographic elevations (e.g. the Urals, Carpathian Mountains, etc.) and also with maximum annual precipitation (e.g. the Baltic coast).

The data of the annual "climatological" and "hydrometric" runoff also agree well. The random error in calculating magnitude of the annual "climatological" runoff is 10-15%.

From the practical point of view it is deemed important to use climatological data in long-term water resources forecasting. Consider the large river forecast (taking the Volga as an example) based exclusively on the data of the standard aerological network [Malinin and Smirnov, 1982].

A basic hypothesis of the procedure is that the amount of water in a large basin stored by the beginning of the flood practically determines the runoff in the successive months until the beginning of a next flood. Hence proceeding from the priority of climatological factors we may write the principal prognostic equation in the following form

$$Q = f\left(\sum_{i=1}^{K} (P - E)_i\right) \quad (6)$$

where K is the number of the months preceding the beginning of the flood.

Thus the essence of the forecast procedure considered consists in calculating the difference "precipitation minus evaporation" for a number of years according to equation (1) and then, in determination of the dependence of the runoff on $\sum^{K} (P-E)_i$ on the statistical basis.

We may consider the period of December to March as that of storing moisture in the snow pack (Ssn) in the Volga basin down to the city of Kuibyshev, and that of September to November as the period of autumn moistening (Sa). As a result the total duration of moisture storage in the basin is assumed to equal seven months, i.e.

$$S_{sn} + S_a = \sum_{i=1}^{7} (P - E)_i$$

The basic calculation period ran to 11 years (1965-1975). The experimental estimation of forecast validity is accomplished by analysing the independent data for 1976-1979.

Thus considerable anomalies of summer moistening have the prognostically valuable information, i.e. they possess long-term hydrological memory. The amount of 30 mm has been taken as a criterion of summer moistening, its account becoming necessary to forecast runoff.

The anomaly of summer moistening was 78 mm in 1972 and in 1974, 31 mm; consequently by the beginning of the flood in 1973 and 1975 the total moisture storage in the basin was 232 mm and 206 mm respectively. It is not difficult to notice that application of these data makes it possible to reduce the difference between the computed and observed values of annual runoff (Table II). But some difference therein is retained and it can hardly be eliminated completely as the non-returnable evaporation losses in the basin are considerably higher for dry years as compared with wet ones.

Year	Annual runoff computed from (7) without summer moistening	Annual runoff computed from (7) with summer moistening	Annual runoff	Difference between columns II and IV	Difference between columns III and IV
1973	195	164	137	58	27
1975	166	154	130	36	24

Table II

Estimation of the role of anomalies of preceding summer moistening for computing the Volga runoff for dry years, in mm per year.

The assessment of the validity of equations (7) and (8) for prognostic purposes has been accomplished using the independent data for 1976-1979. The comparison of the computed and observed values of the runoff is presented in Table III. Note that the tolerated error of the forecast $\Delta\delta_{on}$ equal to 0.647σ amounts to 24 mm for the annual runoff and 17 mm for the spring one.

Year	Annual runoff		Spring natural runoff	
	Q com	Q obs	Q com	Q obs
1976	154	159	82	86
1977	159	149	88	105
1978	206	222	139	124
1979	231	258	167	143

Table III
Comparison of the Volga runoff at Kuibyshev computed by aerological data with the observed runoff.

Thus the results obtained testify to the principal possibility of the long-term annual and spring forecast of the Volga runoff on the basis of the equation of the atmospheric water balance. Figure 2 presents a diagram showing the relationship of values $S_{sn} + S_a$ with the annual (Q_y) and (Q_{spn}) runoff of the Volga at the city of Kuibyshev and the spring natural runoff taking into account the accumulation of moisture in the reservoirs of the Volga-Kama cascade. As easily seen out of the whole set of points on the diagram of the spring runoff there is only one isolated point characterizing the very dry year 1973.

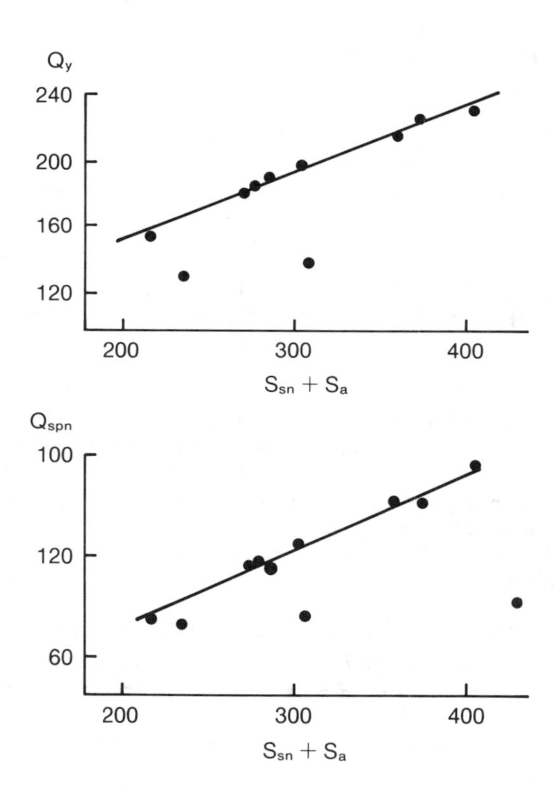

Figure 2 Diagram of relationship of the annual (above) and spring natural runoff (below) of the Volga at Kuibyshev with the magnitude of total accumulation of moisture in the basin by the beginning of a flood

Taking into account the essentially linear relationship between $S_{sn} + S_a$ and the runoff values it is sufficient to use the traditional regression equation. The corresponding prognostic dependences take the following form

$$Q_y = 0.39(S_{sn} + S_a) + 74 \quad (7)$$

$$Q_{spn} = 0.43(S_{sn} + S_a) - 6 \quad (8)$$

correlation coefficient of the annual runoff (excluding the years 1973) being $\tau = 0.95$ and that of the spring one (excluding the year 1973) being $\tau = 0.90$.

Consequently the total storage in the basin almost entirely determines the annual and spring runoff of the Volga.

The further refinement of prognostic dependences (7) and (8) concerns the more complete account of the conditions which influence the formation of runoff volume in dry years. So the main reason of the year 1973 being dry is the severe drought of 1972 over practically all the basin. That is why the part of the autumn moistening and winter deposits of moisture required to recharge the ground water used during the drought was considerably greater as compared to the normal annual conditions, consequently the spring runoff values proved to be much lower than the values which may be obtained from (8). The advantage of the proposed forecasting procedure is that it is based entirely on the data of the standard aerological network and does not require any special research during expeditions.

Note that this procedure can be applied not only to those large rivers that are mostly snowmelt-fed such as the Volga. In [Margues et al, 1980] attention is called to the possibility of forecasting the Amazon runoff on the basis of aerological information known three months in advance.

The examples considered above show a clearly evident but by no means exhaustive illustration of possible applications of meteorological information in hydrological design and forecasting. If one assumes monitoring and water resources forecasting to be one of the most important practical tasks for the immediate future then it is necessary to form automated water balance systems (AWBS) of various levels for its accomplishment: from local ones describing the hydrological cycle of the experimental areas to global ones describing the global hydrological cycle.

The role and position of meteorological information in these AWBS will naturally be various. As is known, the role of the climatic and local (landscape) factors in forming the terms of the water balance does not remain constant when the scales of space-time averaging change. The smaller the area the higher the contribution of the local factors (soil humidity and mechanical composition of the soil, type of vegetative cover, topography, etc.).

When the area increases, the local factors directing their opposing influence on forming the water balance get levelled down and when a certain size is reached their total effect becomes negligible as compared to the climatological factors.

In accord with this, the meteorological data are essentially the main source of information for the AWBS at interregional and global levels.

REFERENCES

Budyko M J, 1971 Climate and life (in Russian): Gidrometeoizdat Publishers, Leningrad, 470 pp.

Volobuev V R, 1983 The Relationship of soil humidity regime and climate of surface air layer (in Russian): Soil Study, No 3, pp 51-62.

Malinin V N, 1977 On assessment of accuracy of water vapour flux divergence computation (in Russian): Works of the Arctic and Antarctic Research Institute, v. 362, pp 40-49.

Malinin V N, 1984 On relationship of atmospheric water balances and underlying surface (in Russian): Geography and Natural Resources, No 4, pp 114-121.

Malinin V N, 1985 On interrelation of terms of atmospheric and land branches of hydrological cycle (in Russian): Works of the State Hydrological Institute, v 296, pp 55-82.

Malinin V N, Smirnov N P, 1982 On construction of prediction scheme of large river runoff on the basis of aerological data (in Russian): Hydrometeorological Support of National Economy, Leningrad Polytechnical Institute Publishers, pp 67-78.

Margues J, Salati E, Santos J, 1980 A divergencia do compodo fluxo de vapor d'aquae as chuvas na regiao Amazonica: Acta ama on, v 10, No 1, 133-140.

Rasmusson E M, 1977 Hydrological application of atmospheric vapour-flux analyses: WMO, No 476, Geneva, 50 p.

33

The application of meteorological information to water services and river basin management

P. D. Walsh, G. A. Noonan and J. M. Knowles, North West Water, Warrington, United Kingdom

INTRODUCTION

The North West Water Authority (NWWA) is one of ten regional Water Authorities in England and Wales whose responsibilities cover the general management of all aspects of the water cycle including water resource development, water supply, sewerage, sewage disposal, river management and flood warning. The Authority covers an area of almost 14,500 sq km and serves some 7 million customers including the main conurbations of Manchester and Liverpool.

Meteorological and hydrological products both in real time and from historic archives are essential ingredients of the information needs of the Authority's business.

At the strategic and investment level where planning and protection of the water environment require effective use of scarce capital investment, knowledge of meteorological variability and extremes, together with assessment of climatological trends, is needed for the planning process.

At the operational level, providing a 24 hour service of water supply, effluent disposal, pollution control, land drainage, flood forecasting and fisheries requires a wide range of current information and forecast products.

Close co-operation with the Meteorological Office ensures

that these and other requirements are effectively and efficiently met and lead to benefits for both organisations.

WATER RESOURCES IN NORTH WEST

The water supplies to the main cities have been developed independently with Manchester deriving its main supplies from the Lake District, in the north of the region, and Liverpool developing supplies from direct gravity and river regulation sources in North and Mid-Wales, (which are located outside the Authority area). The surface water supplied from these sources constitutes over 50% of the total treated water supplied in the region. The Lancashire Conjunctive Use system supplies the central part of the region by the conjunctive use of rivers, two upland reservoirs and boreholes. Figure 1 illustrates the major water supply zones, reservoirs and aqueducts in the NWWA region. The aqueduct systems developed to carry the water to the cities, in particular the Lake District-Manchester aqueducts, form the basis of a water grid which provides supply flexibility. The rest of the supplies are from local surface reservoirs (of which there are over 150) or from groundwater.

The hydrological philosophies developed in the planning and operation of water resource systems in North West Water have been developed over the last 20 years. Most systems are being operated using control (or "rule") curves derived by assessing minimum available runoff volumes from historical records and checked in simulation models. Although the details of the methods may vary, the fundamental approach is based on critical period techniques using non-sequential mass curves of cumulative minimum runoff values, (McMahon and Mein, 1978), though these are sometimes expressed in probability terms.

There is currently a regional surplus of almost 400 Mld (17%) of supply above the demand of about 2600 Mld. This provides the opportunity to utilise water from sources which are cheaper to operate (generally reservoirs) in preference to the more expensive pumped sources, without any reduction in the overall reliability of the system.

Yield Analysis

A technique of non-dimensional presentation of the yield-storage relationship for direct supply reservoirs is presented by Twort et al (1974). The draw to supply the compensation discharge to the natural river are lumped together as the "gross yield" and expressed as a percentage of the average daily inflow (%ADF) and the usable storage as a percentage of average annual inflow (%AAF). ADF and AAF are derived from the Long Term (1941-70) average rainfall less evapotranspiration losses over the catchment area. With this

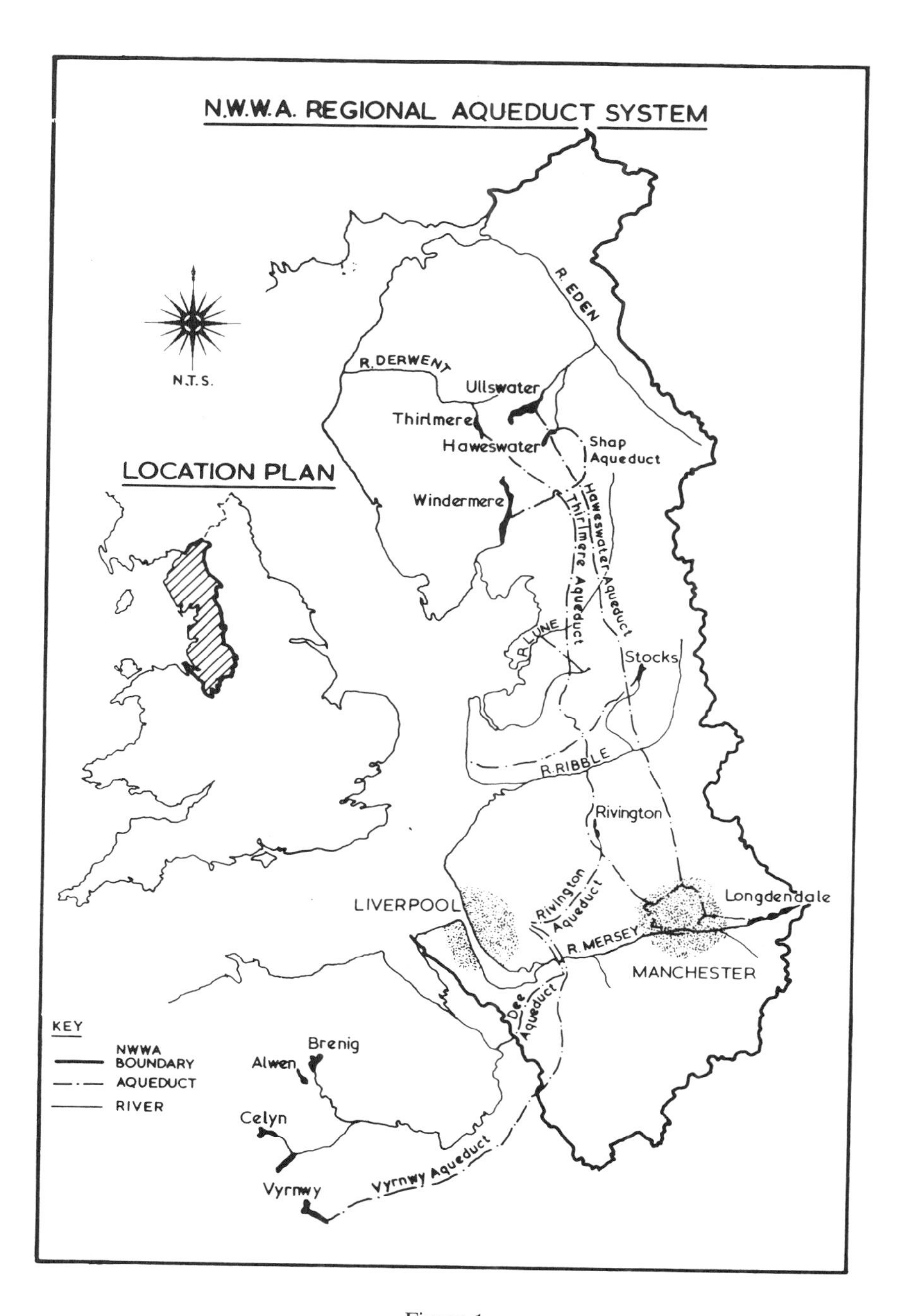
N.W.W.A. REGIONAL AQUEDUCT SYSTEM
N.T.S.
LOCATION PLAN
R. EDEN
R. DERWENT
Ullswater
Thirlmere
Haweswater
Shap Aqueduct
Windermere
Haweswater Aqueduct
Thirlmere Aqueduct
R. LUNE
Stocks
R. RIBBLE
Rivington
Rivington Aqueduct
LIVERPOOL
Longdendale
R. MERSEY
MANCHESTER
Dee Aqueduct
KEY
NWWA BOUNDARY
AQUEDUCT
RIVER
Brenig
Alwen
Celyn
Vyrnwy
Vyrnwy Aqueduct

Figure 1

type of presentation it is then possible to scale the yield-storage graph to another site, assuming that there is consistency in the variability of runoff between sites. This has been demonstrated for the NWWA region by the comparison of cumulative mass curves for various major reservoir inflow records using measured runoff and those based solely on meteorological data (Pearson, 1983).

Derivation of Control Curves

At sites where adequate records of flow data are available the critical period approach of analysing the system inflow sequence is used and a suite of computer programs (Pearson & Walsh, 1981), enable the analysis to be carried out with little human effort. At direct supply reservoirs where the run-off data are inadequate to derive an inflow sequence a different approach is used. To overcome this paucity of data and assist in the rapid derivation of control curves and assessment of reservoir performance a non-dimensional, regional method has been developed (Pearson and Walsh, 1982) using the ADF as the basis for scaling the results of gauged site analyses to sites with no flow data. The strong dependence of these methods on long rainfall records demonstrates the importance of data from meteorological sources.

Reservoir Productivity

It is often useful to estimate the average availability of water from a reservoir taking into account the opportunity to overdraw. Reservoir productivity is the sum of compensation and average supply, expressed as %ADF. The productivity will increase with increasing drawoff capacity and increasing level of development, but with diminishing returns towards a limit of 100% ADF. For non-dimensional presentation the drawoff capacity, in a similar way to gross yield, includes the compensation (Pearson & Walsh, 1982), and again allows effective use of rainfall data as a substitute for direct measurement of runoff.

Operational Management

Analysis of rainfall on a current basis, month by month, and comparison of cumulative rainfall patterns with long term averages provides a guide to current trends of weather over relatively short periods. Operational measures can then be taken when necessary. Drier than average trends can indicate the progression of potential droughts. Assessments are made of how severe the event has been so far compared with previous events in the historic record. These comparisons are usually based on rainfall data in the first instance, due to the longer length of rainfall sequences.

Once system storage levels fall significantly below the control curve, analyses based on current reservoir contents and fixed starting date runoff patterns are used to assess the risks to supply at the current rate of draw from the source. The analysis would also show the levels of supply which the system could support under various probabilities of runoff patterns e.g. minimum historic, 1 in 20 year (5%) 1 in 50 year (2%) etc. This would indicate the severity of water conservation measures needed.

Assessments are also made of the likelihood of refill of sources under different rainfall patterns. Three patterns extending over 6 winter months are used, (i) average rainfall, (ii) 75% average rainfall - this equates to a runoff pattern with a return period of 1 in 5 (20%) and (iii) 62% average rainfall - equating to a runoff pattern with a return period of between 1 year in 20 (5%) and 1 year in 50 (2%). Once there is an 80% chance of the source refilling during the winter the source would again be fully utilised and overdrawn if necessary.

Operation of supply systems depend on control rules to allow maximum use of cheaper water without jeopardising the reliability of the source. Where a number of different sources can be called upon, e.g. rivers or reservoirs, and where different treatment processes may be needed, decisions tend to be made weekly or possibly more frequently. Here timely short and medium range weather forecasts can be of value especially where river sources are used and may be nearing threshold or prescribed flow limits or requiring regulating releases. Delaying changes from cheaper sources, helps to minimise power and treatment costs.

Similar short term or even monthly forecasts are valuable in the timing of particular stages of site work in reconstruction of overflow weirs at impounding reservoirs.

Recently NWW has embarked upon a series of investigations to develop more optimum operating policies using dynamic programming techniques. Initial results show promise of savings in operating costs of 6% (Walker and Wyatt, 1987).

Long Term Changes

Two factors which may affect the long term planning of water supply systems are afforestation and climate change. It is only by collecting meteorological and hydrological data over long periods that effects can be quantified. Afforestation of water supply catchments has been shown (Calder et al, 1978) to reduce the catchment runoff by increasing evapotranspiration through increased interception losses.

Any changes whether in variability or in overall magnitude of rainfall are important to the water industry. On short timescales rainfall intensity is particularly important for sewers, and flooding, whereas changes over months have implications for water resources systems. Over geological timescales the earth's climate shows long term variations but the normal 'design' assumption is that recent decades are typical and can form the basis for sizing future developments. Current beliefs (UNEP/WMO/ICSU, 1985) are that some warming of the climate now appears inevitable due to past increases in concentrations of CO_2 and trance gases in the atmosphere ("Greenhouse Effect"). It is not clear what the consequence of this warming would be, but even if it is no more than increased variability without change in annual average precipitation, the consequences will be significant. A recent report (Wigley & Jones, 1986) shows no evidence so far for long term trends in precipitation mean in England and Wales but indicates that in the North West, in particular, rainfall may be becoming more extreme, with a higher frequency of wet springs and dry summers. Continuing research is required into the causes, consequences and, so far as the water industry is concerned, of the impacts of any climate change.

DESIGN FLOW ESTIMATION FOR RIVER WORKS AND RESERVOIR SAFETY

A variety of engineering works relating to dam spillways, bridges and flood protection/alleviation works require flow estimates for their design and economic appraisal.

In river channel improvement works, assessment and design are related to coping with flood occurrences of between 1 in 10 year and 1 in 50 years. For reservoir safety works extremely rare flood events require estimation.

With a variety of methods available, a general paucity of requisite data, and estimates required at sites without records a major study was commissioned by Natural Environmental Research Council (NERC, 1975) to provide a consistent approach to flood estimation. The Flood Studies Report was developed with major inputs from the Meteorological Office by the Institute of Hydrology, and made the most of all available records of rainfall and flow.

The Report provides generalised methods based on two approaches, firstly the statistical analysis of rainfall and peak discharges and secondly hydrograph synthesis and routing. These general methods can be applied throughout the country using regional factors and catchment characteristics. This approach is now being applied by the Institute of Hydrology in many other parts of the world (Farquharson, Green, Meigh and Sutcliffe, 1986).

FLOOD WARNING

By definition, flooding can take place on both river and tidal flood plains. Even though engineering works might be carried out to alleviate such flooding there are economic and other constraints which limit the protection given by such works. Thus, in practice, it is only possible to alleviate flooding, it can never be eliminated. It follows that flood warning schemes (Noonan, IWEM) are desirable for any area where flooding has occurred in the past. Where there is danger to life or to property owners, shopkeepers, industry and public services can take steps to reduce damage that would be caused by flooding and produce considerable economic benefits.

Flooding can occur from both heavy rainfall and from high tidal surges and waves or, in estuaries, from a combination of these phenomena. In the last decade the improvements in data gathering systems and in computer processing ability has enabled flood warning schemes to be considered and implemented for many flood risk zones which, otherwise would not have been possible.

Meteorological information is used in predicting flooding from both heavy rainfall and from sea conditions. Initially, historical data is used in examining the relationships between meteorological parameters and river and sea conditions so that appropriate models can be designed for use in real time. Secondly, real time and forecast meteorological data is used operationally in the models to give forecasts of possible flooding. In addition, in order to keep fully appraised of the weather position, North West Water duty officers talk to Meteorological Office duty forecasters at least daily and more frequently during actual or potential flooding periods.

Flooding from Rivers

Flooding from rivers is, in the UK, most usually caused by rainfall although snow melt can be important. For long, slow response rivers the classical way of producing a flood forecast for a flood risk zone is to observe an upstream measuring station and correlate the data with levels at the flood risk zone some hours later. This method is usually very satisfactory so long as the time of travel between the upstream station and the flood risk zone is sufficient to give adequate warnings. In many flood risk zones especially on rivers with major tributaries this is not the case and rainfall/run-off models also have to be used. The choice of models for operational flood warning purposes in NW England falls mainly into the first of the following categories:

1. Simple models that can operate continuously and

automatically which would produce high river level forecasts for a specific, relatively short time ahead (for example ISO of Peak Correlation).

2. More sophisticated models requiring greater computer capacity which could, with suitable quantitative forecasts of rainfall, produce forecasts with much longer lead times (for example TFN, ARMA, Muskingum-Cunge, FLOUT, Conceptual).

3. Highly sophisticated conceptual models such as SSAR, Sacramento model etc. which are suitable for very large catchments, and where it is required to forecast the full range of river flows.

Rainfall information for use in real time in the models can be from raingauges and/or from weather radar. Transmission of this information can be by radio, telephone or by using a satellite link.

North West Weather Radar

The first unmanned weather radar installation in the UK specifically designed to monitor rainfall intensity was commissioned in NW England in 1980. It was the prototype used by North West Water and the Meteorological Office for pioneering work in the development of unmanned weather radar installations and their use for flood warning. An additional 13 radars may eventually be built by the Meteorological Office in conjunction with other Water Authorities, to form the UK National Weather Radar Network.

The accuracy of the radar estimate of rainfall was investigated during the project and it was clear that the use of telemetering raingauges to calibrate the radar in real-time significantly improved the accuracy within 75 km of the radar sites on most occasions (Collier, 1986). Studies in the project examined the consequences of any errors in radar rainfall estimates on flow predictions (Collier and Knowles, 1986). Forecasting models that can absorb any occasional inaccuracies in radar estimates through self-correcting techniques, have been developed (Cluckie and Owens, 1987).

The results of the radar development and its incorporation and use in North West Water's flood warning system were discussed at a conference on Weather Radar and Flood Forecasting at Lancaster University in 1985 and published in book form (Kirby and Collinge, 1987).

In NWWA area rainfall patterns and intensities are fed into a computer every 15 minutes which, in conjunction with data from river level gauges, gives forecasts of appropriate river levels for some hours ahead. Duty Officers are on call 24 hours a day, 365 days a year and use this information to alert people, where possible, when there is a danger of flooding.

Rainfall Forecasts

A major step in flow forecasting ability will come when the Meteorological Office introduces, in the not too distant future, their 'FRONTIERS' system of short-period quantitative rainfall forecasting (Browning, 1979). In the FRONTIERS system, rainfall movements as observed by radar are automatically extrapolated by a computer to give forecasts up to six hours ahead. The forecasts are displayed as a series of pictures. An experienced weather forecaster can readily superimpose satellite and other synoptic data onto the picture and use this additional information to adjust, interactively, the automatic forecast. This procedure is likely to be carried out at half-hour intervals and the results transmitted directly to water authority computers for use in models for forecasting flooding. FRONTIERS will make it possible to produce, for the first time, flood forecasts for very fast response catchment areas and, where necessary, improve lead times of existing flood forecasting procedures.

Tidal/Sea State Forecasts

Coastal floods result from high sea levels caused by a combination of astronomical tides and meteorological conditions in which variations of atmospheric pressure and surface winds generate surges and waves. Over the past 30 years or so in the UK the emphasis has been placed on the prediction of surges but it is now increasingly apparent that there is the additional need to forecast wave conditions. The prediction of sea state is effectively in three parts:

(i) The astronomical tide which is the major component of tides. It is independent of meteorological factors and is calculated by the analysis of historical tidal records.

(ii) The surge residual value is obtained by subtracting the predicted astronomical tide from the observed value; it can be positive of negative. In order to predict surge, correlations are made with meteorological parameters such as windspeed and atmospheric pressure, so that an appropriate model can be prepared.

(iii) Waves are caused by wind blowing over the surface of the sea. In addition to locally generated wind waves, there is also 'swell', caused by waves generated from a distant storm.

Numerical sea models developed by the Proudman Oceanographic Laboratory, Bidston (formerly the Institute of Oceanographic Sciences) have been used at the Meteorological Office since 1978 for routine twice daily predictions of storm surges on the coasts of Britain up to 30 hours ahead. These are passed to Water Authorities and others for action. In addition, the Meteorological Office also runs a coarse mesh model twice daily which covers the North Atlantic and provides information on deep water storm waves and swell predictions for up to 36 hours. Offshore waves, in themselves, are of no value in flood forecasting since they are substantially changed before they reach the shoreline. A convenient method of translating offshore waves to inshore waves is to produce mathematical models which take into consideration near-shore topography, depth of the sea on the shore and other relevant parameters for specific flood risk areas on the coast line. The models are run off-line anticipating a wide range of offshore wave conditions, and the results presented in graph or tabular form.

Dissemination of Forecasts

When flood forecasts have been prepared these must be transmitted to the communities that are likely to be affected. In the UK the Police have a duty to pass on such flood warnings. It is of the utmost importance that good communications exist between Water Authorities, Police and Local Authorities so that pre-prepared warning plans can be implemented very quickly after floods are predicted. It is often much easier to reduce the length of time it takes to pass a warning to the public than it is to gain an equivalent improvement in lead time in a flood forecasting model. All organisations involved in communicating flood warnings should know exactly the role that they have to play so that delays in passing messages are kept to a minimum, allowing prompt action to reduce damage as far as possible.

There are two stages in flood warning; the production of the forecast and the dissemination of the warning. Either one is useless without the other. The meteorological input to the first stage is a key element.

SEWERAGE

Over £60 million of the Authority's capital budget is directed toward the improvement of the hydraulic and structural performance of the sewer network in the Region. Procedures outlined in the Sewerage Rehabilitation Manual (Water Research Centre, 1987) have been adopted to ensure this investment is cost-effective.

Currently there is a shift away from localised studies to

production of catchment area plans which allow co-ordinated consideration of performance over the whole sewer drainage area and its consequent impact on receiving rivers.

Fundamental to this approach is the production of a hydraulic model of the sewer system. WASSP, a program in the Wallingford Procedure, (National Water Council, 1981) used in its simulation mode, is usually used. The model allows system performance to be compared with performance criteria and allows identification of reasons why the current system does not meet these. The model also provides a basis for assessing the effect of rehabilitation works. Recently, the use of the model has been extended to cover the performance of the sewer system in relation to water quality criteria.

The model is initially verified using measured rainfall and observed level and flows at points in the sewer system. When modelling observed storms it is necessary to use soil moisture deficit information for the period preceding the storm. This data is obtained from the Meteorological Office.

Two types of rainfall data are significant in the use of WASSP. Firstly high intensity relatively short duration storms are used to assess performance in relation to peak levels and discharges. Secondly, a sequence of rainfall events which have been found to be statistically representative of the long term rainfall pattern is used to assess overpumping requirements, pollutant volumes or to investigate the performance of storm sewage overflows. This latter type of rainfall information is in the form of a Time Series data set (Henderson, 1986).

The relationship between rainfall intensity, duration and return period established for the Flood Studies Report, have been incorporated into WASSP. In design mode, using regional factors, site specific values of rainfall intensity for a particular storm duration and intensity can be derived. These can then be used in WASSP to assess the performance of the sewer under different magnitude of design storms. Similar regional factors are used to apply the generalised sequences of Time Series Rainfall to the particular catchments of interest.

FUTURE INFORMATION DEVELOPMENTS

The requirements of a Water Authority for hydrometeorological information embraces the whole water cycle. Recent developments in Information Technology provide opportunities to make data more widely available to meet information needs which could benefit all aspects of water management.

North West Water has embarked upon a multi-million pound

strategy for communications and telemetry, into which the current flood warning telemetry system will be incorporated. This will allow the development of an integrated system from 'sensor' to 'archive', embracing the need for hydrometeorological data to service all functions, in an immediate mode through links to the Meteorological Office and for planning at a later date.

Almost certainly there will be more rationalisation of existing hydrometric data networks over the next few years, exploiting new technology to make them more efficient and the data more widely available. Data use will be dictated by the information needs of functions within the water industry, which comprise operational processes operating in the natural environment. There will always remain the necessity to collect accurate data in real time and to collate it into long term information, so that our understanding of natural processes and our ability to detect changes in them is not weakened.

It is envisaged that single purpose databases will cease to exist but that they will be integrated onto a common database to which all functions will have access by means of fast retrieval techniques. As these techniques develop increasing emphasis will have to be placed on the best means of presenting data to the user for his individual usage.

REFERENCES

Browning, K A, 1979 The FRONTIERS PLAN: A strategy for using Radar and Satellite Imagery for very-short-range precipitation forecasting. Met Mag 108, 161-184.

Calder, I R, Newson, M D and Walsh P D, 1982 The Application of Catchment, Lysimeter and Hydrometeorological studies of Coniferous Afforestation in Britain to Land Use Planning and Water Management. Proceedings or the International Symposium on Hydrological Research Basins and their use in Water Resources Planning, Berne.

Cluckie, I D, and Owens, M D 1987 Real-time Rainfall Runoff Models and Use of Weather Radar Information. In: Weather Radar and Flood Forecasting, ed. Kirby and Collinge, Wiley and Sons.

Collier, C G 1986 Accuracy of Rainfall Estimates by Radar Part 1: Calibration by Telemetry Raingauges. Part 2: Comparison with Raingauge Network. Journal of Hydrology, 83.

Collier, C G and Knowles, J M 1986 Accuracy of Rainfall Estimates by Radar Part 3: Applications for Short Term Forecasting. Journal of Hydrology, 83 : 237-249.

Farquharson, F A K, Green, C S, Meigh, J R and Sutcliffe, J V 1986 An analysis and Comparison of Flood Frequency Curves from many Different Regions and Countries of the World. Int. Symp. on Flood Frequency and Risk Analysis, Louisiana, USA May 1986 (Also available from Institute of Hydrology, Wallingford, UK).

Henderson, R J, 1986 Rainfall Time Series for Sewer System Modelling. WRc Engineering, Swindon, UK.

IWEM (Institute of Water and Environmental Management) 1987 Water Practice Manual No 7. "River Engineering - Part 1, Design Principles", Chapter 4.

Kirby C and Collinge, V R 1987 Weather Radar and Flood Forecasting. J Wiley and Sons Ltd.

McMahon, T A and Mein, R G 1978 Reservoir Capacity and Yield. In: Developments in Water Science, No. 9, Elsevier, Amsterdam.

National Water Council, 1981 Design and Analysis of Urban Storm Drainage, The Wallingford Procedure, Volumes 1 to 4. NWC London, (now published by Hydraulics Research, Wallingford, UK).

NERC, 1975 Flood Studies Report. Natural Environmental Research Council, UK.

Noonan, G A, 1986 An Operational Flood Warning System. Journal of IWES 40: 437-453.

Pearson, D and Walsh, P D, 1981 The Implementation and Application of a Suite for the simulation of complex water resource systems in evaluation and planning studies. In: Logistics and Benefits of using Mathematical Models of Hydrologic and Water Resource Systems (Proc. Pisa Symp., Oct 1978). IIASA Proc. Vol 13.

Pearson, D and Walsh, P D, 1982 The Derivation and Use of Control Curves for the Regional Allocation of Water Resources. In: Optimal Allocation of Water Resources (Proceedings of the Exeter Symposium, July 1982) ed. Lowing M J, IAHS Pub No 135.

Pearson, D 1983 Control Curve Report. NWWA, UK.

Twort, A C, Hoather, R C and Law, F M, 1974 Water Supply.

2nd Edition, Edward Arnold, London.

UNEP/WMO/ICSU, 1985 Conference statement on: International assessment of the role of Carbon Dioxide and of other Greenhouse Gases in Climate Variations and Associated Impacts. Sponsored by UNEP/WMO/ICSU at Villach, Austria, October 1985.

Walker, S and Wyatt T, 1987 The use of a Dynamic Programming Technique In Optimising the Operation of a Regional Water Resource System (UK). In press.

Walsh, P D and Lewis, A M, 1987 The Potential of Radar and Automated Data Capture in Information Systems for Water Management. In: Weather Radar and Flood Forecasting, ed. Kirby and Collinge, Wiley and Sons.

Water Research Centre, 1986 Sewerage Rehabilitation Manual, 2nd Edition. WRc, Swindon, UK.

Wigley, T M L and Jones P E, 1987 England and Wales Precipitation: A discussion of Recent changes in Variability and an Update to 1985. Journal of Climatology 7, (3).

34

Use of meteorological data and information in hydrological forecasting

J. R. Moore, Institute of Hydrology, Crowmarsh Gifford, Wallingford, England

INTRODUCTION

The use of meteorological data and information in hydrological forecasting will be illustrated through two applications: short-term flow forecasting for flood warning and long-term water resource forecasting for drought management. First, the complementary use of weather radar and telemetered raingauge data for short-term flood forecasting is considered with emphasis on possible future developments. The need for improved weather radar calibration procedures, algorithms for the detection and correction of anomalous radar values, and the development of flood forecasting models capable of exploiting distributed grid-square weather radar data is discussed in the context of providing more robust real-time flood forecasts.

Second, the use of meteorological data and information for forecasting in the longer term (say, from one to nine months into the future) is illustrated with reference to an operational system for managing pumped storage reservoirs supplying water to the London area during a drought. Meteorological information in the form of the UK Meteorological Office's categorical probability forecast of the forthcoming month's rainfall is employed together with long-term rainfall records as input to a novel procedure for assessing the reliability of the reservoir system to supply water at various times in the future. The procedure forms part of a decision support system employed operationally by Thames Water for drought management.

These two forecasting applications serve to highlight the differing data needs of short-term and long-term hydrological forecasting.

FLOOD FORECASTING AND WEATHER RADAR DATA

Research on improving methods for short-term forecasting of flow for flood warning purposes has traditionally concentrated on the development of improved models of the rainfall-runoff and channel flow routing processes. Over the last decade attention has turned to developing updating techniques which exploit the ability to use real-time data on flow to correct the model flow forecast. Techniques based on error prediction, parameter adjustment and state updating have been reviewed by Reed (1984) and Moore (1986). Current research in the UK, however, is more concerned with the recognition that forecast performance is data limited and that the use of weather radar, and the spatial coverage of rainfall data it provides, has yet to be fully exploited for flood forecasting purposes. An impetus to this research has been the formation of a NERC Steering Committee on Hydrological Applications of Weather Radar with the aim of fostering research into the improved use of weather radar for hydrological application, including flood forecasting. Research under forecasting applications is being directed on two broad fronts, one concerned with improvement of the radar product itself through research on improved calibration using additional local raingauges, and the other with the development of models capable of exploiting the grid-square rainfall data that weather radar provides. The former will be considered first.

Calibration of weather radar

Development of calibration procedures by the UK Meteorological Office has been based on the premise that a radar network supported by a sparse set of calibrating raingauges would be sufficient to provide quantitative estimates of rainfall over a large area. Specifically Collier et al (1983) state that "Methods using a large number of raingauges are likely to be too expensive for real-time calibration as the raingauges would have to be of the telemetry type" and go on to "point out how real-time radar estimates of surface rainfall might be improved using data from only a small number of telemetry raingauges". Radars which form the UK network typically use five calibrating raingauges and are judged to provide quantitativestimates of rainfall out to a range of 75 km, while providing only estimates of qualitative value beyond this up to a maximum theoretical range of 200 km. The original premise was no doubt influenced by the potential to replace telemetering raingauge networks by a sparse network of radars and calibrating raingauges. In practice in the UK, water authorities have been reluctant to dispense with established networks of telemetering raingauges, partly on account of existing investment in such systems but also because of a growing awareness of what conventional weather radar systems can and cannot do. Having now reached a stage where both systems are commonly available for water authority use, the logical next step is to exploit the dual systems by developing local calibration procedures. The potential exists to supplement the existing five calibrating gauges per radar with, for example, 50 or more telemetering raingauges commonly available within a water authority region, to obtain a more precise and robust estimate of the rainfall field.

The existing real-time calibration system is used by the UK Meteorological Office is presented in detail for the Hambledon Hill radar in Collier et al (1983). Error ratios of raingauge to radar grid square values where the two are coincident are used to factor neighbouring grid-square radar values. The neighbourhood is defined by calibration domain maps which specify the area over which a particular raingauge calibration factor applies. Different domain maps are used for different synoptic conditions and the procedure is automated through a synoptic type identification algorithm based on the temporal variability of the calibration factor. To reduce the effects of sampling variability of the radar data the calibration procedure is applied to data aggregated in space and time from the basic 2 km 5 minute resolution to 4 km 1 hour. The raingauge value is regarded as truth despite sampling rainfall at a point rather than over the 16 km^2 area of the coincident radar grid square. As well as causing problems when rainfall fields exhibit large spatial variability the procedure also introduces a delay of one hour before calibration is applied.

Current research in the UK aims to improve upon this calibration procedure by exploiting the space-time correlation structure of the square lattice data that weather radar provides together with point gauge estimates of rainfall available from the UK water authorities' networks of telemetering raingauges. An example of the spatial correlation present in the radar field is depicted in Figure 1. This shows the spatial correlation function (Ripley, 1981, p 79) obtained from 2 km 5 minute data from Chenies radar in London for the rainfall field caused by Hurricane Charlie. Note the strong dependence at low spatial lags (a lag of one corresponds to 2 km). Such short range dependence, together with similar dependence in time, can be incorporated in a space-time model of the rainfall field. For example, if $x_t^{(i,j)}$ denotes a mean adjusted rainfall for grid square (i,j) at time t then this may be related to rainfall in neighbouring

(a) 2km grid-square radar data field
(maximum 5 minute rainfall intensity is 32.3 mm/hr)

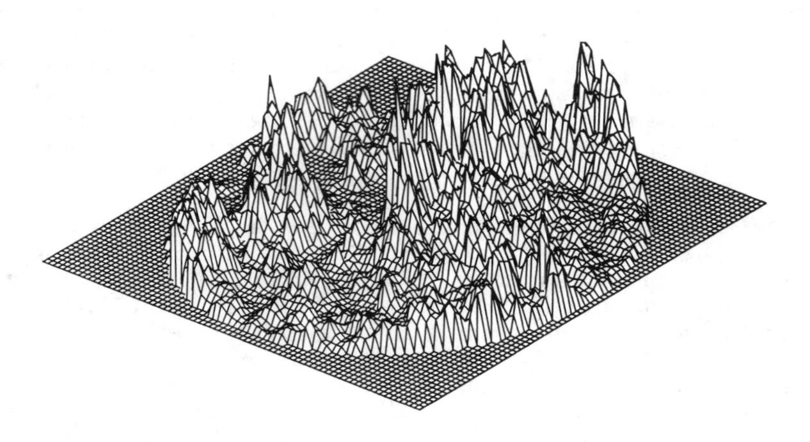

(b) Spatial correlation function

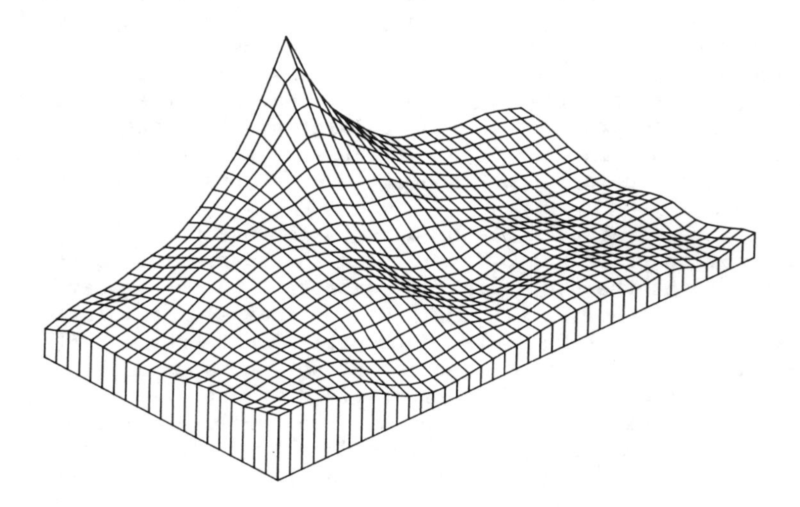

Figure 1

Spatial correlation function obtained using the 2 km grid-square radar data field available to a range of 76 km from Chenies radar. The storm concerned is Hurricane Charlie and the data is for a time resolution of 5 minutes at 2000 hr 25 August 1986. Left to right is for lags 0 to 20 in a west to east direction. Front to back is for lags -20 to 20 in a south to north direction. Since the function is symmetric only the right half-plane is presented.

grid squares through a linear dependence function

$$x_t^{(i,j)} = \sum_{\tau \in T} \sum_{l,k \in S} a_\tau^{lk} \; x_{t-\tau}^{(i+l,j+k)} + \;\; {}_t^{(i,j)} \qquad (1)$$

where ${}_t^{(i,j)}$ denotes model error. The spatial summation is over the neighbourhood S defined through a finite set of positive, zero and negative spatial lags (l,k) and the temporal summation ranges over a finite set of non-negative lags denoted by T. The parameter set $\{a_\tau^{lk}\}$ may be constrained to have certain symmetry properties, for example, $a_0^{l,k} = a_0^{l,-k}$. Rainfall $x_t^{(i,j)}$ may be related to measurements of rainfall provided by weather radar, $y_{1t}^{(i,j)}$, and by a raingauge measurement $y_{2t}^{(i,j)}$ if present in the (i,j)th square, so that

$$\begin{aligned} y_{1t}^{(i,j)} &= x_t^{(i,j)} + \epsilon_{1t}^{(i,j)} \\ y_{2t}^{(i,j)} &= x_t^{(i,j)} + \epsilon_{2t}^{(i,j)} \end{aligned} \qquad (2)$$

where $\epsilon_{1t}^{(i,j)}$ and $\epsilon_{2t}^{(i,j)}$ denote errors of radar and raingauge measurements respectively. Extension of the model to include spatial and temporal differencing (Ord and Rees, 1979) may prove appropriate.

The above model could form the basis of a calibration procedure, for example by obtaining the conditional mean of $x_t^{(i,j)}$ given a set of neighbouring (in time and space) radar and raingauge measurements, y_1 and y_2. It might also form the basis of a rainfall forecasting procedure. However, it is not straightforward to specify the covariance structure of the errors and of the true rainfall, even for basic, simple forms of the model, and practical implementation is problematic. The spatial and temporal correlation structure will differ with synoptic type, for example exhibiting stronger dependence in space and time at higher lags for frontal storms than for convective storms. This might require developing different linear dependence functions for different synoptic types. Also covariate information might need to be introduced. For example, in regions of orographic influence contour data might be used to define an orographic enhancement field.

Other possibilities for adjusting the radar data field to accord with telemetered data might be considered as being more straightforward. A surface could be fitted to raingauge to radar ratios (or to the logarithm of the ratios) according to a criterion chosen to achieve close correspondence between ratio and surface interpolation values whilst preserving the smoothness of the surface. The surface could be adjusted in real-time. Pattern recognition techniques could be used to identify domains within which the error ratio field is judged homogeneous. An average calibration factor used to adjust the domain's radar values could be calculated as the geometric mean of the error ratios for gauges within the domain in order to avoid bias introduced by the skewed distribution of error ratio values. Techniques of optimal linear interpolation (Gandin, 1965) and Kriging (Matheron, 1971) also have a role to play. Cokriging could be used as a formal way of introducing height into a scheme for spatial interpolation of calibration factors. Scope for further research to develop improved calibration procedures is clearly very wide, but must be sensibly restricted to algorithms simple enough to be implemented in real-time within a small computer.

A problem closely associated with calibration is the detection of anomalous values in the radar field. If such values are undetected and used with a rainfall-runoff model for flood warning then false warnings may be initiated and distrust of the forecasting system may ensue. A major cause of anomalous radar grid values is 'bright band' where the presence of melting snow in the radar beam causes a large electromagnetic return signal falsely indicating high intensity rainfall. Recently an automatic bright band detection algorithm has been developed by the UK Meteorological Office (Smith, 1986) based on using data from a number of radar beam elevations to estimate the height and intensity of the bright band and to apply a correction. The algorithm basically exploits the fact that bright band causes a maximum return signal at a range dependent on the height of the bright band and the beam elevation. Other techniques for detecting anomalous data depend on examining the smoothness of the spatial correlation function of the rainfall field. Grid squares having anomalous values will have an unusually high influence on the spatial correlation function allowing such outlier values to be detected. Krajewski (1987) has formally developed such a detection technique based on using the influence function of the spatial correlation function for a homogeneous gaussian field. Such techniques, however, do not consider the temporal dependence structure of rainfall in addition to the spatial dependence in order to detect anomalous values. It would be possible to develop detection algorithms exploiting space-time dependence as part of the linear dependence function approach previously described and presented as equations (1) and (2).

Weather radar data and flood forecasting models

The previous discussion has concerned itself with the importance of not simply using raw weather radar data as input to flood forecasting models. Not only would the forecasts be less accurate but more importantly they would lack the property of robustness, being prone to produce occasional anomalous flood warnings. Techniques which combine telemetering raingauge and grid-square radar data to increase the accuracy of weather radar data can also be developed to detect anomalous radar values and correct for them prior to input to a flood forecasting model. The calibrated and corrected weather radar grid data can then be used with greater confidence as input to flood forecasting models in a number of ways. Moore (1987) identifies three approaches to use radar grid-square data in forecasting models:

(1) to form a basin or sub-basin areal average rainfall input to a lumped rainfall-runoff model;

(2) as input to a grid-square flood forecasting model;

(3) to define a spatial probability distribution of rainfall over a basin used as input to a probability-distributed rainfall-runoff model.

The rainfall-runoff model may be used in isolation for flood forecasting, for forecasting lateral inflow to a channel flow routing model, or as part of an overall river network flood forecasting system.

The first approach is illustrated in Figure 2 where sub-basin areas of the River Dee basin are defined approximately to coincide with the sides of the 2 km radar grid of the Llandegla radar installed as part of the Dee Research Programme (Cole et al, 1975; Harvey and Lowing, 1976; Central Water Planning Unit, 1977). Sub-basin areas may be defined with respect to flow gauging stations, tributaries, or hydrologically homogeneous response units.

One type of grid square flood forecasting model is illustrated in Figure 3. Whilst such models may be expected to maximise the benefit of distributed weather radar data they do present problems of parameter estimation since parameters controlling the runoff response and travel time characteristics of each grid square must be determined. This can be made feasible by exploiting information contained in contour maps of the river basin concerned. Isochrones, defined as lines joining points of equal time of travel to the basin outlet, may be estimated using contour maps and an assessment of flow velocities via hillslope and channel paths. Effective soil moisture storage depth controlling the production of saturation overland flow may be estimated for each grid square by introducing some form of functional dependence of depth with slope which can be measured from the contour map. Digital image analysis techniques together with a digital terrain model may also be used to automate isochrone and slope calculations.

In the third approach geometric information on rainfall distribution is considered to be of only secondary importance

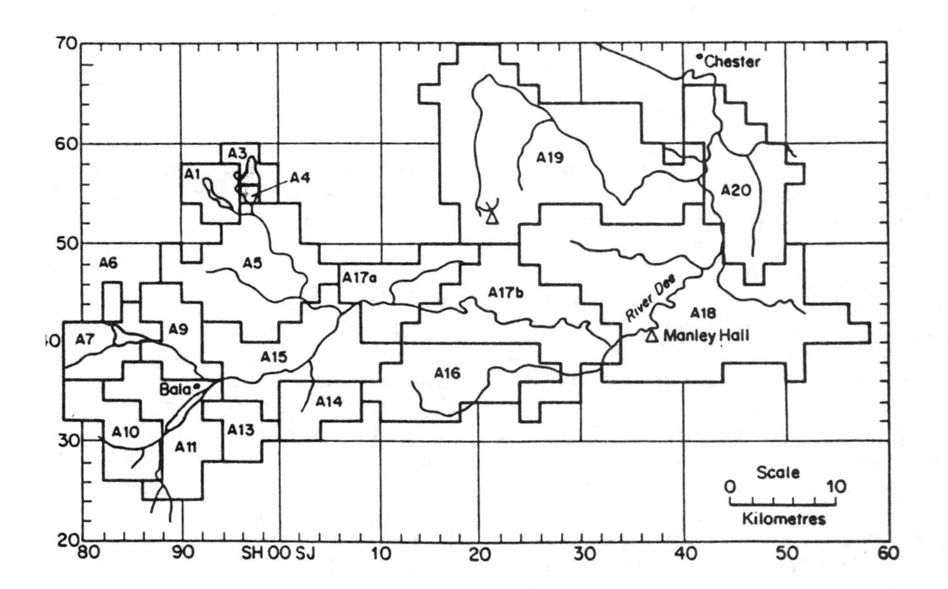

Figure 2 Grid-square representation of the River Dee basin used to calculate sub-basin average rainfalls from weather radar data.

(a) River basin with superimposed weather radar grid

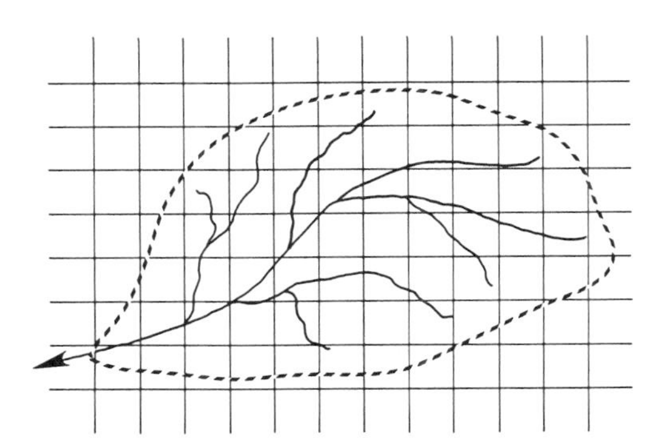

(b) Block representation of runoff response for grid square (i,j)

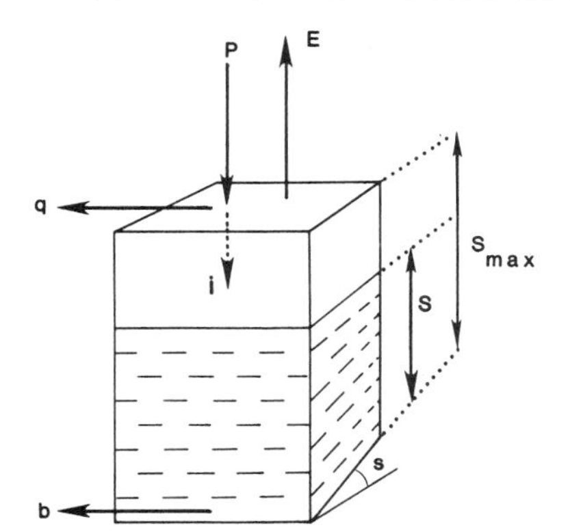

P: weather radar estimate of rainfall

E: evaporation

q: direct runoff

b: baseflow

S, S_{max}: water storage and maximum storage capacity

i: infiltration

s: average slope of block

(c) Isochrones for grid square (i,j)

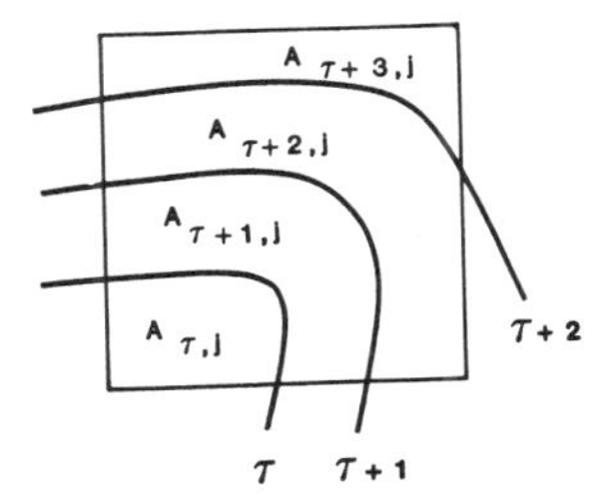

Figure 3 Simple grid-square model for flood forecasting

relative to the spatial probability distribution function of rainfall over a basin in terms of its effect on streamflow response. Then the radar grid data can be used to identify and estimate an appropriate parametric (or non-parametric) spatial probability distribution function of rainfall. A rainfall-runoff model can be constructed which employs such a density function as input and further can be developed to account for spatial variability in the runoff response over the basin in a probabilistic way (Moore, 1986, 1987). Figure 4 shows an example of a probability distribution function estimated from radar grid data, and Figure 5 shows the nonlinear relation between rainfall and runoff that results if rainfall and soil moisture storage capacity controlling runoff generation by a saturation excess mechanism are considered to be exponentially distributed over the basin; p and c denote the mean rainfall and mean storage capacity for the basin respectively and p would vary from one time interval to the next as estimated from the radar grid data. Models based on applying the probability-distributed principle to the rainfall input and the runoff response mechanism to account for hydrological spatial variability over the basin can be developed as practical flood forecasting tools with the aid of grid-square weather radar data.

Research is required to investigate the relative merits of these three approaches to using grid-square radar rainfall data in flood forecasting models. Indeed models of the grid-square and probability-distributed type require further development before an assessment would be even worthwhile. Operational use of weather radar in flood forecasting models is currently confined to the calculation of basin areal average rainfall for input to simple lumped models (North West Authority et al, 1985 Collier and Knowles, 1986) and to river network models (Haggett, 1986).

FLOW FORECASTING FOR DROUGHT MANAGEMENT

Flow forecasts for longer periods of time, say from one month to a year into the future, are employed for managing water resource systems during a drought. A case study will be presented next which serves to illustrate how meteorological data in the form of long sequences of historical rainfall data can be used to produce scenarios of future flow conditions. When combined with a water resource model and a probability forecast of the next month's rainfall provided routinely by the UK Meteorological Office, an assessment can be made of the reliability of a water resource system to supply water at given times in the future. Management discussions can be made on the basis of reliability assessment.

```
   LH
ENDPOINT  COUNT
   0.       99   **************************************************
  15.       39   ********************
  30.       32   ****************
  45.       12   ******
  60.        6   ***
  75.        4   **
  90.        5   **
 105.        1
 120.        4   **
 135.        0
 150.        3   *
 165.        1
 180.        1
 195.        0
 210.        2   *
```

Figure 4 Histogram of hourly rainfall in 1/10 mm derived from a 12 x 10 5 km grid west of Birmingham airport at 13.00 hrs 14 July 1982.

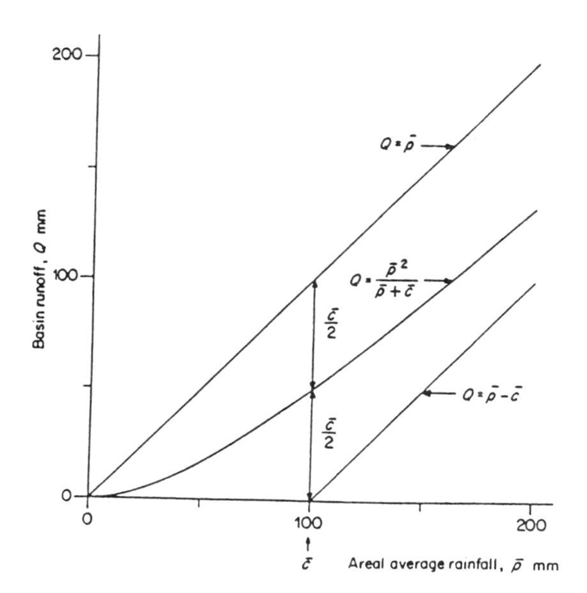

Figure 5 Rainfall-runoff relationship for an exponential probability-distributed model of rainfall and runoff production.

Risk assessment procedure for London's water supply reservoirs

A system of pumped storage reservoirs in the Thames basin replenished by river abstractions supplies 58% of the water supply needs within the basin. Water resource models (Sexton et al, 1979) developed to simulate reservoir system behaviour depend on flow data being available at key abstraction points. The availability of long records of daily rainfall from gauges within the basin dating back to the last century means that long sequences of flow can be generated using rainfall-runoff models for the basins draining to the abstraction points. A conceptual rainfall-runoff model developed specifically to represent the complex hydrological response of sub-basins of the Thames has been developed by Greenfield (1984) and calibrated for the basins concerned. The water budgeting calculation involved in the model requires evaporation data in addition to basin areal average rainfall. A standard annual profile of Penman potential evaporation daily values is used for long-term simulations, calculated as a basin average using a reciprocal-distance weighting scheme. Daily basin average rainfalls are calculated using a procedure based on standardisation of daily measurements at each gauge by the gauge's long-term average annual rainfall (Moore et al, 1986a).

Scenarios of future flows starting on the current day are obtained by the following procedure. The rainfall-runoff model is initialised at some time in the past and rainfall from the initialisation date to the current time used to obtain a simulated flow sequence up to the current day. Flow simulations in the recent past, preferably during recession conditions, are compared with observed flows and an adjustment made to the water content of the conceptual stores of the model to achieve correspondence between observed and model flows. The conceptual model thus encapsulates information on the current hydrological status of the basin. Next, the historical record of basin areal average rainfall is used to provide scenarios of equi-probable future rainfall by selecting from each year of the record the period corresponding to the forecast period required. For example, a 6 month forecast made on 1 July 1987 would require the extraction of 97 rainfall sequences available for the months July through to December from each year of a historical record dating back to 1890. Each rainfall sequence is used as input to the rainfall-runoff model with its water storage contents initialised on the current day to the previously adjusted values. The resulting equi-probable flow scenarios are then used as input to the water resource system model to derive equi-probable scenarios of reservoir levels and imposed demand restriction levels. The scenarios are then analysed statistically to derive (i) the reservoir level on a given day of the forecast period that has a specified probability of not

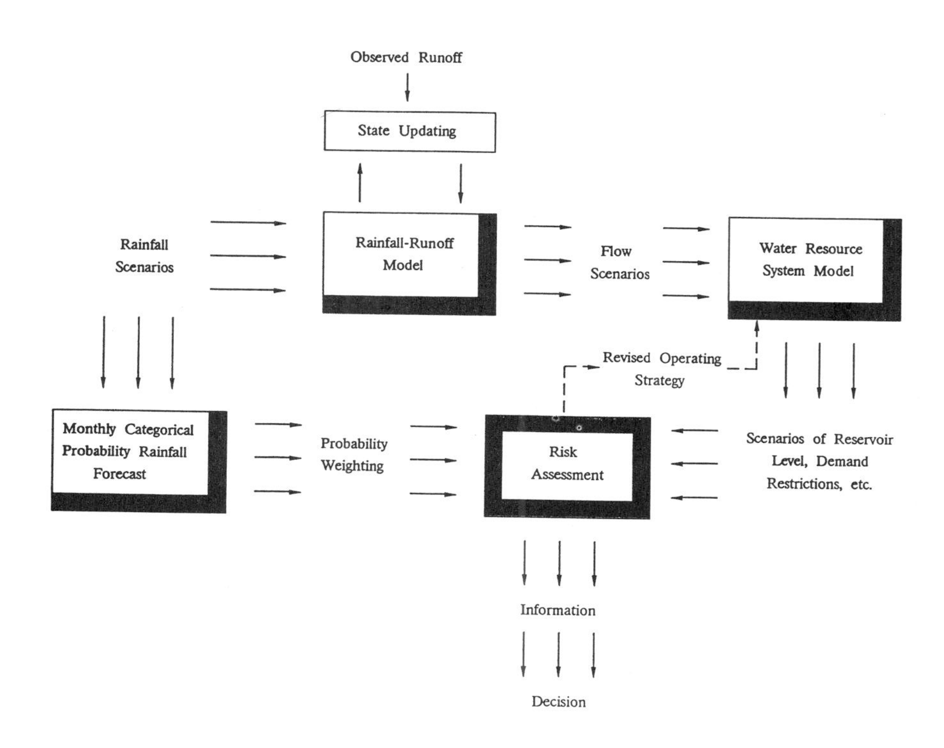

Figure 6 The risk assessment procedure for drought management.

being reached, and (ii) the probability that a given level of demand restriction lasts for more than a certain number of days over the forecast period. Figure 6 illustrates the overall risk assessment procedure including the option for the decision-maker to modify the operation of the water resource system model, observe the change in reliability, and use such information to support a decision.

Incorporation of monthly rainfall forecasts

The basic principle underlying the risk assessment procedure is that each rainfall scenario extracted from the historical record may be used to represent an equi-probable sequence of future rainfall. Flow and reservoir level sequences derived from such sequences through transformations in the form of rainfall-runoff and water resource system models may in turn be regarded as equi-probable. However, if additional information is available which indicates that a rainfall sequence for a particular past year is more likely to resemble rainfall over the forecast period, then such information could be incorporated by giving the sequence greater weight when estimating risk probabilities. Such information is available in the form of monthly rainfall forecasts provided by the UK Meteorological Office. These forecasts are expressed in terms of probabilities that rainfall will be "Below Average", "Average" or "Above Average" for the forthcoming month. Each category represents one-third of all possible rainfall sequences. The basis of these categorical probability rainfall forecasts is described in detail by Maryon and Storey (1985) and Folland and Woodcock (1986); essentially, the forecast procedure involves the application of multivariate analysis techniques to mean surface pressure anomalies and sea surface temperatures over the Europe and North Atlantic area.

A simple use of these probabilities in the risk assessment procedure is as follows:

(1) use the historical record to determine the rainfall amount defining the two category boundaries for the month concerned;

(2) identify for each year of the historical record the category to which the monthly rainfall belongs;

(3) to each of the n years assign a weight, w_i, calculated as the forecast probability of rainfall falling in the category divided by the sum of these probabilities for every year of record;

(4) rank the n reservoir levels on a particular day of the forecast period derived from each rainfall sequence, keeping the weights attached;

(5) sequentially sum the weights to obtain the probability of exceedence of the r'th ranked storage, S_r, as $p_r = \sum_{i=1}^{r} w_i$, so that $\{p_r\}$, r=1,2,...,n defines an empirical distribution function of storage for the day concerned;

(6) calculate the storage levels having specified risks of not being reached on that day by interpolation of the empirical distribution function.

Steps (4), (5) and (6) are repeated for each day of the forecast period to produce the final risk assessment information shown in Figure 7.

In practice a refinement of this procedure is used to overcome the possibility of applying very different weights to historical rainfall sequences which, although they have similar monthly totals for the month of the rainfall forecast, fall either side of a category boundary. A piecewise linear smoothing function (Moore et al, 1986a) has been developed which preserves the forecast probability of rainfall falling within each of the three categories. A plotting position formula applied to the ranked historical rainfalls for the month concerned is used to enter the function and return weights to be attached to each year.

Decision support system for drought management

The risk assessment procedure including the option to employ monthly rainfall forecast information forms part of a microcomputer-based decision support system for managing London's water supply reservoirs (Moore et al, 1986b; Moore et al, 1987). An important feature of the system is an automatically updated archive of hydrometric data maintained through a real-time communication link with a second computer dedicated to real-time data acquisition via telemetry. Fifteen minute resolution data from eleven raingauges in the Thames basin are automatically aggregated in time and space to update a long-term archive of daily basin average rainfalls extending back to 1890. Powerful data and model management facilities coupled with a menu-and form driven dialogue user interface makes the system very user-friendly. Graphical displays such as the one shown in Figure 7, are used extensively to present information on water resource reliability in an easily assimilated form. The system was first used operationally by Thames Water's Regulation and Monitoring Division in March 1987.

(a) Equi-probable scenarios

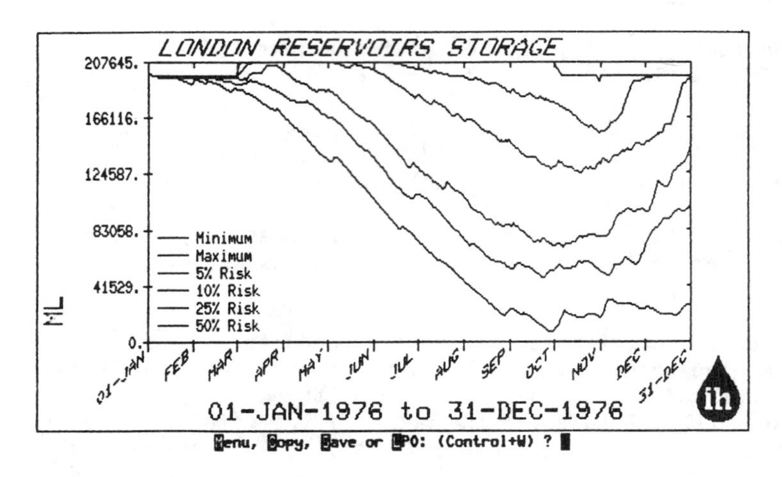

(b) Monthly rainfall forecast weighted scenarios (the rainfall forecast for January 1976 is assumed to be 60% below average, 30% average, and 10% above average).

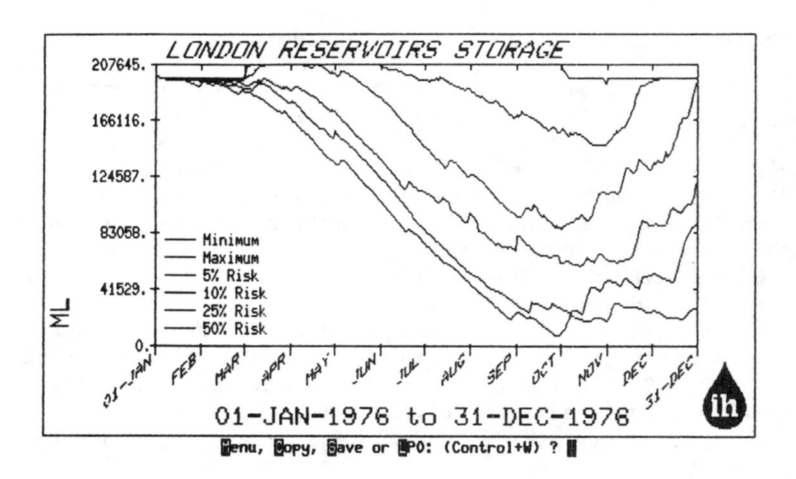

Figure 7 Risk assessment of the London reservoir system for the drought of 1976.

DISCUSSION

The two forecasting applications considered, one for flood warning and the other for drought management, serve to highlight the differing data needs of short-term and long-term hydrological forecasting. At short lead times the natural lag of a basin's response to rainfall is exploited to use recently telemetered rainfall and weather radar data to forecast flows a few hours or so ahead. Also the dependence structure of rainfall in time and space may be used as the basis of a statistical forecast of rainfall to extend the useful forecast lead time. Models based on a linear dependence structure in time and space have been introduced as having potential when used with weather radar and raingauge data to provide more accurate and robust estimates of current rainfall and to serve as the basis for short-term rainfall forecasting. It is clear that the main concern in short-term flood forecasting is to obtain the most accurate forecast of future flow conditions, and in this regard the recognition that forecast performance is data limited argues strongly for the more effective use of weather radar data in flood forecasting models.

As the forecast lead-time increases it becomes more relevant to consider possible scenarios of future conditions rather than attempt to forecast what the flow will be at a particular time in the future (Jones and Moore, 1985). The main data requirement then shifts from accurate high resolution spatial and temporal data available in real-time towards long-term records which can be used reliably as the basis of a risk assessment approach to the forecasting of conditions over the next week, month or year. Only during sustained periods of no rain will weekly forecasts be expected to correspond to actual flow conditions. The equi-probable scenario approach employed in the case study of London's water supply reservoirs serves to illustrate how information on the possible range of future conditions of flow and reservoir storage level can be useful in the management of reservoir systems during drought. The confidence that can be attached to the risk estimates obtained are, however, directly related to the number of rainfall scenarios used and hence to the length of historical rainfall records available (Moore et al, 1987). Introduction of additional meteorological information in the form of monthly probability forecasts of rainfall presents a novel way of attaching weights to the rainfall scenarios extracted from previous years according to how likely they are to be similar to the forthcoming month's rainfall. The alternative risk assessment it provides, however, is probably best regarded as complementary to the equi-probable assessment as part of an overall decision support system for drought management.

ACKNOWLEDGEMENTS

Past contributions from colleagues at the Institute of Hydrology to the work described are gratefully acknowledged, as is the financial support of Thames Water Authority and the Flood Protection Commission of the Ministry of Agriculture Fisheries and Food.

REFERENCES

Central Water Planning Unit, 1977 Dee Weather radar and real-time hydrological forecasting project. Report by the Steering Committee, HMSO, London, 172 pp.

Cole, J A, McKerchar, A I, Moore J R, 1975 An on-line flow forecasting system, incorporating radar measurements of rainfall, as used to assist the short-term regulation of the River Dee in North Wales; Application of Mathematical Models in Hydrology and Water Resources Systems. Proc Bratislava Symp, LAHS Publ No 115, 57-66.

Collier, C G & Knowles, J M, 1986 Accuracy of rainfall estimates by radar, Part III: Application for short-term flood forecasting, Journal of Hydrology, 83, 237-249.

Collier, C G, Larke, P R and May, B R, 1983 A weather radar correction procedure for real-time estimation of surface rainfall, Quart. J. R. Met. Soc., 109, 589-608.

Folland, C R & Woodcock, A, 1986 Experimental monthly long-range forecasts for the United Kingdom, Part 1. Description of the forecasting system. Meteorological Magazine, 115 (1371), 301-318.

Gandin, L S, 1965 Objective analysis of meteorological fields, Leningrad, USSR, 1963. Trans. Israel Programme for Scientific Translation 242 pp.

Greenfield, B J, 1984 The Thames Water Catchment Model, Internal Report, Technology and Development Division, Thames Water, 13 pp.

Haggett, C M, 1986 The use of weather radar for flood forecasting in London, Conference of River Engineers 1986, Cranfield 15-17 July, Ministry of Agriculture, Fisheries and food, 11 pp.

Harvey, R A & Lowing, M J, 1976 The development and implementation of a real-time flow forecasting system for the River Dee, Institute of Hydrology.

Jones, D A & Moore, R J, 1985 Strategies for flow forecasting, Institute of Hydrology, 32 pp.

Krajewski, W F, 1987 Radar rainfall data quality control by the influence function method, Water Resources Research, 23(5), 837-844.

Maryon, R H & Storey, A M, 1985 A multivariate statistical model for forecasting anomalies of half-monthly mean surface pressure, Journal of Climatology, 5, 561-578.

Matheron, G, 1971 The theory of regionalised variables and its applications, Centre de Morphologie Mathematique.

Moore, R J, 1985 The probability-distributed principle and runoff production at point and basin scales, Hydrological Sciences Journal, 30(2), 273-297.

Moore, R J, 1986 Advances in real-time flood forecasting practice. Symposium on Flood Warning Systems, Winter meeting of the River Engineering Section, Inst. Water Engineers and scientists, 23 pp.

Moore, R J, 1987 Towards more effective use of radar data for flood forecasting. In: Collinge, V K and Kirby C (eds), Weather Radar and Flood Forecasting, 233-238, J Wiley.

Moore, R J Jones, D A, Black K B and Parks, Y, 1986a Real-Time Drought Management system for the Thames basin, contract report to Thames Water, 56 pp, Institute of Hydrology.

Moore, R J, Black, K B, Jones, D A and Bonvoisin, N, 1986b User guide to the Real-Time Drought Management System for the Thames Basin, contract report to Thames Water, 71 pp, Institute of Hydrology.

Moore, R J, Jones, D A, Black, K B, 1987 Risk assessment and drought management in the Thames basin. Int. seminar on Recent Developments and Perspectives in Systems Analysis in Water Resources Management, WARREDOC Spring Meeting, Perugia, Italy, 1-3 April 1987, 16 pp.

North West Water Authority, 1985 North West Weather Radar Project, Report of the Steering Group, 73 pp.

Ord, K and Rees, M, 1979 Spatial processes: recent developments with applications to hydrology, in Lloyd, E H, O'Donnel, T, and Wilkinson, J C (eds), The Mathematics of Hydrology and Water Resources, 95-118, Academic Press.

Reed, D W, 1984 A review of British flood forecasting practice, Institute of Hydrology Report No 90, 113 pp.

Ripley, B D, 1981 Spatial Statistics, J Wiley

Sexton, J R, Cook, D J, & Jones, A E, 1979 Water Resources Model: an introduction, 8 pp, Thames Water.

Smith, C J, 1986 The reduction of errors caused by bright bands in quantitative rainfall measurements made using radar, Journal of Atmospheric and Oceanic Technology, 3, 129-141.

35

The role of carbon dioxide and other greenhouse gases in climate variation, and the mechanisms for informing decision makers

A. Gilchrist, Meteorological Office, UK

PART I - THE CLIMATIC EFFECTS

Carbon dioxide and other greenhouse gases

It is well known that the concentration of carbon dioxide has been increasing at least through the modern industrial era. Keeling's measurements from Mauna Loa, Hawaii, show that the concentration in 1958 was about 315 ppmv and that now it is in excess of 345 ppmv i.e. a rise of ≈ 10% in 30 years. Measurements of CO_2 trapped in glacier ice indicate that in the 18th century, the concentration was around 275 ppmv (SCOPE Report 1985 Chap 3). Thus the concentration has increased by ≈ 25% since pre-industrial times.

Although relatively inert chemically, carbon dioxide is a radiatively active gas, which absorbs infra-red radiation principally at about 15 μm but also more weakly at 7.6 μm (where it overlaps water vapour bands) and at 10 μm (where it partly overlaps the 9.6 μm ozone band). Radiation in these wavelengths leaving the earth's surface is absorbed by the gas and partly re-radiated back to earth. The net effect is a relative warming of the earth's surface. It is to be noted that because the 15 μm band is partially saturated, the absorption does not increase linearly with increased carbon dioxide concentrations. Augustsson and Ramanathan (1977) concluded that the temperature rise at the surface should depend approximately logarithmically on the concentration.

Carbon dioxide is not the only potentially important greenhouse gas. Attention has been drawn to others whose concentration is increasing, particularly methane CH_4, nitrous oxide N_2O, chlorofluorocarbons CFCs and ozone. Present concentrations and rates of increase are (SCOPE Report, 1985):

CH_4	1.65 ppmv	1.2% /annum
N_2O	.30 ppmv	0.3% /annum
$F_{11}(CFCl_3)$	.168 ppbv	5.7% /annum
$F_{12}(CF_2Cl_2)$	.285 ppbv	6.0% /annum
tropospheric O_3		0.25% /annum

The concentrations of these gases are much less than that of carbon dioxide but their absorption is much greater because it is mainly in the atmospheric window. It is estimated (SCOPE Report, 1985) that if concentrations continue to increase at present rates, these additional trace gases will have the same effect as 140 ppm of carbon dioxide by the year 2030.

General Circulation Model Results

In estimating the climatic effects of increased carbon dioxide, modellers have usually considered the equilibrium climate with either doubled or quadrupled amounts from a nominal 300 ppmv modern value. I shall not try to review all the results available but will draw attention to some salient features.

At the time of the NAS Second Assessment of Carbon Dioxide and Climate, 1982, most model integrations for a doubling of carbon dioxide produced global surface temperature rises of 2-3°C. The main exception was the GISS model of Hansen et al which was reported as giving a warming of ≈ 4°C; the main reason for this was thought to be the method of dealing with interactive cloudiness.

More recently, four independent models have obtained global warming of around 4°C viz:

Hansen et al (1984)	4.2°C	11% inc in ppn
Washington & Meikl (1984)	3.5°C	7.1% inc in ppn
Wetherald & Manabe (1986)	4°C	8.7% inc in ppn
Wilson & Mitchell (1986)	5.2°C	15% inc in ppn

In these integrations, the models had a number of common features: principally, (1) a realistic description of the earth's topography was used (2) the annual variation of the solar cycle was included (3) the ocean was represented by a mixed layer slab of the depth required to reproduce the annual variation of SSTs (4) cloud was interactive, and estimated by the model itself.

Thus the estimated warming in recent integrations has been almost twice that of earlier values. There is evidence that the primary reason for this is the use of model-generated cloud which, in all cases, led to a reduction of average cloudiness in a high-CO_2 climate. However, Somerville and Remer (1984) have pointed to the possibility that the cloud properties themselves might change, so as to reduce the warming.

The Uncertainties

There are many uncertainties involved in predicting the climate change due to greenhouse gases. The most important are

A. Future concentrations of greenhouse gases

(i) Carbon dioxide. These depend on amount and type of fossil fuels burned in the future. Estimates therefore vary widely. The SCOPE Report gives a range of 380-500 ppm for the year 2030 based on fuel use changing from the present value of 5 GtC/yr to GtC/yr (lower bound) or 20 GtC/yr (upper bound).

(ii) Other greenhouse gases. Mostly these gases are either manufactured or are dependent on man's activities. Their future concentrations are therefore highly dependent on political decisions, and the assumption of a constant rate of increase could be wildly wrong.

B. General Circulation Models

Estimates of the climate change due to an effective doubling of carbon dioxide. The main causes of uncertainty are thought to be:

(i) Ocean circulation. The ocean will delay a warming but if the circulation is also altered, the effects could be even greater. Investigation of this depends upon the availability of reliable ocean models, and is therefore just beginning.

(ii) Cloudiness. The main factor in the recent large warmings obtained by models is believed to be changing cloudiness. However, model algorithms for cloud have involved a significant amount of tuning and therefore the confidence that can be placed on estimated changes are low.

(iii) Sea-ice. The general circulation models use, within them, models for the creation and dissipation of sea-ice. The feedback effects of ice in the magnitude of the global warming is relatively large. The models, however, are greatly simplified being purely thermodynamic and take no account of such important dynamical factors as the effects of wind and waves.

(iv) Regional estimates of climate change. The uncertainties noted already apply to global values of general circulation model estimates. The estimates for a particular region are even more uncertain.

C. Uncertainties in the impacts

(i) Vegetation. Other things being equal, increased carbon dioxide increases the rate of growth of vegetation, and this may counteract negative effects on growth by climatic factors (e.g. reduced availability of moisture)

(ii) Sea-level rise. With increasing warmth sea-level tends to rise for two possible reasons: the expansion of sea-water already in the sea, and the increase in the volume of sea-water due to the melting of land-based glaciers. However, it is not inconceivable that the higher moisture content in high latitudes, consequent upon an atmospheric warming, might lead to a higher snowfall, and the growth of glaciers.

(iii) Regional effects. Even if there is global warmings and increased precipitation globally, some places might, on average, be colder and drier, and at present there is no consensus among models on these regional variations.

PART II - ADVISING OTHERS ABOUT THE CLIMATIC EFFECTS OF GREENHOUSE GASES

Summarizing the scientific position as regards the climatic effects of greenhouse gases, we note

(1) there is a consensus among scientists that they could be very significant, but

(2) there is no consensus on their magnitude, detailed distribution or, indeed, whether or not they will be beneficial to mankind as a whole.

Many questions arise, which are especially difficult because of the uncertainties that have been noted. Whom should we, as meteorologists with an understanding of the science, inform? How should we inform them so as to evoke the correct response?

What, in view of the uncertainties, is the correct response we seek?

It is already evident that, dependent on who is informed and the light in which the information is conveyed, the response may vary from deep concern and alarm to boredom and indifference.

First of all, to look again at the matter of who should be told. It is often assumed that the people we should now be telling belong to those mysterious sets of individuals whom we characterize as 'decision makers', or 'policy-makers'. At the Villach Conference for example, the Working Group on Socio-Economic impacts concluded that scientists and policy-makers should begin an active collaboration to explore the effectiveness of alternative policies and adjustments. But whom do we recognize as policy-makers in this sense, and are they, in fact, the best people to explore the effectiveness of policies? For example, it is then often assumed that members of the European Parliament are policy-makers as far as the EEC are concerned, and similarly for MPs in the UK. However, the power of the latter is much less than that of US Congressmen and the analogy is quite misleading. In the UK most government departments have a Chief Scientist who has the responsibility for advising Ministers on scientific matters. I would guess that in a scientific question, the relevant Chief Scientist and his staff are the most important individuals to convince about a correct course of action. The point I wish to make is that the people who should be influenced differ from country to country, and it is not possible to generalize across the spectrum of nations.

The next question was that of who should be exploring the effectiveness of alternative policies. This is difficult, and, in my view, probably requires collaboration between scientists and specialists on the evaluation of policies, who may be scientists, or economists or sociologists, or agriculturists etc. At the Villach Conference, an approach to this problem was discussed by Dr. Clark. Only when the consequences of different courses of action have begun to be understood, can we reasonably expect 'policy-makers' to make their individual contribution to the debate.

We have so far considered how the information about greenhouse gases should be dealt with in order to ensure that responses are effective. Only a small group of individuals may be involved. There is the more general problem of keeping the general public informed. They have a right to know and mostly such knowledge as they have is gleaned from newspapers, periodicals and television or radio programmes. Unfortunately, these sources of information are often grossly inadequate and biased. One reason is that the writer or producer often seems intent, not on maximising the information content, but on entertaining an audience and on homing in on a simple message, even when the simplicity is achieved as a result of an incorrect or unbalanced interpretation of the science. It must also be said that sometimes it is due to the forcefulness with which individual scientists present their views. It is obvious the scientists are selected by the media more for their articulateness and their willingness to speculate than for their scientific ability and integrity. As a result a great deal of misinformation and uninformed comment is given wide publicity. It is fortunate that the public memory is short.

There is no easy was of avoiding these problems. I suggest, however, that, as scientists, we have to try to base our actions on establishing

(1) What are the scientific facts?

(2) Who are the scientists who have the depth of scientific knowledge combined with the insight and understanding required to provide informed opinions going beyond the established scientific facts?

The first of these is difficult enough, but fortunately there are a few publications which set out the scientific position very well. The second is even more difficult, and the corollary, that having selected the right individuals their advice should be heeded before that of others is obviously a counsel of perfection - but that is no reason not to aim for it.

REFERENCES

Augustsson, T, Ramanathan, V, 1977 A radiative-convective model study of the CO_2 climate problem. J T Atmos. Sci., 34, 448-451.

Somerville, R C, Remer, L A, 1984 Cloud optical thickness feedbacks in the CO_2 climate problem. J. Geophys. Res., 89, 9668-9672.

36

Role of meteorology in activities related to the protection of the environment

E. Mészáros, Institute for Atmospheric Physics, H-1675 Budapest

1. INTRODUCTION

Atmospheric processes interact with different pollutants released into the atmosphere by human activities. Regular and irregular air movements disperse gases and particles around emission sources. During this transportation air pollutants can be physically and chemically transformed as a function of the state and dynamics of the atmosphere. After a certain time, called residence time, man-made species, in the same way as similar natural components, leave the atmosphere to be deposited into other media such as soil, water, plants etc., of our environment.

It follows from this discussion that the concentration or deposition of a certain pollutant over a given area or at a certain site will depend, besides emission parameters, (e.g. rate and height of emission), on atmospheric phenomena. In other words this means that meteorological information is needed to determine the distribution of air pollution on different scales. More precisely: meteorological models are necessary to connect emission parameters with the concentration/deposition field. The main aim of this lecture is to discuss in a simple way how to use meteorological information and products to solve the problem mentioned above. The practical usefulness of such an operation for the protection of the environment is outlined.

Pollutants are not only passively transported in the air, but at the same time they influence atmospheric composition. The activity of meteorologists to record these changes and to foresee the climatic consequences of such modifications is also briefly summarized. It is obvious that information on possible changes in climate can be used for long-range planning of agricultural production, water management and other human activities.

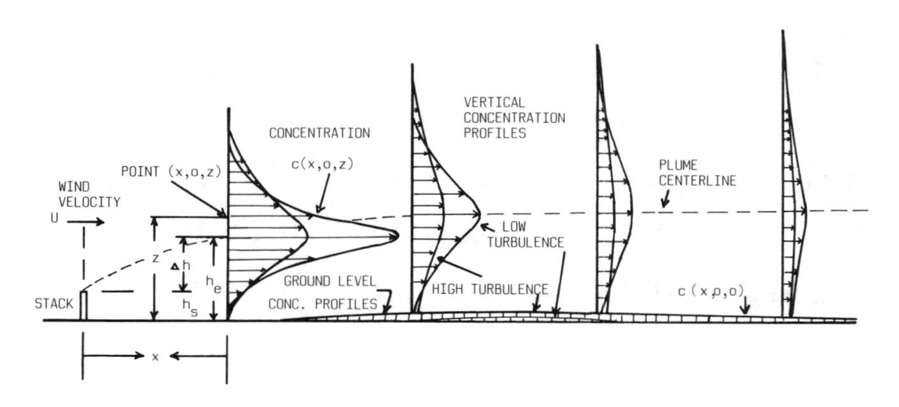

Figure 1

Distribution of the concentration of air pollutants in the vertical xz plane emitted by a high point source according to Strom (1976)

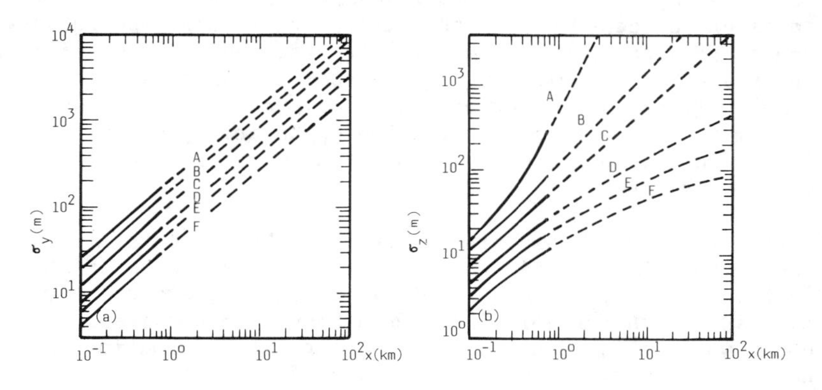

Figure 2

Variation of σ_y and σ_z as a function of x according to Pasquill and Gifford (taken from Högström, 1978).

A: Extremely unstable B: Moderately unstable
C: Slightly unstable D: Neutral
E: Slightly stable F: Moderately stable

2. LOCAL AIR QUALITY MANAGEMENT

The role of meteorology in activities related to local air quality management is demonstrated by the relatively simple Gauss model[1] widely used in air pollution practice. By applying this approach among other things we assume that the wind blows in direction x of our coordinate system. (The center of the stack base is the origin of the system), and the field of meteorological parameters, including turbulence, is homogeneous. In this case at the effective plume height (see Figure 1) gaseous materials emitted are transported in direction x by regular movements, while they are dispersed by turbulence in directions y and z. Due to the characteristics of turbulence the concentration is distributed in directions y and z around the centerline of the plume according to the Gauss function, well-known from statistical mathematics. Under these conditions the concentration C of the pollutant can be calculated in the surface air, (the surface is flat and the ground reflection of the molecules is supposed "perfect"), by the following simple equation (see e.g. Strom, 1976).

$$c(x,y,o) = \frac{Q}{\pi \sigma_y \sigma_z u} \exp\left[-\frac{1}{2}\left[\frac{y}{\sigma_y}\right]^2\right] \exp\left[-\frac{1}{2}\left[\frac{h_e}{\sigma_z}\right]^2\right] \qquad (1)$$

where Q is the emission by unit time, σ_y and σ_z are the standard deviations of the concentration distribution of Gaussian form in directions y and z, u is the wind speed, while h_e is the effective stack height:

$$h_e = h_s + \Delta h \qquad (2)$$

In equation (2) h_s is the physical height of the stack and Δh is the plume rise which can be determined by taking into account the emission and stack characteristics (see Appendix I). Equation (1) can be further simplified if we want to calculate the variation of the concentration only in the wind direction (y = 0). In this case

$$c(x,0,0) = \frac{Q}{\pi \sigma_y \sigma_z u} \exp\left[-\frac{1}{2}\left[\frac{h_e}{\sigma_z}\right]^2\right] \qquad (3)$$

In equations (1) and (2) the concentration of the pollutant, the rate and height of the emission and the meteorological parameters characterized by u, σ_y and σ_z are included. These latter two parameters are functions of the vertical stability of the atmosphere and x as given in Figure 2.

Equations (1) and (2) are very important tools for

[1]Note that the Gauss model can be derived from the general advection/diffusion equation as discussed by Rote (1980).

meteorologists since, by using these equations and meteorological information, the following practical problems can be solved (for numerical examples see Appendix II):

a. What will be the concentration distribution of a given pollutant if a new power plant or works is established in a certain area? Will the concentration be higher at any point of the area than the ambient air standard?

b. What is the minimum reduction in Q necessary to have concentrations lower than the ambient air standard?

c. What is the minimum stack height to have concentrations lower than the ambient air standard for the area[1].

Meteorologists are able to answer these questions if the wind and the vertical stability of the atmosphere are measured for example by radiosonde observations. More detailed information is obtained by measuring wind and temperature profiles on appropriate meteorological towers. If the concentration field averaged for a longer time is calculated, the average values of meteorological parameters should be taken into account. In this case the standard deviation of meteorological data around averages will be a measure of the variation of the concentration. In other words, this means that concentrations calculated will correspond to the averaging time of meteorological data. If we make the calculations for a time period shorter than 1 year or 1 day, the annual variations or daily changes in meteorological elements must be taken into consideration.

Finally, we note that the model outlined above can be applied for multiple and areal sources (e.g. urban pollution) as well as for irregular surfaces. Also, the sedimentation velocity of coarse dust or flying ash particles can be taken into account. However, the mathematical and physical complications involved do not change the principle of the operation.

3. LONG-RANGE TRANSPORT OF AIR POLLUTANTS

The model presented above can be applied for solving local scale (below 20 km) air pollution problems. If we want to calculate the transport of air pollutants on regional (about 20-500 km) or continental (500-3000 km) scales, an other approach should be used. The first point to be considered is that the effects of turbulence on the transport can be neglected on these scales. The second difference is that the

---

[1] It should be noted that while local air pollution decreases with increasing stack height, high stacks are very dangerous from the point of view of larger scale pollution.

transformation and removal of pollutants, neglected in Section 2, should be taken into account.

The principle of the procedure is as follows. Along the wind trajectory determined for a certain isobaric surface (e.g. 850 hPa) an air column (box) of unit base moves. The height of the column is equal to the mixing height determined from radiosonde observations. In the box the concentration pollutants is uniform. The air column is transported over the surface with known emission characteristics. In this way the material input into the box can be calculated for any time interval. On the other hand, the decrease of the concentration in the box due to transformation and deposition is also determined for each time period.

Long-range transport models have mainly been elaborated for sulfur compounds. On the basis of the above discussion the temporal variations of sulfur dioxide (C_1) and sulfate (C_2) concentrations in the box will be (Eliassen, 1987):

$$\frac{dC_1}{dt} = \frac{Q'}{H} - k_1 C_1 - k_2 C_1 - k_3 C_1$$

$$\frac{dC_2}{dt} = k_1 C_1 - k_4 C_2 - k_5 C_2 \qquad (4)$$

where Q'is the emission by unit surface and time, H is the mixing height, k_1 is the rate constant of the transformation of SO_2 to SO_4, k_2 and k_4 represent the dry deposition of sulfur dioxide and sulfate, respectively, while k_3 and k_5 are the same parameters for wet deposition. It should be noted that the coefficients k_1-k_5 vary as a function of meteorological parameters. However their average values are around 10^{-4} -10^{-5} s^{-1}.

Long-range transmission models can be used to determine the origin of the sulfur deposited in a certain region (country). The fate of the sulfur emitted in the area can also be estimated (see Appendix III). Such calculations are necessary to estimate trans-boundary air pollution transport. In other words, atmospheric scientists on the basis of meteorological and other information can give advice to decision-makers as to how to organize international air quality management in the future.

Long-range air pollution modelling can also be applied in ecological studies. Atmospheric deposition values provide input data for studying the effects of different pollutants on aquatic and terrestrial ecosystems. The problem of acid rain is a good example in this respect. Without details we note that ecologists try to simulate the effects of air pollution as a function of atmospheric deposition. Since deposition is connected with the emission field in atmospheric models, meteorological information and products should be used to determine damages in ecosystems in the case of a given source configuration.

Meteorological services are also able to help regional and continental air quality management by carrying out background air pollution measurements in the monitoring network of WMO (BAPMoN: Background Air Pollution Monitoring Network). At the regional and continental stations of this network, air and precipitation chemistry data are produced which make the determination of dry and wet deposition possible (Mészáros, 1985). By comparing deposition values to emission data the difference of pollutant "exportation" and "importation" over a certain region can be estimated. Data measured in BAPMoN are also useful to check the results of long-range transport calculations.

4. THE ROLE OF METEOROLOGY IN GLOBAL POLLUTION STUDIES

Owing to the interaction of sources and sinks some anthropogenic gases have an atmospheric residence time of several years. Therefore, they are well mixed in the entire troposphere causing pollution problems of global scale. From the meteorological point of view the most important environmental consequence of these gaseous pollutants is the modification of atmospheric composition. Since composition controls short-wave and long-wave radiation transfer in the atmosphere, modifications in composition can cause changes in climate. Because the effect of anthropogenic carbon dioxide and other radiatively active gases on climate variations will be discussed in another paper in this volume, the role of meteorology in this very serious environmental problem is only briefly outlined here.

First of all, meteorological services and institutions have to monitor changes in atmospheric composition on a global scale. This so-called climate-related environmental monitoring is done at the baseline stations of the WMO BAPMoN, the number of which is over 10 at present (Pueschel, 1986). The observations carried out at these stations provide important information concerning changes in atmospheric composition. The well-known carbon dioxide measurements at Mauna Loa Observatory, (Hawaii, USA), constitute a good example in this respect.

The second role of atmospheric scientists in this field is the estimation of possible climatic consequences caused by anthropogenic modification of atmospheric composition. While this problem is much more complicated than is discussed here, we note that this work is done by means of appropriate climate models in which meteorological information and products among other input data, for example possible emission inventories, are used. Since human life and more generally the state of the biosphere are extremely climate dependent, such modelling activities are very important for decision-makers responsible for planning human activities and social behaviour in the future.

5. CONCLUSION

The role of meteorology in environmental management is most essential. While meteorological activity in itself is not sufficient for the protection of the environment, meteorological information and products constitute the basis of several important decisions, from local scale emission control to planning of future human activities.

REFERENCES

Eliassen, A, 1978 The OECD study of long range transport of air pollutants: long range transport modelling. Atmospheric Environment 12, 479-487.

Högström, U, 1978 Dispersal models for sulfur oxides around chimneys and tall stacks. Sulfur in the Environment (ed J O Nriagu) 123-169. Wiley and Sons, New York.

Mészáros, E, 1985 The importance of deposition measurements in the study of atmospheric sulfur and nitrogen cycles: an application of BAPMoN data. Special Environmental Rep. No 16, 604-622, WMO-No 647.

Pueschel, R, 1986 Man and the composition of atmosphere. A contribution to GEMS, WMO, 80 pp.

Rote, D M, 1980 Gaussian plume models-Theoretical aspects. Atmospheric Planetary Boundary Layer Physics (ed A Longhetto) 211-228. Elsevier, Amsterdam.

Strom, G H, 1976 Transport and diffusion of stack effluents. Air pollution, Vol I, Third edition (ed A C Stern) 401-501. Academic Press, New York.

APPENDIX I

Determination of the effective stack height

Many approaches can be found in the literature to calculate the final plume rise. A very simple formula was used by American workers of the Tennessee Valley Authority (see Strom, 1976):

$$h = 114\ C_1 F^{1/3}\ u^{-1} \qquad \text{(m)}$$

where

$$C_1 = 1.58 - 0.414\ \sigma\theta/\sigma z \qquad \text{(no unit)}$$

and

$$F = g\ V_s r_s^2\ /Ts-T/\ \ T \qquad (m^4 s^{-3})$$

while u is the wind speed (ms^{-1}).

In these latter two formulae $\sigma\theta/\sigma z$ is the potential temperature gradient (°C/100 m), g is the acceleration due to gravity (9.81 ms^{-2}), V_s is the emission velocity of the effluent from the stack (ms^{-1}), r_s is the radius of stack outlet (m), while T and Ts are the temperature of the ambient air and of the effluent at stack outlet (°C), respectively.

It is to be noted that the concentration at the plume centerline (at a height h_e) is given by

$$c(x,0,h_e) = \frac{Q}{2\pi\sigma_y\sigma_z}$$

which means that the concentration is inversely proportional to the wind speed and the strength of the turbulent diffusion (see also Fig 1).

APPENDIX II

Numerical calculation by the Gauss model

Suppose that we want to establish a new coal-fired power plant. According to the plans the emission of SO_2 and h_e are 10^5 mg s^{-1} and 100 m, respectively. According to meteorological observations carried out in the area the average wind direction and speed at this height are NW and $1 ms^{-1}$, while atmospheric vertical stability is neutral on an average (curves D in Fig 2).

Question A:

What will be the concentration distribution of SO_2 in the surface air along the direction of the average wind?

The first step to be done is the determination of σ_y and σ_z for different distances (x) from the stack by means of Fig 2.

The second operation is the calculation of C as a function of x by using equation (3) and the above meteorological information.

The results obtained are tabulated in the second column of Table II/1. As it can be seen, the concentration increases very strongly between 500 m and 1000 m to reach a maximum at 2000 m from the stack. After that it slightly decreases with increasing distance. Suppose that in the area considered the maximum allowable SO_2 concentration (air standard) is 0.50 mg m^{-3}, so we can state that the maximum concentration calculated is about two times higher than the air standard. We can solve this problem in three ways:

a. to place the power plant at another site,

b. to reduce the emission,

c. to increase the stack height

Since the increase of the stack height is not desirable because of large-scale environmental problems (e.g. acidification) there are only two possibilities.

The first possibility is to establish the power plant at another site economically still acceptable. Suppose that at that site the average wind speed is 5 ms^{-1} and the air is generally slightly unstable (curves C in Fig 2). In this case the concentration distribution in direction x is given by the third column of the table. We can see that changes in meteorological parameters modify drastically the distribution of SO_2. Higher wind speed and stronger turbulence result in lower concentrations in surface air. Thus, at the new site the maximum concentration is lower by a factor of two than the air standard for the area.

Finally suppose that the site of the power plant cannot be changed. In this case the following question arises:

Question B:

What is the maximum allowable emission for having concentrations below the air standard? The answer is very simple. Since the concentration is linearly proportional to the emission, we can say that a decrease by a factor of two is necessary to obtain this condition.

Table II/1

Concentration of SO_2 in mg m^{-3} at different distances from the stack (x) in the case of Q = 105 mg s^{-1} and h_e = 100 m.

x (m)	u = 1 ms^{-1} stability: D	u = 5 ms^{-1} stability: C
500	1.7×10^{-4}	0.03
1000	0.11	0.25
2000	1.02	0.17
5000	0.64	0.05
10000	0.29	0.01
20000	0.15	4.1×10^{-3}

APPENDIX III

Long-range transport calculations

The European Economic Commission has an environmental program called EMEP (Environmental Monitoring and Evaluation Program). The aim of EMEP is partly to monitor air pollution in the air over different European countries and partly to evaluate transboundary air pollution transport by means of appropriate atmospheric models. Table III/1 give an example of the results of calculations carried out for sulfur by the Meteorological Synthesizing Centre, West of EMEP (Norway) for Hungary in 1984.

It can be seen that in 1984 65% (226 x 10^3 t) of the total deposition was due to emission from Hungarian sources, while an important part of the sulfur deposited was emitted in other countries. On the other hand, considering the Hungarian emission in 1984 (552 x 10^3 t) we can state that 59% (552-226/552) of the Hungarian emission was removed from the air outside Hungary.

Table III/1

Sulfur deposition (in 10^3 tonnes/year) in Hungary during 1984 due to the emission of some neighbouring countries.

Austria	5
Czechoslovakia	23
GDR	6
Hungary	226
Italy	10
Poland	14
Yugoslavia	30
Total	348

Note: Hungarian sulfur emission in 1984: 552 x 10^3 t.

37

Education of users to increase the effectiveness of agrometeorological activities[1]

W. J. Maunder, New Zealand Meteorological Service

INTRODUCTION

Some of the impacts of severe weather and climate events on national and regional economies are now well reported. For example, in discussing "Europe's Big Freeze", Business Week (2 February, 1987) stated:

> 'The arctic weather threatens to push the Continent's first-quarter economic growth, like its temperatures, below zero. Such a weak start is the last thing Europe needs right now. ... There are also fears on the Continent that frigid weather may revive inflation. Drastic shortages are pushing up prices of fresh fruit and vegetables. French market vendors now ask $2 a lb. For in-season leeks, double the usual price. The Kiel Institute's Trapp estimates that the German consumer price index could jump 2% in January over December.'

Nevertheless, it is often very difficult to assess the increased economic and social benefit in agrometeorology that may arise from an improvement in the *use* of weather and climate *information*. Indeed in a simple manner one may consider that the economic outcome of an agroclimatic

---

[1] Based in part on extracts form the author's book The Uncertainty Business: Risks and Opportunites in Weather and Climate, Methuen, London, 1986; New York, 1987

sensitive process under management is influenced by four factors (McQuigg and Thompson, 1966). First, weather events (that is, what actually occurs or occurred at a specified place); second, weather information (that is, what is reported to have occurred, or what is forecast to occur, or what an analysis of climate information tells us about what has occurred in the past); third, non-weather events (such as the actual price structure and current government policy); and fourth, non-weather information (such as the expected price structure, or the expected government policy). In all cases, the economic and, in some cases, the social and political outcome of an agroclimatic sensitive process under management, is subject to some degree of uncertainty, principally because at the time the most appropriate alternative is chosen, the decision-maker does not know the actual value of either the 'weather information box' or the 'non-weather information box'.

Weather and climate information may of course constitute a number of separate items. It may, for example, consist of (1) the weather existing at the present moment, (2) the weather and the climate that is expected to exist at a specified time in the future, and/or (3) analysis and interpretation of the records of the weather and the climate that has existed in the past, including the recent past. All three types are of value (see: Maunder, 1970, 1981, 1986) but the specific type of weather or climate information required is (or should be) determined by the kind of problems which are confronted, and the associated decision-making involved.

For example, an analysis of the past weather and climate is useful for assisting in *planning* the location and design of many types of facilities, such as irrigation schemes, water storage systems, and agri-business enterprises. On the other hand, relevant real-time or near real-time weather and climate information is essential for any *operational* type decisions such as marketing, agricultural production forecasts, and analysis of agri-business trends. Further, present weather and short-period forecasts of future weather are most useful in making *operational* decisions such as the scheduling of irrigation, estimating fertiliser demand, or the forecasting of agricultural production and the "futures" prices of agricultural products.

Applications of meteorological and climatological information, have been considered by a number of investigators including those at a special symposium on The Business of Weather held over 20 years ago in Chicago. Hallanger (1963) made some very pertinent remarks to businessmen at that time which are possibly even more relevant today than they were in 1963. He stated:

> . . . the key point in the realisation of the potential value of weather information to your activity (is that) . . . the meteorologist, working as a team with your people, must become familiar with your operation. Only then is he in a position to identify the true weather problems. Only then can he provide the appropriate weather information in the most useful form. Only then will the return exceed the cost by the greatest amount.

Thus, the potential value of weather information to an economic activity (including farming and related agri-business activities) and to a particular operation will be realised only when the qualified meteorologist and/or climatologist working with the co-operation and backing of management, develops the information most suited to the specific needs of the activity.

REALISING THE VALUE OF WEATHER AND CLIMATE INFORMATION

In most countries, changing social patterns, trends in population growth from rural to predominantly urban areas, the accelerating demand for energy, the changing impact of markets, and the emergence of more sophisticated and often weather - and climate-sensitive systems, have created a need for more rational and effective responses to atmospheric events.

In this regard it is relevant to ask three important questions as they apply to any country or region: first, what is the current and potential value of weather and climate services; second, how will prospective advances in the atmospheric sciences improve the usefulness of such services; third, what are the potential benefits or losses associated with the deliberate or inadvertent modification of both the weather and the climate?

Until the early 1960's, questions like these, or attempts to answer them, were of little concern to world governments or to the meteorological community. In the last decade however, several factors have contributed to a need for much more serious 'meteorological bookkeeping'. But can these questions be answered? In some cases the answer is no, while in others the answers can only be given imprecisely. Nevertheless, attempts to answer such questions must be made for two main reasons: first, both scientific and technical advances have brought about increasing understanding of the atmosphere; second, complex sensing, communications, and data processing systems are capable of producing a virtually unlimited flow of data.

In assessing the value of weather and climate information, the concept of the 'weather and climate package' is useful.

What, for example, is the value of weather forecasts published in newspapers, the value of agricultural weather forecasts to farmers, the value of forecasts of extreme weather to catchment authorities, or the value of the published and unpublished climate data? Moreover, how does the value of *this* weather and climate information compare with the value of *other* weather and climate information which could be provided with present technology and finance, and the potential information which could be provided if technology and finance were provided? Regrettably answers to most of these questions are not known. In particular, very little is known about how efficient the present 'weather and climate package' is, or whether it is economically desirable to provide more weather - and climate-related information. Further, if a better 'weather and climate package' was made available, would the users of meteorological and climatological products know how to use it, and how would these customers use the 'package' to their advantage?

It is clear that the decision-making involved in managing a weather - and climate-sensitive aspect of an economy or a business must take these factors into account. The decisions involved are made by many people with a wide spectrum of responsibility, experience, and training (both in the meteorological and non-meteorological sense). But it must be clearly realised that the potential value of the flow of weather and climate information in this context, can only be realised when it affects the decisions made by managers of weather - and climate-sensitive processes.

COMMUNICATING THE WEATHER PACKAGE

Information about the weather and the climate is a key factor in many socio-economic aspects of weather and climate enterprises. Indeed, in several countries specialist information companies can provide users with any kind of weather and climate information they require - at a price. Most of the information marketed by such companies is highly time dependent but whether the question relates to, say the most optimal routing or a ship from Calcutta to New York, or weather and soybean production in Brazil, information about the past, present and future weather and climate conditions is available to assist the decision-maker.

The presentation of such weather and climate information is, or course, a highly specialised and important form of communication. Specifically, the weather and climate package can be viewed in three phases; first, the *preparation* of the package - the technical task of the meteorologist or climatologist; second, the *presentation* of the package through radio, television, newspapers, journals, or videotex systems; and third, the *use or application* of the weather and climate package by the consumer. In all cases education and marketing is the key: first, education of the meteorologist and the climatologist in the techniques of the media; and second, marketing of the weather and climate package to the consumer in the best possible way. To do these things, meteorology and climatology must broaden its vision by actively encouraging research in the social, economic, and marketing aspects of the profession. But of prime consideration should be the realisation that meteorology, and particularly the communication of weather forecasts and weather information, embraces many things besides the physical sciences.

WEATHER AND CLIMATE INFORMATION AND DECISION-MAKING

The need for a more rational and effective response to atmospheric events, and the associated requirement for better and more useful information about the atmosphere, has resulted from an increasing awareness that information about the weather and climate can play a very important role in the decision-making processes associated with the management of weather - and climate-sensitive enterprises. Included in these enterprises are aspects of regional, national, and international economies concerned with agriculture, energy, trade, insurance, and prices.

At the international level, the problem is considerably more complex, but it is evident that all economies - whether these are monitored by international organisations, national governments, or private companies - are sensitive to weather and climate variations, the most obvious examples being food and energy aspects of economies. In addition, it is important to note that since it is the relatively 'short-term' variations from the 'near average' weather and climate that most governments and international agencies have to contend with on a day-to-day and week-to-week basis, the recognition that 'short-term' variations are important - economically, socially, and politically - could well help to avoid problems such as those which first arose in 1973, when grains were 'traded' between a number of countries to an unnecessary degree, described by one writer as 'the great grain robbery' (see Trager, 1975). Of course the 'long-term' variations are also important, but to most if not all governments it is *this* seasons 'crop' which is of paramount importance and not what *may* happen to the crops next decade.

Although it is often suggested that the provision of better information (both weather and non-weather) will lead to better decisions, a great deal depends on the way in which the information, and specifically in this case weather and climate information, is presented to, and used by, the consumer. Clearly, it is the presentation of the 'package' to the consumer that is critical to the success of the marketing strategy of both government funded and 'private' meteorological services, and there is strong evidence that this aspect of the 'weather business' will become even more important in the future.

In this regard, a recent review of assessing user requirements and the associated economical values for both short-range weather forecasts and current weather information made by Murphy and Brown (1984) is noteworthy. In commenting on the nature of user requirements they emphasise three factors:

> First, it should be recognized at the outset that requirements will vary from user to user, even for individuals involved in essentially identical activities. . . . Second, the degree of sophistication of users -more accurately, the degree of sophistication of their information-processing and decision-making procedures - has important implications for the manner in which user requirements can and should be met . . . Third, studies of user requirements should be sufficiently detailed to ensure that it is possible to differentiate between activities that are weather sensitive and activities that are weather information sensitive.

In discussing the third factor Murphy and Brown correctly point out that many operations that are affected by weather conditions may not be sensitive to weather information, either because of the presence of constraints on the options available to the user, or because the state of the art of observing and forecasting the weather is such that information with the required accuracy or with a sufficient lead-time cannot be provided. They further note that many studies of user requirements and economic values have not been sufficiently detailed to distinguish between these two conditions and, as a result, have led some investigators to misjudge the requirements for and overestimate the values of meteorological information.

NATIONAL BUSINESS AND GOVERNMENT

When the economic planning processes in many countries are examined, it is evident that many top-level decision-makers (in both the government and private sectors) appear to be unaware of the potential value of weather information (past, present and future). Important questions arise from these matters, notably: (1) what is the role of the meteorologist in national economic planning and in advising top-level decision-makers, (2) what is the function of a government meteorologist or national meteorological service, and (3) what is the function of a consulting (or private) meteorologist or company in these processes?

It is obvious that the answers to these questions will vary from country to country, particularly where both government and private meteorologists are active. For example, in the United States, more than 200 professional meteorologists and climatologists are employed in over 30 consulting meteorological companies. The role of the 'public service' meteorologist/climatologist in the United States may therefore be quite different from that in most countries where 'private' meteorologists/climatologists are few or non existent. The situation in the United States is discussed by Epstein (1976) who, speaking for the National Oceanographic and Atmospheric Administration (NOAA), stated: "The industrial meteorologist represents an indispensable link in our ability to bring the best meteorological services to the nation. We have important jobs to do, and we do them best when we work together".

INTERNATIONAL RESPONSES AND CONCERNS

In considering the atmosphere as a resource, and in particular as an elite resource, there are three important factors to be considered: first, monitoring and understanding its variability, secondly, predicting its variability, and thirdly, assessing the impact on consumption and production of this variability. Furthermore, because the *impact* of short-term variations in the available atmospheric resources can expect to continue to increase in importance because of the growing demands for food and energy, any change in the climate - for the worse - could well result in food shortages and energy supply problems that will be far more critical in some areas than will occur if the climate remains *normal*. Appropriate international meteorological planning must therefore be evolved if we are to live within the limit of our atmospheric resources. But, if this is to be accomplished, the international politician and the planner must become more weather orientated, for only then will optimum use be made of the climate resources of the 1990's and beyond. Central to this meteorological planning is the need for a much more comprehensive monitoring and analysis of the world's weather and climate, both to detect and predict changes, and to understand the consequences of such changes.

An excellent example of good international meteorological planning is the lead role taken by the World Meteorological Organization (WMO). WMO is a unique international organisation which, among other things, provides an open forum for discussions on a wide range of meteorological and climatological matters. Most importantly it allows the small nations of the world (both developing and developed) to have a considerable influence on how this United Nations specialised agency is organised. Indeed from an organisational point of view - particularly in regard to standards of observations, exchange of information, formats etc., WMO is very much respected among the various specialised agencies of the United Nations. But perhaps the greatest contribution of WMO is through its Secretariat and Officers, and through its various technical committees and commissions, who collectively are very much aware of the economic, social and political implications of international meteorology and climatology.

INFORMATION FOR DECISION-MAKERS

To assist decision-makers, national meteorological services, and several private meteorological companies prepare and disseminate a wide variety of weather and climate forecasts - ranging in time from a few hours to a few months. For example, the Swedish Meteorological and Hydrological Institute (SMHI) is developing a new meteorological information system called PROMIS-90, which covers the whole spectrum of predictions from very short-range to medium-range. In a paper discussing some aspects of PROMIS-90, Liljas (1984) notes that in order to establish a cost-effective and developing weather service the Swedish weather service has to work, both operationally and with developmental tasks, after the principle that 'weather information has no value until it is used with success in weather sensitive plannings and decisions'. Liljas (1984) further comments that the requirements for very short-range forecasts are increasing, but that high costs are involved in the development of a meteorological information system for the whole spectrum of predictions from the very short-range to the medium range. However, Liljas also correctly emphasises the gap that exists between the large technical and economic efforts that are devoted to the collection and processing of data, and our lack of knowledge of what and how the weather-dependent society would really like to know.

Clearly, to provide a better service for decision-makers it is necessary to identify those activities directly or indirectly affected by the weather and the climate. National and consulting meteorological services must also provide decision-makers with 'real-time' weather and climate information, appropriately weighted by economic activities and areas, and this information needs to be available on a time-scale which provides sufficient lead-time for decision-making. A significant question in this regard is: Is there a difference between weather and climate? In a real-time sense the answer must be no; it is also evident that the time-frame is becoming less important as the ability to monitor *all* aspects of the atmospheric environment in real-time becomes a viable and economic reality. In this regard it is relevant to note two important decisions of the Ninth Session of the WMO Commission for Climatology held in Geneva in December 1985. First, in terms of monitoring the weather and the climate in real-time, the Commission for Climatology endorsed the re-affirmation by the WMO Commission for Basic Systems of the principle of the 'free' exchange of World Weather Watch (WWW) data; second, the Commission for Climatology noted that many data requirements for climate application and impact studies are met by the operational exchange of data through the WWW system.

VIDEOTEX SYSTEMS : AN EXAMPLE OF INFORMATION-EDUCATION

The 1980's have brought significant changes in the way weather and climate information is communicated, and the development of videotex systems in many countries offer a considerable marketing challenge to meteorologists and climatologists in the provision of the past, present, and future weather and climate information. Indeed, it is, evident that as we move into the late 1980's communication of information will be paramount in many climate-dependent decision-making activities. This is particularly so, as the various videotex systems become more common.

Videotex systems are specially designed to provide information to specific groups. More specifically, it is evident that systems capable of providing information to groups such as wheat growers in Australia, grain storage companies on the Canadian Prairies, kiwifruit growers in New Zealand, international soybean buyers in St Louis, hydro-electric authorities in Switzerland, television companies in the United States, or wool marketing companies in the Southern hemisphere, will be the *norm* well before the end of the present decade. The type of information that can be provided (and already is in some cases) to these specialised customers include the whole range of weather and climate products, including - at least in theory - every weather and climate forecast, and every piece of weather and climate data available in every national meteorological service. That such a range of information will be available to any customer at the touch of a button, creates a formidable marketing and educational challenge to meteorologists and climatologists, but this will only be realised if a bold and imaginative approach is used. But clearly, there are already developments in a number of countries which will mean that *communicating* weather and climate information will finally breakthrough the communications barrier.

WEATHER AND CLIMATE : THE CHALLENGE OF OPERATIONAL DECISION-MAKING

While it is comparatively easy to make assessments of the general relationship between weather and climate factors and some aspects of production or consumption, or in a few cases prices, the more precise relationships necessary for operational decision-making are much more difficult to formulate. Moreover, even with a perfect weather-and climate-economic model, major problems still exist in the acceptance by decision-makers of this new aid. In particular, the acceptance and use of commodity weighted weather and climate information, and forecasts of production, resulting from this information, clearly offers a challenge to both the meteorologist and the climatologist, as well as the user of weather and climate information, and a necessary first step is to place greater emphasis on the meteorological and climatological aspects of planning and development.

The question remains - where do we go from here: the issues are clear. First, the impacts of weather and climate on productivity and consumption should be assessed and presented in terms of production figures, costs, or other similar measures which can be used directly by decision-makers - including economists, agriculturalists, planners, and politicians. Secondly, national meteorological services or their equivalent should actively encourage personnel who have a background that will allow them to become 'development' or 'application/marketing type' meteorologists and climatologists. Indeed, one could and indeed must comment, following Bernard (1976), that the purely physical and mathematical approach of conventional meteorologists and climatologists results in their being too impervious to the scientific and technical applications of meteorology and climatology for socio-economic progress. However, it must also be stated that these comments *have* been acted upon by a number of national meteorological services including Canada, Japan, France, Sweden, United Kingdom, United States, and New Zealand; indeed, it is now clearly recognised that the days of the 'purely physical and mathematical approach of conventional meteorologists' are rapidly becoming ' a thing of the past'.

Clearly, the opportunities provided through *real* operational decision-making in the weather and climate business have (or will) become key issues in the meteorological or climatological scene. In this regard the 'lead' role of the World Meteorological Organization should be emphasised. For example, at the Ninth Session of the WMO Commission for Climatology held in Geneva in December 1985, it was agreed that to improve the usefulness of application activities, national Meteorological Services should develop capabilities for the extensive implementation of professional knowledge for using and interpreting the complexity of the data/information package. It was also noted that for these user-oriented activities, meteorologists have to understand the problem *from the users' point of view*; further, with a view to achieving mutual understanding, professional dialogues should be established and maintained continuously. Similarly, the Commission agreed that for many operational purposes new approaches are needed to tailor short-term weather forecasts to meet the requirements of specific users, and that both governments and specific users need to be better informed as to the relevance and value of both weather and climate services.

WEATHER AND CLIMATE : THE INFORMATION OPPORTUNITY

The sensitivity of the world's commodity markets to weather and climate information is a clear indication that, in the economic world, weather and climate sensitivity is a reality. There is also realism in the very difficult areas of disaster relief, and agricultural and energy policies. However, the real sensitivity - in economic, social and political terms - of nations, sectors of nations, and commodities to weather and climate variations and changes, has to be better understood. Indeed, the need for such understanding offers the most important challenge to the meteorological and climatological community (see Mather et al., 1981), and a necessary step is to educate both the *producers* of weather and climate information, and the potential *users* of these products, in the *specific* applications of weather and climate information to problems.

The connections between the difficulties of weather (and climate?) forecasting and the equally complex problems which politicians and social scientists face, point clearly to an even more difficult problem when one tries to link the meteorological and climatological system *with* the economic, political and social system. It can of course be done, indeed it must be done, and Gordon McKay had some very pertinent comments to make in an editorial in Climatic Change (Vol. 2, No.1, 1979):

> Most applications of climatology have advanced because of recognised value rather than through theoretical conjecture. Where value can be demonstrated clearly, the product will be demanded. Our challenge is to produce practical information that can be readily understood and integrated in a smooth and timely fashion into the planning process. The chances of success in this regard are improved when the planning process is understood - they are much improved when the user is convinced and involved.

These comments underline the difficulty facing applied climatologists and applied meteorologists in convincing the potential decision-maker of a weather or climate sensitive operation, that there *is* much more to meteorology and climatology than tomorrow's forecast or the average rainfall. The key factor is the need for closer involvement with the user to ensure viability, relevance and real benefits from new information. Indeed, the users' interest is, says McKay, 'in more useful information, not in answers to complex problems that they do not understand or complex answers they have to suspect. Usefulness is the prime criterion'.

REFERENCES

Bernard, E A, 1976: Costs and structure of meteorological services with special reference to the problem of developing countries. WMO Technical Note, No 146.

Epstein, E S, 1976: NOAA Policy on industrial meteorology. Bulletin American Meteorological Society, 57: 1334-1340.

Hallanger, N L, 1963: The business of weather: its potential and uses. Bulletin American Meteorological Society, 44: 63-67.

McQuigg, J D, and Thompson, R G, 1966: Economic value of improved methods of translating weather information into operational terms. Monthly Weather Review, 94: 83-87.

Mather, J R, Field, R T, Kalkstein, L S, Willmott, C J, and Maunder, W J, 1981: Climatology: The impact of the seventies and the challenge for the eighties. Weather and Climate, 1: 69-76.

Maunder, W J, 1970: The Value of the Weather, Methuen and Co, London, 388 pp. Also published as University Paperback No. 347.

Maunder, W J, 1981: The economic climate : fact or fiction? Proceedings of the Eleventh New Zealand Geography Conference, pp. 187-192.

Maunder, W J, 1986: The Uncertainty Business : Risks and Opportunities in Weather and Climate, Methuen and Co, London, 420 pp., Also published by Methuen Inc, New York, 1987.

Murphy, A H, and Brown, B G, 1984: Short-range weather forecasts and current weather information: User requirements and economic value. Proceedings Nowcasting II Symposium, Norrkoping, Sweden.

Trager, J, 1975: The Great Grain Robbery, Ballantyne Books. (Revised edition of The Amber Waves of Grain published in 1973).

38

Agrometeorological crop monitoring and yield forecasting

G. F. Popov, Food and Agriculture Organization of the United Nations (FAO) Rome, Italy

1. INTRODUCTION

The evolution of crop production over the past 30 years has shown striking progress in most developed countries, both in yield and in global production, although this increase has been counteracted somewhat by a reduction in areas under cultivation. In developing countries, however, increase in production has been much slower, and in most of them anyway the population has grown drastically, thanks to better living conditions. In addition to this, climatic variability has produced some extreme periods of drought, especially over the last decade, and often in several production areas at the same time.

As a result of this evolution, it has become increasingly necessary to forecast the size and quality of harvests -particularly for cereal crops - which remain the basic source of food in most parts of the world. Forecasts are very important in helping the producing countries to know in advance what percentage of the harvest can be exported. They are also very important for food-importing countries, in order to predict in advance the size of their national harvest, their exporting capacity or alternatively, the percentage of the country's food consumption which will have to be procured abroad. The same remarks apply, of course, to smaller areas within a given country.

Finally, it is imperative that the international and donor agencies know the food procurement possibilities and requirements in advance, so as to organize emergency food assistance programmes in the best possible manner.

In the last 20 years, good progress has been made towards this goal, through the use of agrometeorological information.

2. AGROMETEOROLOGICAL MODEL

2.1 Objectives

An agrometeorological model was designed in FAO (Agrometeorology Unit, 1976), for crop monitoring and forecasting, based on a cumulative weekly or ten-daily crop water balance, which at a given moment of the crop growing cycle gives an index expressing the degree of satisfaction of the crop water requirements. This index is strongly correlated with the yield and gives a very good idea of the yield to be expected.

The method has been successfully utilized for the purpose of the Early Warning System in many countries in Africa. Over recent years it was also utilized to follow up the summer crops or rice, maize and sorghum in South Asia. It relies on a minimum amount of actual data and makes use of some climatological information which may be assembled before the "operational" phase, i.e. the cropping season.

Finally, the method is designed in such a way that it provides a first qualitative monitoring of crop conditions by successive steps. The precision of these assessments will improve towards harvest time.

2.2 Basic principles of the method

The method is based on a cumulative water balance established over the whole growing season for the given crop and for successive periods of one week or ten days.

The water balance is the difference between precipitation received by the crop, and the water lost by the crop and the soil. Water retained by the soil should also be taken into account in the calculation. The method does not directly involve temperature, which conditions growth of the crop. However, it appears that the temperature intervenes indirectly in the balance. The effect of air temperature will be noted in the length of the growing cycle and in the calculation of potential evapotranspiration.

2.3 Steps of calculation of the water balance

The different steps of the calculation of the cumulative water balance are detailed hereunder, with an example (Figure 1).

STATIONS		MAY			JUNE			JULY			AUGUST			SEPTEMBER			OCTOBER			NOVEMBER			DECEMBER		
		1	2	3	1	2	3	1	2	3	1	2	3	1	2	3	1	2	3	1	2	3	1	2	3
MATAM 1978	PN	1	1	2	10	17	23	36	44	49	68	70	64	50	41	31	20	14	9						
	Po	0	0	0	0	23	11	33	46	17	41	10	6	23	21	29	0	14							
	do	0	0	0	0	2	1	2	5	2	5	1	1	3	2	2	0	2							
	PET				65	62	59	53	51	52	46	43	47	43	45	49	49	48	54						
	Kcr							0.3	0.4	0.5	0.8	1.0	1.0	1.0	0.6	0.5									
	WR							16	20	26	37	43	47	43	27	25				Σ WR 284 mm					
	Po-WR							17	26	-9	4	-33	-41	-19	-6	4									
	Rs							17	43	34	38	5	0	0	0										
	S/D							0	0	0	0	0	-36	-19	-6										
	I							100	100	100	100	100	87	80	78	78									
ZIGUINCHOR 1978	PN	1	3	6	23	40	62	118	121	124	176	180	176	130	120	111	85	35	26						
	Po	0	0	2	46	35	115	104	100	202	218	56	149	81	87	74	34	72							
	do	0	0	1	3	4	6	6	5	7	9	5	7	5	7	6	3	5							
	PET	64	61	62	52	48	44	41	38	41	35	34	38	37	39	40	41	41	45						
	Kcr				0.3	0.3	0.4	0.4	0.5	0.7	0.8	1.0	1.0	1.0	0.9	0.6	0.5								
	WR				16	14	18	17	19	29	28	34	38	37	35	24	20			Σ WR 329 mm					
	Po-WR				30	21	97	87	81	173	190	22	111	44	52	50	14								
	Rs				30	51	60	60	60	60	60	60	60	60	60	60	60								
	S/D				0	0	97	87	81	173	190	22	111	44	52	50	14								
	I				100	100	100	100	100	97	94	94	94	94	94	94	94								
NIAMEY 1978	PN	11	16	19	20	26	32	50	61	70	69	70	67	45	40	16	12	7	2						
	Po	1	0	64	71	18	10	28	35	44	54	51	90	37	40	26	21	2	0						
	do	1	0	4	4	1	2	2	2	4	6	2	4	2	6	2	3	1	0						
	PET	75	76	82	69	64	61	59	56	59	49	46	52	46	51	54	58	58	64						
	Kcr			0.3	0.3	0.4	0.4	0.5	0.7	0.8	1.0	1.0	1.0	0.9	0.6	0.5									
	WR			25	21	26	24	30	39	47	49	46	52	44	31	27				Σ WR 461 mm					
	Po-WR			39	50	-8	-14	-2	-4	-3	5	5	38	-7	9	-1									
	Rs			39	60	52	38	36	32	29	34	39	60	53	60	59									
	S/D			0	29	0	0	0	0	0	0	0	17	0	2	0									
	I			100	100	100	100	100	100	100	100	100	100	100	100	100									
ABECHE 1978	PN	6	8	10	6	8	12	40	48	53	77	80	75	30	20	17	8	6	3						
	Po	0	2	21	18	21	15	72	74	107	15	11	83	7	45	0	6	0							
	do	0	2	2	1	2	2	6	3	8	4	2	8	2	5	0	1	0							
	PET	67	67	72	64	61	58	55	57	52	43	41	45	44	55	50	56	58	54						
	Kcr							0.3	0.4	0.5	0.8	1.0	1.0	1.0	0.6	0.5									
	WR							17	20	26	34	41	45	44	28	25				Σ WR 280 mm					
	Po-WR							55	54	81	-19	-30	38	-37	17	-25									
	Rs							55	60	60	41	11	49	12	29	4									
	S/D							0	49	81	0	0	0	0	0	0									
	I							100	100	100	100	100	100	100	100	100									
DORI 1978	PN	7	9	10	14	20	25	49	50	52	63	65	61	40	32	24	10	4	1						
	Po	4	11	14	23	18	58	51	48	134	9	29	28	15	32	6	0	3							
	do	1	1	2	1	2	3	2	3	5	1	2	5	3	4	1	0	1							
	PET	75	78	80	68	63	59	59	57	59	48	47	50	47	50	52	55	59	59						
	Kcr						0.3	0.4	0.5	0.8	1.0	1.0	1.0	0.6	0.5										
	WR						20	24	29	47	48	47	50	28	25					Σ WR 318 mm					
	Po-WR						38	27	19	87	-39	-18	-22	-13	7										
	Rs						38	60	60	60	21	3	0	0	7										
	S/D						0	5	19	87	0	0	-19	-13	0										
	I						100	100	100	100	100	100	94	90	90										

Figure 1: Example of water balance calculation

2.3.1 Normal precipitation (PN) - ten day or weekly precipitation total calculated from long term series of climatological data for the station concerned.

2.3.2 Actual precipitation (Pa) - represents the total precipitation which falls in each ten days/week.

2.3.3 Number of days of precipitation in the ten days/week - shows the distribution of rain during the period.

2.3.4 Potential evapotranspiration (PET) - the potential evapotranspiration taken as reference in this work is the maximum quantity of water which may be lost by a uniform cover of dense short grass when the water supply to the soil is not limited, as defined by Penman (1948). When no parameters are available for the Penman calculation, the Thornthwaite formula may be used or good measurements made with a Pan A evaporation tank.

2.3.5 Crop coefficients (K_c) - cultivated crops, particularly annual crops, pass through several stages from emergence to maturity. The crop coefficient is the ratio between the evapotranspiration for a given crop during a given phenological stage and potential evapotranspiration PET as defined above. For the application of the method to different crops, suitable adaptation of the crop coefficients should be carried out.

For this reason, it is essential that the inventory of background data for crop monitoring includes very precise information on the variety of crops grown in the various areas, the normal length of their growing cycle and the date of sowing (Figure 2).

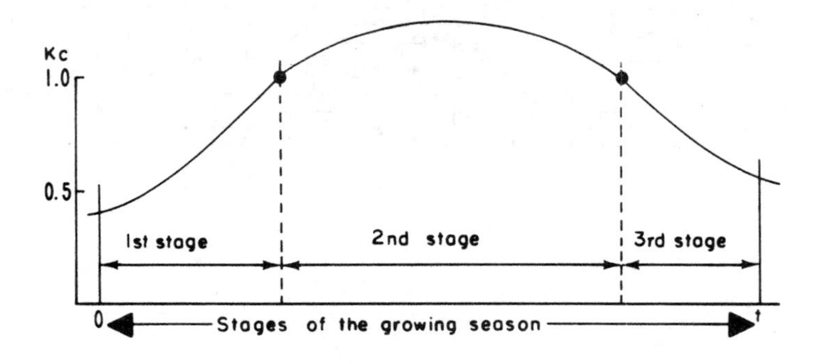

Figure 2

2.3.6 Water requirements of the crop (WR) - is arrived at by multiplying the potential evapotranspiration for the period by the respective crop coefficient for the same period. Since potential evapotranspiration is calculated from climatological values and since the crop coefficient is "pre-set" according to the normal length of the growing period, it is possible to calculate at the beginning of the season the total water requirements of the crop for the season by adding the successive water requirements ten days by ten days (or week by week).

2.3.7 Difference between actual precipitation and crop water frequirements (Pa - WR) This expresses the quantity of water available to the crops, without taking into account the water stored in the soil. From this it can be seen that the effect of a given precipitation may vary according to the crop development stage.

2.3.8 Water reserves in the soil (RS) - this term expresses the water stored in the soil, which can readily be used by the crop. In other words, it is the water content between the field capacity and the permanent wilting point. The amount of water usefully stored in the soil will depend on the depth of the soil exploited by the roots of the crop and the phyto-chemical characteristics of the soil.

2.3.9 Surplus or deficits of water (S/D) - this step indicates the surpluses or the deficits regarding the water storage capacity of the soil. Surplus refers to any quantity of water above the water retention level. Deficits refer to any water requirement under the zero level of the water storage capacity.

2.3.10 Index (I) - the Index (I) indicates in percentage the extent to which the water requirements of a crop have been satisfied in a cumulative way at any stage of its growing period.

The Index is calculated as follows: it is assumed that at the beginning of the growing cycle, sowing takes place when ample water is available in the soil. The Index is thus assumed to be 100 and will remain at 100 for the successive ten day periods until a deficit appears. If, for example, the water reserves fall to 0 and a deficit of 20 mm appears, then the quotient between the water deficit 20 and the total water requirement, say 400 mm (as calculated previously), is made and gives a value of 0.05. This corresponds to 5% of the water requirements which are not satisfied and the Index goes from 100 to 95. The calculation is continued until the end of the growing season, taking into account the fact that the index number starts in the first ten day period (or week) at

100 and thereafter can only remain at 100 or go down. The Index at the end of the growing season will reflect the cumulative stress endured by the crop and will usually be closely linked with the final yield of the crop, unless some other harmful factors (pests and diseases, for example) have significant effects.

For the quantitative estimates of yields, a direct relationship between the water requirement satisfaction index (I) and the crop yield is used. This yield can be expressed either in absolute figures (Kg/ha) or in relative figures (percentage of an optimal crop yield).

2.4 Other factors affecting yields

While the present method demonstrates the utility of calculating cumulative water balances for short periods of seven or ten days to show the yield losses due to water stresses in the plant during its growing cycle, it is also evident that other factors may contribute to the reduction of yields. These elements may be physical, such as strong winds or floods causing waterlogging, or biological, such as locusts, birds, fungi or insects.

For this reason, the establishment of a final forecast of the yields will depend in many cases on the water status of the plant, but should also take into account all the other causes. This is why information as complete as possible on all aspects of crop development is important for the establishment of a good crop monitoring and forecasting system in the country.

2.5 Applications and experience

2.5.1 General

The method of crop monitoring developed in FAO in 1976 after the catastrophic drought which struck the Sahelian area during the period 1969-73, has been tested in many countries and is now in regular use in the following countries: Cape Verde, Senegal, Mauritania, Mali, the Gambia, Burkina Faso, Niger and Chad, which utilize the method through the AGRHYMET programme, executed by WMO in association with FAO. In the first operational phase of the AGRHYMET programme, FAO is continuing the monitoring already carried out over the past 10 years.

In addition, FAO carries out some direct assistance programmes in this field in Nepal, Tanzania, Pakistan, Bangladesh, Bhutan and Zambia. WMO is also assisting Botswana along the same lines as the FAO method. A new FAO project is just starting to assist the SADCC countries (Southern Africa Developing Countries Conference), namely Angola, Botswana,

Lesotho, Malawi, Mozambique, Swaziland, Tanzania, Zambia and Zimbabwe. This assistance is part of a Regional Early Warning System for the SADCC countries.

The present method is also in use in various countries which have received some FAO assistance in this regards, namely Ethiopia, Togo, Turkey and Somalia. Other technical assistance projects in this field are in preparation for other countries such as Zaire, the Sudan, Madagascar, Rwanda, Burundi, Haiti and Nigeria. Finally, the method has been tested experimentally in some other countries like India and Thailand.

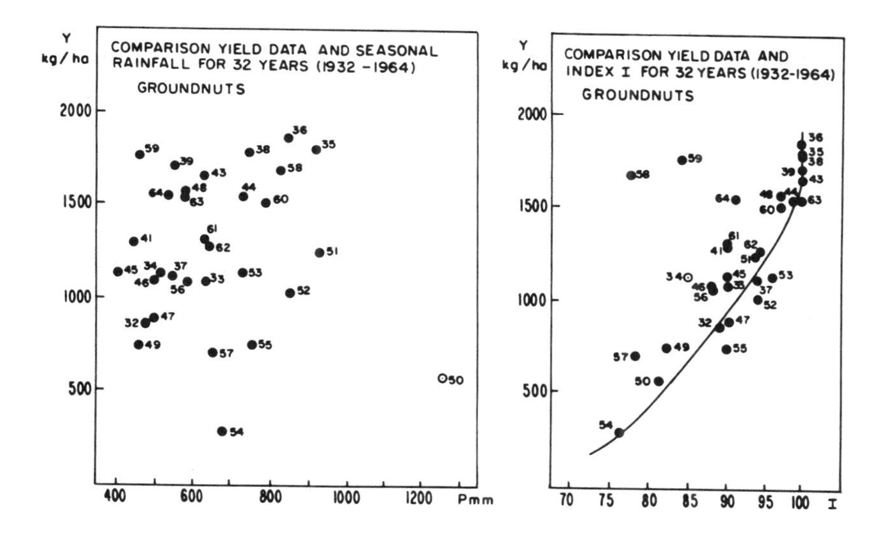

Figures 3 and 4

2.5.2 Senegal

The Centre of Agricultural Research in Bambey, Senegal, has experimented for many years with various food and cash crops. In particular, daily rainfall observations, sowing and harvest dates of yields of groundnut are available for a period of 32 years (1932 to 1964). The analysis of the results, illustrated in Figures 3 and 4, show that:

- there is no correlation between total seasonal rainfall and final yield of the crop (Figure 3);

- the distribution of rains over the season is usually more important than its total seasonal amount;

- minimum yields (about 700 kg/ha) occur in both the driest years (300-400 mm/year), and the wettest (1200-1300 mm/year);

- a good correlation exists between the water requirements satisfaction index (as calculated by the method) and the yield in Kg/ha (Figure 4).

2.5.3 Bangladesh The example taken for Bangladesh is interesting because the crop concerned is rainfed rice and because it comes from a wet, tropical climate which should not present drought constraints for crop growth.

Monsoon areas in Asia often show excessive rainfall amounts over the growing season. However, dry spells of two or more ten day periods can occur during the growing season and be harmful to crops. In this particular year, dry spells occurred during the growing cycle and in addition the monsoon rains ended early. Statistical production data confirmed the yield reduction pattern.

3. ORGANIZATION OF A CROP MONITORING AND FORECASTING SYSTEM BASED ON AGROMETEOROLOGICAL INFORMATION

The system, as it has been proposed, can be started in many cases with the present infrastructures and does not require any sophisticated equipment. The first step of such a system is to form a committee at national level comprising the various governmental services involved in agricultural production and covering the fields related to the factors influencing agricultural production. This formal committee should have an equivalent at the technical level, which would constitute an operational working group responsible for the conception and the day-to-day operation of the system in specific regions or at national level, according to the possibilities offered.

The number of institutions concerned may vary from one country to another, according to the organizational schema at national level or agricultural research and production. Normally, a working group is created with the participation of specialists from the following services: agricultural statistics, plant production, plant protection, meteorological service, soil surveys, land improvement, agricultural research, etc.

4. CONCLUSIONS

The method presented here does not claim to replace other forms of crop assessment based mainly on statistical sampling. Instead, it constitutes a useful complement, allowing an early assessment of the crop situation based on the causes of possible modification in crop behaviour leading to production losses. It is hoped that the introduction of such a method in national crop forecasting systems will improve their general food situation assessments.

On the other hand, any agrometeorological information which has accumulated can serve as a solid base for the determination of agro-ecological zones, actual and potential land use, agricultural planning, etc.

Finally; the application of agrometeorological crop monitoring and forecasting can be a beginning or a means to strengthening close collaboration between specialists of different disciplines, all of which are essential to the realization of the important objectives of agrometeorology.

39

Agrometeorological information for use in agricultural production, including weather forecasting

B. A. Callander, Agricultural Meteorology Section, Meteorological Office, Bracknell, UK

1. WEATHER DATA RELEVANT TO AGRICULTURE

The weather variables with the greatest impact on agriculture are:

i) Rainfall

ii) Sunshine (including photoperiod)

iii) Temperature and Humidity (including soil temperature)

iv) Wind

Of these, rainfall is normally the single most important, because it exhibits such wide fluctuations from year-to-year and from place-to-place. Often a crop can tolerate extreme values of temperature, sunshine or wind as long as it is not accompanied by extremes of rainfall.

1.1 Observed data

1.1.1 Standard observations

Meteorological observations, when they form part of a national or regional network, should be made under standard conditions of exposure. The WMO Guide to Agricultural Meteorological Practices (WMO, 1981 - hereafter called 'The Guide') gives detailed advice on instrument site and exposure in order to make observations that conform to standard practice.

Such 'Standard Observations' are designed to be representative of the weather conditions over a large surrounding area. That is, they measure the macroclimate. They do not measure the microclimate that surrounds the individual plant or animal. For example on a sunny day the maximum temperature may be reported as 32°C, but this relates to air temperature as measured in a screen and young plants, being so near the ground where the Suns's energy is being absorbed, would experience temperatures far in excess of this.

1.1.2 Microclimates

It would be impractical for meteorological services to collect observations of the microclimates that exist for a given macroclimate. Consequently it may be necessary to estimate conditions within a crop from standard observations only. A great deal of research has been done in this field in order to relate the microclimate over and within crops to the macroclimate above them. Rudolph Geiger was one of the first in the field of microclimatology and his book (Geiger, 1965) contains a wealth of observational data though with few formulae. Goudriaan (1981) draws together much of the more recent research with emphasis on algebraic formulae suitable for inclusion in computer models.

1.1.3 Spatial variability

Standard observations are designed to measure the macroclimate, but the macroclimate itself changes with distance. Unless you are fortunate to have your own meteorological site nearby you will often need to use data from a station or stations of the national network that may be several kilometres, or even tens of kilometres distant. How accurate will these data be for your own site? The main factors which cause differences in macroclimate between areas are:

i) Topography (including slope and aspect)

ii) Soil type and ground cover

iii) Proximity to the sea or large bodies of water

How representative a particular observing station is will therefore depend on how different these factors are between your site and that of the observing station. The one general rule is that rainfall is the most variable element, particularly where the mechanism of rainfall is convective ascent of air (e.g. thunderstorms) rather than widespread dynamic ascent (e.g. frontal rain). Beyond this there are few universal guidelines and it would be misleading in this paper to give sample figures for the rate at which measurement accuracy diminishes with increasing distance from the station, because of the wide variation of the above factors in different areas of the world. National meteorological services normally carry out studies in this field to help them plan the optimum observing network, so reference should be made to them for such information.

1.1.4 Quality Control

An important feature of any observing network is quality control of the data. There are attractions to operating your own on-site weather station, particularly if the nearest national network station is several kilometres away. But even with modern equipment considerable effort has to be devoted to sensor maintenance and to checks of the data in order to ensure that the recorded information is as free from errors as possible. Automatic weather stations which are controlled by a microcomputer may perform some simple checks of their measurements as they are made, but data collected by a national Meteorological Service should be subject to more thorough quality control, such as comparing a station's observations with near neighbours and with long term averages. Where this is done properly it adds significantly to the value of the data, particularly where long series of records are involved. It may be preferable to use records from a non-local station which are reliable, to records from a local station whose reliability is unknown.

1.2 Agrometeorological Products

So far we have talked of individual observations, but usually the influence of weather on agriculture is due to a combination of two or more variables, or to the mean value of a particular variable or variables. Dry matter production depends on accumulated intercepted radiation; evaporation depends on sunshine, temperature, humidity and wind speed; successful crop-spraying may require a particular 'window' of wind speed, temperature and no rainfall.

Any agriculturally-useful number or index which is derived from a combination of the basic weather observations will be called an agrometeorological product. A statistical quantity derived from a sample of measurements of a single entity, such as a mean temperature or a rainfall return period will also be termed an agrometeorological product.

The reason for making the distinction between variables and products is that while weather observations and hence weather records are mainly of single variables, the users of

Table 1

Presentation of agroclimatic information - from Section 3.2.4 of the Guide to Agricultural Meteorological Practices, WMO (1981)

(*a*) *Air temperature*
- (i) Temperature probabilities;
- (ii) Chilling hours;
- (iii) Degree days;
- (iv) Hours or days above or below selected temperatures;
- (v) Interdiurnal variability;
- (vi) Maximum and minimum temperature statistics;
- (vii) Growing season statistics;
- (viii) Frost risk;

(*b*) *Precipitation*
- (i) Probability of specified amount during a period;
- (ii) Number of days with specified amounts of precipitation;
- (iii) Probabilities of thundershowers; hail;
- (iv) Duration and amount of snow cover;
- (v) Date of beginning and ending of snow cover;
- (vi) Probability of extreme precipitation amounts;

(*c*) *Wind*
- (i) Wind rose;
- (ii) Maximum wind, average wind speed;
- (iii) Diurnal variation;
- (iv) Hours of wind less than selected speed;

(*d*) *Sky cover, sunshine, radiation*
- (i) Per cent possible sunshine;
- (ii) Number of clear, partly cloudy, cloudy days;
- (iii) Amounts of global and net radiation;

(*e*) *Humidity*
- (i) Probability of specified relative humidity;
- (ii) Duration of specified threshold of humidity with time;

(*f*) *Free water evaporation*
- (i) Total amount;
- (ii) Diurnal variation of evaporation;
- (iii) Relative dryness of air;
- (iv) Evapotranspiration;

(*g*) *Dew*
- (i) Duration and amount of dew;
- (ii) Diurnal variation of dew;
- (iii) Association of dew with vegetative wetting;
- (iv) Probability of dew formation with season;

(*h*) *Soil temperature*
- (i) Mean and standard deviation at standard depth;
- (ii) Depth of frost penetration;
- (iii) Probability of occurrence of specified temperatures at standard depths;
- (iv) Dates when threshold values of temperature (germination, vegetation) are reached.

meteorological information usually require it in the form of products. So the message to a user is, wherever possible, avoid accessing basic weather records. Always aim to obtain the data in as processed a form as possible.

National Meteorological Services normally calculate weather statistics on a routine basis. If the service has a strong hydrological or agricultural interest then it may also provide some of the agrometeorological products listed in Section 3.2.4 of the Guide and shown here in Table 1.

The important point for the non-meteorological user is this: find out what products are available before asking for basic data.

1.3 Forecast Information

1.3.1 Forecasting of Weather

A primary function of most national meteorological services is to provide weather forecasts. Relatively few provide forecast services that are specifically tailored to agriculture, and not simply adaptations of the general weather forecast.

One basic feature of forecasts is that the further ahead they look, the less accurate they tend to be. This may seem obvious but it is a fact often ignored by users. Forecasts are updated at regular intervals and users should always attempt to access the latest forecast. In the UK a 5-day farming forecast is issued on a Sunday and updated on a Wednesday. Yet often farmers are caught out by the weather on a Thursday because they are still planning their activities on the basis of the Sunday forecast.

Meteorological Services constantly monitor the accuracy of their forecasts. With faster and bigger computers and improved understanding of the atmosphere it is technically possible for every Met Service to continually improve their forecasts, though lack of resources may prevent them from doing so. Figure 1 demonstrates how the UK Met Office model has steadily improved over the last 10 years. 'Pressure Change Correlation' is now around 90% for a 24 hour forecast but falls to around 70% for 120 hours ahead. By comparing the 24 hour and 72 hour traces we can see that the 3-day forecast is now as good as the 1-day forecast was 10 years ago. Figure 2 shows the relative performance of the model in the Northern and Southern hemispheres and in the Tropics.

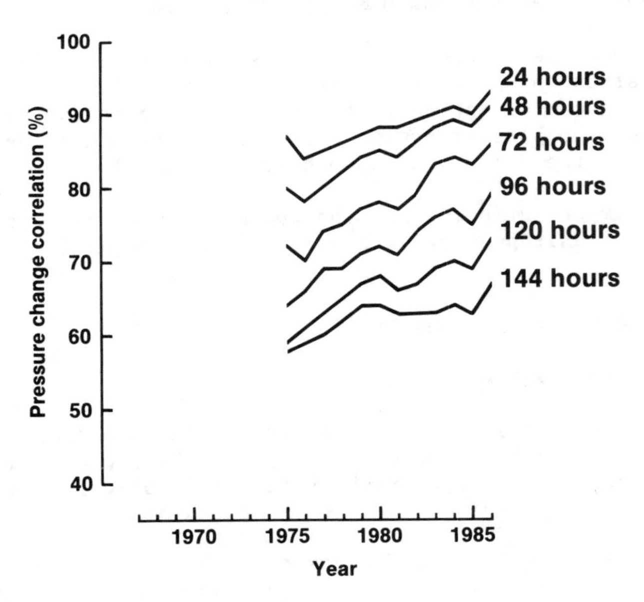

Figure 1
Improvements in accuracy of Met Office forecasts for the North Atlantic area, tested by comparing predicted with actual changes in Mean Sea Level Pressure (MSLP)

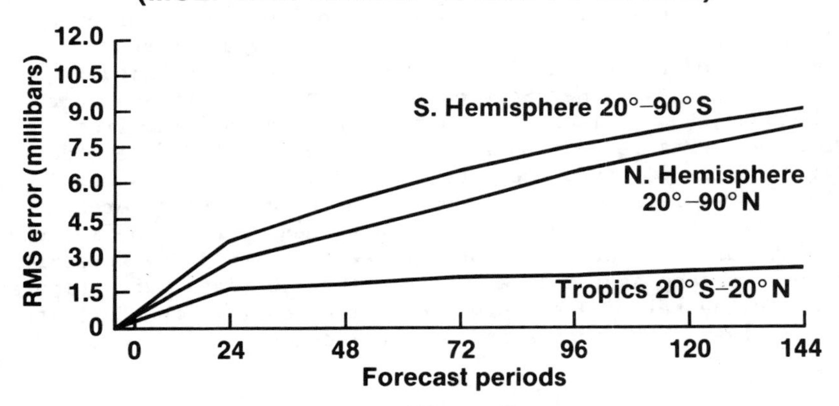

Figure 2
Current accuracy of Met Office forecasts of global Mean Sea Level Pressure (MSLP)

Forecasts generated by a human forecaster are produced by developing and moving forward weather systems (i.e. fronts, centres of low pressure, ridges etc.) according to well-tried rules. The coming weather is then inferred from the position and strength of the various systems. By knowing the typical weather sequences of his own area and some of these basic meteorological rules it is possible for the non-meteorologist by intelligent observation of the clouds and weather to successfully predict the weather over the next few hours.

Numerical forecasting models which run on computers start from fields of initial values of meteorological variables on a regular grid (temperature, pressure, wind speed etc.) and predict how these fields will change over time. When the human forecaster sees plots of these fields he has to decide how they relate to the weather systems that he - and you and I - are familiar with, and then to translate these into terms of sunshine, temperature, wind and rain. Thus the traditional human forecaster deals mainly with pictorial data while the computer deals with numerical data.

1.3.2 Agrometeorological forecasting on the basis of forecast weather

Chapter 4 of the Guide gives many examples of the way in which forecast values of weather variables can be combined to produce, in effect, forecast products that are of more direct value to agriculture. Figure 3 is one example. Instead of issuing forecasts of the individual elements that affect evaporation, all the factors are combined into a number that relates directly to the amount of irrigation that needs to be applied.

Water loss forecast - An example:

"Free-water loss during the past 24 hours average 0.6 cm. Expected free-water loss today is 0.6 cm and tomorrow is 0.8 cm.
Rainfall probability will remain low for the remainder of the week and crops will begin to suffer from moisture stress unless supplementary water is applied".

References: WMO Technical Notes Nos 21, 83, 92, 97, 126

Figure 3
An example of an agrometeorological forecast from the Guide to Agricultural Meteorological Practices (WMO, 1981)

Preparation of this type of forecast requires agrometeorological knowledge and experience, and not all Meteorological Services have the necessary staff. An alternative is for agrometeorologists within the national agricultural extension service to compile agricultural weather forecasts on the basis of raw forecast data received from the National Meteorological Service. A recent survey of operational agrometeorological services in WMO Region VI shows that most weather-based disease warnings are issued by a national agriculture services, not the Meteorological Service.

Because it is difficult for a human forecaster to assimilate all the information present in weather charts, or to provide the required level of detail in his forecast (which usually has to cover a large area), there are advantages in deriving forecast products entirely automatically. To produce an automated forecast product (AFP) it is necessary to be able to access, by computer, the data files containing forecast fields. The forecast products is then calculated automatically from these data according to formulae programmed into the computer. The advantage of AFPs is their relatively low cost of production.

At the time of writing this is a relatively new area of forecasting, and not many Services are researching it. One main reason is that agriculture requires forecasts of values at or near the ground. Numerical models are designed to predict the state of the atmosphere as a whole, and the values they predict for variables at the surface are subject to errors because of the large effect that small scale features, such as topography, type of crop, local soil moisture status, can have on the microclimate.

In practice this difficulty can be solved to a large extent by developing regression equations between observations at a particular site and variables predicted by the model. For example, observed minimum temperature could be statistically related to model predictions not only of temperature, but also of wind speed and cloud cover at the neighbouring model grid points. This technique of statistically relating actual observations to model predictions is called Model Output Statistics. Once the equations are derived they can be used to produce a station-specific forecast from the model forecasts. However the regression equations tend to be unique to each station so an extensive amount of development work is required before such a scheme is operational

1.3.3 Agrometeorological forecasting on the basis of recent weather

It is not necessary to use forecast weather in order to make agrometeorological forecasts. Current soil moisture status depends on past rainfall and evaporation; the growth stage of a crop depends on the total number of day-degrees accumulated since sowing, and predictions can be made of the growth stage likely to be reached in the next week or month on the basis of accumulated temperature to date, and of climatological averages for the season. Disease warnings based on current weather are often called 'forecasts' though they may contain no forecast weather data.

Chapter 5 of the Guide describes a range of methods of agrometeorological forecasting covering crop development, soil moisture and crop yield. Most of these are purely empirical models - statistical correlations between the variable of interest and those weather variables known, or thought likely, to affect it. The statistical approach is often the most practical way of modelling very complex processes; its major disadvantage is the difficulty of successfully applying the derived formula to other climatic regions.

Physical models of crop growth are under development in various countries. These require much more information about the physics of the underlying processes, so require more detailed validation data. However in the longer term these models offer greater prospect for successful application in a range of climates.

2. SOURCES OF METEOROLOGICAL INFORMATION

2.1 Basic Data

First of all I underline again the point that you should avoid, wherever possible, referring to basic weather observations. However, there will be occasions where, because of the nature of your work, or because the basic data have not been processed, it will be necessary to refer to unprocessed weather data. For example, imagine that research leads you to believe that a particular crop is sensitive to extreme daytime values of vapour pressure deficit followed by low night-time temperatures. To find out how often such conditions are encountered you would have to look at records of basic observations because this information could not be obtained directly from any simple statistics. Where would you look in order to get this information? Before answering this question it is necessary to explain that national or regional observing networks contain different kinds of station.

2.1.1 Synoptic Stations

Every country maintains a number of stations which make hourly or three-hourly observations of weather, and which feed these observations directly to their national Meteorological Service. The national Service not only uses these data in support of their day-to-day operational services, but also passes the data onto the WMO Global Telecommunications System (GTS). GTS, which allows the free exchange of real-time meteorological data between countries, is an essential part of weather forecasting, because the large computer models used in numerical weather prediction need regularly to 'know' the state of the atmosphere before they can carry out their calculations of how it will develop over the next few days. A global forecast model will require data from several thousand surface stations at the start of a forecasting run.

Stations which observe weather at such regular times and transmit the data for immediate use are synoptic stations.

2.1.2 Climatological stations

There are also stations, often run by non-meteorologists, which make daily or twice-daily observations of certain weather parameters such as maximum and minimum temperature, daily wind run, hours of sunshine, soil temperature and rainfall. Such stations are climatological stations. If certain biological observations are also made they may qualify as agricultural meteorological stations (Chapter 2 of the Guide gives full specifications for agricultural meteorological stations). Climatological stations recording only one weather element may also be called rainfall, anemograph (wind) or sunshine (or radiation) stations. Normally their data are sent to the national Meteorological Service at monthly intervals though some report daily or even in real-time to their regional or national centre.

Both climatological and synoptic data are valuable for research work, but for operational services which require near real-time data, it is normally only synoptic stations which can supply the data quickly enough.

2.1.3 Directories of synoptic stations

WMO-No 9 'Weather Reporting' (WMO, 1986a) lists all currently-available synoptic observing stations according to Region. Figure 4 is a sample entry from WMO-No 9 showing the station parameters that are normally listed. Each national Meteorological Service maintains records of its own synoptic observations and the data can be accessed through your own Service. In many cases the data can be provided on computer tape or disk.

Through GTS your service also has access to synoptic data from other countries, particularly those within its own Regional Association. Your Service may store some of these foreign synoptic data particularly if they are used for

REGION III - SOUTH AMERICA / AMERIQUE DU SUD

COLOMBIA / COLOMBIE

INDEX NUMBER		NAME	LAT.	LONG.	ELEVATION HP	ELEVATION H/HA	PRESSURE LEVEL	SURFACE OBSERVATIONS 00	03	06	09	12	15	18	21	OBS.H OBS.S	UPPER-AIR 00	06	12	18	OTHER OBSERV. AND REMARKS
80144	P	QUIBDO/EL CARANO ...	05 43N	76 37W		33		23	.	.	.	X	X	X	X	H10-24	.	.	.	.	A;METAR;CLIMAT(C);EVAP; SPECI;SUNDUR
149	P	MANIZALES/LA NUBIA .	05 02N	75 28W		2080		.	.	.	.	X	X	X	X	H11-23	.	.	.	.	A;METAR;SPECI
175	P	TRINIDAD/TRINIDAD ..	05 26N	71 39W		217		.	.	.	.	X	X	X	.	H12-20	.	.	.	.	
210	P	PEREIRA/MATECANA ...	04 49N	75 48W	1342	1338	STATION	23	.	.	.	X	X	X	X	H10-01	.	.	.	.	A;METAR;CLIMAT(C);SPECI; SUNDUR
211	P	ARMENIA/EL EDEN	04 30N	75 43W	1219	1204	STATION	.	.	.	.	X	X	X	X	H11-23	.	.	.	.	A;METAR;SPECI
214	P	IBAGUE/PERALES	04 26N	75 09W		928		.	.	.	.	X	X	X	X	H11-23	.	.	.	.	A;METAR;CLIMAT(C);EVAP; SPECI;SUNDUR
219	P	GIRARDOT/SANTIAGO VILLA	04 17N	74 48W	293	286		.	.	.	.	X	X	X	X	H11-23	.	.	.	.	A;METAR;CLIMAT(C);SPECI
222	P	BOGOTA/ELDORADO	04 42N	74 08W	2548	2547	STATION	X	X	X	X	X	X	X	X	H00-24	RW	.	RW	.	WTR;A;METAR;CLIMAT(CT); EVAP;SPECI;SUNDUR
234	P	VILLAVICENCIO/ VANGUARDIA	04 10N	73 37W	431	423		23	.	.	.	X	X	X	X	H10-24	.	.	.	.	A;METAR;CLIMAT(C);SPECI; SUNDUR
241	P	GAVIOTAS	04 33N	70 55W	167	165		X	.	.	.	X	X	X	X	H10-24	.	.	RW	.	WT;A;METAR;CLIMAT(CT); EVAP;SPECI;SOILTEMP
251	P	BUENAVENTURA/ COLPUERTOS	03 55N	77 05W		10		X	.	.	.	X	X	X	X	H10-24	.	.	.	.	METAR;CLIMAT(C);EVAP
252	P	BUENAVENTURA	03 53N	77 04W	15	12		.	.	.	.	X	X	X	X	H11-23	.	.	.	.	A;METAR;CLIMAT(C);EVAP; SPECI;SUNDUR
259	P	CALI/PALMASECA	03 33N	76 23W	969	962	STATION	X	X	X	X	X	X	X	X	H00-24	.	.	.	.	A;METAR;CLIMAT(C);EVAP; SPECI;SOLRA;SUNDUR
300	P	GUAPI	02 35N	77 54W		54		.	.	.	.	X	X	X	X	H11-17	.	.	.	.	
308	P	POPAYAN/MACHANGARA .	02 28N	76 36W		1730		.	.	.	.	X	X	X	X	H11-23	.	.	.	.	A;METAR;SPECI
315	P	NEIVA/LA MANGUITA ..	02 58N	75 18W	443	449		23	.	.	.	X	X	X	X	H10-24	.	.	.	.	A;METAR;CLIMAT(C);EVAP; SPECI;SUNDUR
318	P	SAN VICENTE DEL CAGUAN.	02 09N	74 48W		300		.	.	.	.	X	X	X	.	H12-19	.	.	.	.	A;METAR;SPECI
322	P	SAN JOSE DEL GUAVIARE	02 34N	72 38W	155			.	.	.	.	X	X	X	.	H12-19	.	.	.	.	A;METAR;SPECI
336	P	TUMACO/LA FLORIDA ..	01 49N	78 45W	6	3		.	.	.	.	X	X	X	X	H11-23	.	.	.	.	A;METAR;SPECI
337	P	TUMACO/EL MIRA	01 34N	78 41W	5	16		X	.	.	.	X	X	X	X	H10-24	.	.	RW 1)	.	WT;AGRIMET;METAR; CLIMAT(C);EVAP;SOILTEMP; SOLRA;SUNDUR
342	P	PASTO/ANTONIO NARINO	01 25N	77 16W	1826	1796	STATION	.	.	.	.	X	X	X	X	H11-23	.	.	.	.	A;METAR;CLIMAT(C);EVAP; SPECI
346	P	PITALITO	01 52N	76 03W		1320		.	.	.	.	X	X	X	.	H12-18	.	.	.	.	A;METAR;SPECI
354	P	FLORENCIA/CAPITOLIO	01 36N	75 32W		244		.	.	.	.	X	X	X	X	H11-23	.	.	.	.	A;METAR;SPECI

====================

1) TEMPORARILY DISCONTINUED / TEMPORAIREMENT SUSPENDUES

Figure 4
Page from WMO-No.9 "Weather Reporting" giving paremeters of synoptic stations in Colombia.

operational purposes. In general Meteorological Services maintain much more complete records from their own national network than from foreign stations.

2.1.4 Directories of climatological stations

There is no equivalent to WMO-No 9 for climatological stations yet available, though one for Latin America is currently being prepared for publication. However, each national service should maintain a list of all climatological stations which report to it. This list should include climatological stations which, though now closed, have provided data in the past.

It is of course possible to obtain data direct from a synoptic or climatological station for they will have records of all their own observations. However, they will rarely be in computer-compatible form and will not have been subject to the fullest quality control.

To assist research workers who need to access raw data, WMO (1965) has produced a 'Catalogue of Meteorological Data for Research'.

2.1.5 Specialised Data

Solar radiation and radiation balance on a global basis has been published since 1974 though the size of the collation task means that information is normally 18-24 months in arrears (SCHCNE, 1974 et seq).

Many meteorological services provide daily weather reports but these are usually in the form of analysed synoptic charts. These are useful if the weather sequence on one or two days is required, but it would be labour-intensive to extract numerical data from these for more than a few days. The European Centre for Medium-Range Weather Forecasting publishes daily analysis for the globe, though again the published information is 18-24 months in arrears. (ECMWF, 1982 et seq).

Satellite information is published by the US Department of Commerce (NTIS, 1979 et seq) and by the European Space Agency, (ESA, 1978 et seq) but this is of a very specialised form, is qualitative rather than quantitative, and requires skilful interpretation.

As a footnote to this section it is appropriate to mention the Data Rescue Project of the World Climate Data Programme. Recognising the steady deterioration of many original manuscript data records, this project aims to place on microfilm as much meteorological data as possible as soon as possible, particularly in Region I. More information on this project can be obtained from WMO in Geneva.

2.2 Meteorological summaries and products

2.2.1 Climatological summaries

Virtually every meteorological service publishes climatological statistics at least annually, often monthly and sometimes weekly. Because the aim is usually to make these statistics as complete as possible publication awaits receipt, followed by quality control, of all the data from the observing network including climatological stations. As a consequence the information may not be available for weeks or months after the end of the period.

The WMO World Climate Programme is operating an experimental monthly bulletin called 'Climate System Monitoring'. (WCP, 1984 et seq). This is a monthly publication based on weather observations transmitted on GTS, and includes not only summaries of weather conditions but also assessments of crop prospects. In the UK this is being received two or three months after the end of the month in question. In cooperation with WMO the United States publishes 'Monthly Climatic Data for the World' (NCDC, 1948 et seq).

Periodically particular countries or organisations publish extensive summaries of climate over the world. The US Dept. of Commerce publishes 'World Weather Records' in five volumes (USDC, 1985). The equivalent UK Meteorological Office publication (Meteorological Office, 1980) is in six parts.

2.2.2 Agrometeorological Information

In many cases meteorological information has been processed much further than simply producing climatological statistics. The extent to which this is done by a national Meteorological service depends on the emphasis it places on services to agriculture.

Relatively simple processing of the basic weather records can give agriculturally useful information such as:

Lengths of growing season

Accumulated temperature or sunshine likely to be available during the growing seasons.

Geographical limits for profitable crop production (on the basis of known relationships between crop growth and climate).

Evaporation, soil moisture and irrigation demand

Rainfall probability, including return periods

Work days (for example, how many days are normally available for crop spraying)

Research results that link crop response to weather have been summarised in a number of reports by the WMO Technical Commission for Agricultural Meteorology. Those available to 1986 are listed in 'Publications of the World Meteorological Organisation' (WMO 1986b). At the ninth session of the Commission for Agricultural Meteorology (CAgM, 1986) reports were received on the agrometeorology of the potato and sugar cane crops. This session also appointed rapporteurs to compile reports on the agrometeorology of the following crops: Coffee, Citrus, Grass and grasslands, Trees (as a component in a mixed trees/crops agricultural system), Yam, Chickpeas and Cassava.

Agrometeorological surveys of particular areas have been carried out by WMO in collaboration with other agencies (see WMO, 1986b).

2.2.3 CLICOM

The realisation that agrometeorological products are of much greater use to agricultural users than basic meteorological data is central to the concept of CLICOM (WCP, 1986). Developed internationally under the auspices of the World Climate Programme, CLICOM is not simply a data base or a computer or a program, but a whole system designed

a) To gather together all available climate data in a specific region in standard format

b) To provide readily and widely available computer hardware and software which will process the information into useful agrometeorological products.

When a country or region acquires its first CLICOM system it is necessary for existing meteorological records to be keyed into the computer (or loaded directly if the data are already stored in a form compatible with CLICOM). After this is done, it is a simple matter for other users to make copies of the data. They are free to augment the data set as they wish. Because the data base follows a standard format, all of the programs written for CLICOM - and new ones are continually being produced - will operate on any CLICOM data base. Thus CLICOM allows a user to turn an archive of basic weather observations into more directly useable agrometeorological products; the only major effort required is the initial entry of the data.

Forecast runs are made from both 00 GMT and 12 GMT data times. Bulletins are coded either in WMO GRID (FM47-V) or WMO GRIB (FM92-VIII Ext). Further details can be supplied on request.

1. Global NWP products in GRID code (available about 0500 and 1700 GMT)

Elements:	Sea level pressure, surface wind (10 m), surface temperature (1.5 m). Maximum wind and tropopause. Heights, winds and temperatures at 850, 700, 500, 400, 300, 250, 200, 150, 100 mb. Relative humidities at 850, 700 and 500 mb.
Areas:	There are 16 areas. In each hemisphere there are two polar areas covering latitudes 70 to 75 and 80 to 90 degrees and four mid-latitude areas spanning 20 to 67.5 degrees in quadrants. A further four areas spanning 15N to 15S cover the tropical belt.
Resolution:	The mid-latitude areas are at a resolution of 2.5 degrees latitude by 5 degrees longitude; the tropical belt is at 5 degrees latitude by 5 degrees longitude. In the polar regions there is reduced longitudinal resolution.
Forecast Times:	Fields are available at 6-hour intervals from the analysis (T+0) to a forecast time of T+48. In addition, sea level pressure and 500 mb height are available at T+60, T+72, T+96 and T+120.

Table 2

Selected list of numerical forecast products available from Bracknell on GTS. For a complete list see Meteorological Office (1987)

2.2.4 Agrometeorological advisers

In a number of countries close links have been established between the national Meteorological Service and the agricultural extension service. Where these links exist agricultural field workers can talk directly to a meteorologist who not only has access to central records of weather data but can also act as an interpreter and advisor on those data.

2.3 Forecast information

2.3.1 Weather forecasts

Forecasters are usually available on radio, TV, newspapers or direct from the national Service. Often these forecasts are for the general use and are not phrased in terms specific to agriculture.

In some countries forecasts for a specific agricultural application may be available through a national agricultural organisation.

Forecast weather data is also transmitted on GTS. Table 2 is an example of some of the data sent from the UK Met Office computer at Bracknell; other major numerical weather prediction centres such as Washington and Moscow also send out forecast values on GTS. Output from the European Centre for Medium Range Weather Forecasting, on the other hand, is available only to those countries contributing to its operation.

While it is possible, with the cooperation of the national meteorological service, for any organisation to receive forecast data from GTS, the interpretation of these data does require meteorological expertise. Consequently any exploitation of forecast meteorological data for agricultural purposes should involve both meteorologists and agrometeorologists. Nevertheless in some areas the forecast data on GTS may be the only source of numerical forecast information available.

2.3.2 Agrometeorological forecasts (without forecast weather)

Yield of a nationally important crop is of strategic importance to a country, and often Governments, national organisations or financial institutes will attempt to predict the national harvest weeks, or even months in advance. The basis is often the statistical methods discussed in Section 1.3.3, though satellites can also provide the means to estimate crop prospects over large geographical areas.

While total national yield is important at Government levels, the day-to-day management of a crop is the concern of the farmer, who requires tactical information. Often these needs are met by regular bulletins of soil moisture status, disease risk, or crop growth stages issued by a national or regional agricultural service. Because this information usually implies some form of management input to a crop, commercial companies supplying, for example, irrigation equipment or agrochemicals may also operate a service providing agrometeorological forecasts based on current and past weather.

REFERENCES

CAgM, 1986 Provisional final report, WMO Commission for Agricultural Meteorology, Ninth session, Madrid, 17-28 Nov 1986.

ECMWF, 1982 et seq Daily global analyses, operational data assimilation system. European Centre for Medium Range Weather Forecasts, Shinfield Park, Reading.

ESA, 1978 et seq Meteosat Image Bulletin. MEP/Data Service, European Space Operation Centre, Darmstadt, Germany.

Geiger, R, 1965 The climate near the ground. Translation of 4th German edition of Das Klima der bodennahen Luftschicht by M N Stewart and others. Harvard University Press, Cambridge, Mass, 1965.

Goudriaan, J, 1977 Crop micrometeorology: a simulation study. Centre for Agricultural Publishing and Documentation, Wageningen, The Netherlands 1977.

Meteorological Office 1980 Tables of temperature, relative humidity, precipitation and sunshine for the world. Met O 856a, HMSO, London, 1980.

Meteorological Office, 1987 Quarterly report on Numerical Products from Bracknell, October-December 1986. Met O 2b, Met Office, Bracknell, UK.

NCDC, 1948 et seq Monthly climatic data for the world. National Climatic Data Center, Asheville, NC.

NTIS, 1979 et seq Key to Meteorological Records Documentation, National Technical Information Service, US Department of Commerce, Sills Building, 5285 Port Royal Rd, Springfield, VA 22161.

SCHCNE, 1974 et seq Solar radiation and radiation balance

data (The World Network). USSR State Committee for Hydrometeorology and Control of Natural Environment, Leningrad, USSR.

USDC, 1985 World Weather Records 1961-1970 (Five volumes) US Department of Commerce/National Climatic Data Center, Asheville, NC.

WCP, 1984 et seq Climate System Monitoring (CSM) Monthly Bulletin. WMO World Climate Programme, Geneva, Switzerland.

WCP, 1986 CLICOM project (Climate data management system). WMO World Climate programme WCP-119, WMO/TD - No 131, Geneva, Switzerland.

WMO, 1965 Catalogue of Meteorological Data for Research. WMO-No 174 TP 86 Geneva, Switzerland. Part I. Published synoptic and climatological data. Part II. Meteorological stations with observational series extending over 80 years or more. Part III (1972). Meteorological data recorded on media usable by automatic data-processing machines. Part IV (1979). Sources of additional data needed for research on climatic change.

WMO, 1981 Guide to Agricultural Meteorological Practices. WMO-No 134, Geneva, Switzerland.

WMO, 1986a Weather Reporting - observing stations. WMO-No 9, Volume A, Geneva, Switzerland.

WMO, 1986b Publications of the World Meteorological Organisation. Geneva, Switzerland, 1986.

40

Introduction to the Telematic Irrigation Report

S. Dervaux and R. Specty, Météorologie Nationale, Strasbourg, France

HISTORY

In Alsace the climatic situation is particularly suited to the development of irrigation guidelines:

- winter precipitation does not always fill the reserves
- the rainfall deficit during the summer (from June 20 to August 20) varies between 130 and 170 mm
- maximum water available is often low (lower than 80 mm)

A study carried out by the Experimentation and Information Service at the National Institute of Agronomical Research shows a notable increase in crops (harvest increase of 83% for maize, 80% for rapeseed and 59% for wheat) and a levelling-off for the latter (for maize standard deviation of harvest without irrigation is 18 quintals (qx)/hectare (ha), with irrigation 5.6 qx/ha), 1 qx/ha is about 544 lb per acre, see Figure 1 on page 462.

Since 1964, the written guidelines had been based on the results of weekly water budgets which the farmer would receive by post. The potential evapotranspiration (PE) was estimated using the PENMAN formula; the farmer would be sent details of crop factors.

The water budget was gathered from 25 rainfall stations and from 3 meteorological stations measuring radiation, wind force and vapour pressure.

In the final version of the written report the farmer would, at the beginning of the irrigation season receive forms which allow him to keep a personal daily record of his fields: the coefficients used to calculate the maximal evapotranspiration (ME) would be given to him during the phenological stages, see Figure 2 on page 462.

The farmer would then carry out his weekly record starting from PE and rainfall totals which he would receive week by week. He would be advised to obtain a raingauge, as the wide differences in rainfall amounts is not sufficiently accounted for by the 25 Alsatian checking stations.

THE USE OF TELEMATIC

In 1982, the Department of Telecommunications launched a programme to develop the "Telematic". This was done by setting up the Teletel network, using the Transpac network and particularly involved the distribution of the Minitel terminal to the general public. This was to allow access to data banks (the first being the electronic directory).

The change to the telematic system gives us several advantages:

- the water budget can be made on a daily basis and without delay

- when the "ready-to-use" reserve is exhausted the evapotranspiration drops below the ME due to water stress. The water budget established by a computer can account for this actual evapotranspiration (AE).

- data of a more detailed nature can be given as can, in particular the length of time in which the "ready-to-use" reserve will be exhausted, in case of negligible rainfall.

- and especially this interactivity allows the joint gathering of quantitative results.

INTRODUCTION TO THE PROGRAMME

The first page, see Figure 3 on page 463, gives to the farmer the present state of his fields (he can keep a check on 8 fields): the proportion of the "ready-to-use" available and the time taken for the reserve to become exhausted in the case of negligible rainfall.

The farmer can obtain more accurate information of each of his fields and enter the information necessary for the calculation, see Figure 4 on page 463.

- The actual phenological stage: once this has been obtained it can determine the k factors used

- Last rainfall count: the farmer can at any moment gain access to the record of rainfall for his field and change it if it differs appreciably from his own check measured on his raingauge

- Record of last irrigation: irrigation and the rates of rainfall can be checked following the same procedure

- Present state of the reserve (represented by a funnel bearing a variable water level) showing the amount of water available to the plant, the rate of irrigation required and the number of days before the reserve is exhausted

Finally two types of checks are available, a table, see Figure 5 on page 464, in which all the parameters used in the water budget are recorded and a histogram, see Figure 6 on page 464, on which is figured the minimal state of the reserve over 5 day periods; the farmer must ensure that this figure does not go lower than the "survival" reserve.

At the start of the campaign, the farmer himself builds up a file of his fields. For each one he must enter:

- the type of crop

- the meteorological station to be used for the PE and rainfall

- the maximum available water; if the former is not known to the farmer the latter can be deduced by the depth of root-growth by the soil-type and by the density of stones (this part has been carried out with the help of the National Institute of Agronomical Research)

HOW THE SYSTEM WORKS

The formula used is:

ResD+1 = ResD + Rain + Irrigation - AE

where ResD is the state of the reserve at the day D

The maximum available water is represented by a reserve

which is divided into two parts:

the first is called the "ready-to-use" reserve (about two thirds of the maximum).

the second is called the "survival" reserve; when it is reached a water stress is supposed to reduce the evapotranspiration.

Rain: rainfall is measured in 25 checking stations and can be modified by the farmer.

Irrigation: the dates and water rates are indicated by the farmer.

AE = ME   if ResD + Rain + Irrigation are greater than the "survival" reserve

or

AE = ME * (ResD+Rain+Irrigation)   in the opposite case ("survival" reserve)

with ME = k * PE   where k is the crop factor varying through the phenological stage

Rainfall and PE are measured at 25 checking stations. The data which allow them to be calculated come from a number of sources:

- the synoptic meteorological network which transfers the figures from the 3 Alsatian meteorological checking stations (extreme temperatures, rainfall, sunshine, radiation, wind, vapour pressure)

- the automatic weather stations which measure the same parameters and which are checked daily by a computer via the public telephone network

- the inspectors of the climatological network who enter their figures through Minitel on the host computer

All the figures are collected daily; the PE is then estimated and an overall calculation for the fields can thus be made.

If then, on consulting the computer, the farmer enters new information, the calculation is immediately modified for that particular field.

It must be noted that this method for establishing the water budget is a compromise amongst numerous approximations:

- the figures for the reserve are generally not known precisely and they should vary with root-growth
- the PE is calculated on a daily basis by the PENMAN formula which is more suitable for a 10 day period
- the cultural factors should depend, not only on the phenological stage but also on the PE figure
- during the days of rainfall the calculation of evapotranspiration is perhaps underestimated
- the rates of irrigation are not always known precisely
- the rainfall rates should be measured more accurately in the field

OTHER ITEMS

The meteorological forecast over 5 day periods is adapted to the needs of the farmers and presented in sections:

Sky conditions: sunshine, clouds
Forecast phenomena: rain, storms, frost
Temperatures
Wind

The phytosanitary guidelines established by the Department for the protection of Crops (attached to the Ministry of Agriculture).

The stage of maturity of the maize according to the temperature.

The supplying of agrometeorological data, such as PE, rainfall temperatures and heat units over a daily or weekly period. Finally an electronic mail box allows communication amongst farmers, agricultural advisers and the host centre.

LIMITING FACTORS IN THE DISTRIBUTION OF THE PROGRAMME

This system of guidelines in irrigation was conceived to be used autonomously by farmers but has, during its operational phase, revealed that regular monitoring is needed. This problem must be resolved with the assistance of agricultural advisers; they are acquainted both with the new computer methods and the needs of farmers.

Irrigation is frequently used in Alsace and its application is often according to personal judgement. Certain farmers have used irrigation for more than ten years and believe they are employing it correctly.

Even if they recognise climatic differences, it seems that they have the tendency to inverse them, by underestimating hydric needs for rather humid years and by overestimating them for dry years.

It is easier for use to introduce the farmer to this method when he is in the process of obtaining irrigation materials then it is when he is a traditional irrigating farmer, preferring to trust in his past experience.

It is worth noting that often (before taking on the materials) farmers set up temporary irrigation fields to assess their needs.

The second obstacle to this method is the inferior capacity of equipment. During the irrigation period certain farmers cannot achieve proper water distribution despite the constant use of their materials.

They no longer need to keep a running check, and subsequent updating after a rainy period is not always considered necessary.

Finally, it is interesting to note that an increasing number of farmers equip themselves with tensiometers and are not therefore, interested in the method of running checks. A comparative study of the two methods is under way. The first results, obtained last year, show a high similarity between the two methods.

RESULT AND CONCLUSION

The service outlined here is still being extended; the numbers have expanded from 50 farmers in 1985 to 100 in 1986. This number seems to be maintained this year and one can hope that with the forthcoming period of maize irrigation, new farmers will again show some interest in the coming months.

It is worth bearing in mind that the national proportion of consumers of the telematic services amongst the farmers is 1 in 40 whilst in Alsace, counting only those using irrigation and this particular service, the proportion is 1 in 20.

The number of inquiries about the report made by the farmers using irrigation has grown from 1300 in 1985 to 2050 in 1986 and the first 3 months of 1987 already show a total of 920 calls despite a surplus rainfall. At the same time the meteorological forecast rose from 2500 in 1985 to 5500 in 1986 to 3300 during the months of April, May and June 1987.

In conclusion it seems that this type of distribution of information, updated on a daily basis, personalised and available on demand 24 hours a day, corresponds to the needs of farmers even if at first some of them are unsettled by the use of a computer.

	MAIZE			PEA	
	not irrigated	irrigated		not irrigated	irrigated
1967	10.7	76.3	1967	68.8	80.5
1968	71.1	78.2	1968	52.2	51.0
1969	38.3	62.5	1969	80.8	80.9
1970	44.6	69.8	1970	44.6	69.3
1971	39.0	69.4	1971	47.3	77.9
1972	52.2	73.2	1972	29.5	48.9
1973	57.3	78.2	1973	36.3	61.7
1974	38.2	74.9	1974	41.8	77.1
1975	40.8	76.5	1975	49.3	78.7
1976	15.0	82.2	1976	37.4	62.0
Average	40.7	74.1		48.8	68.8
Stand. dev'n	18.0	5.6		15.5	12.3

Figure 1

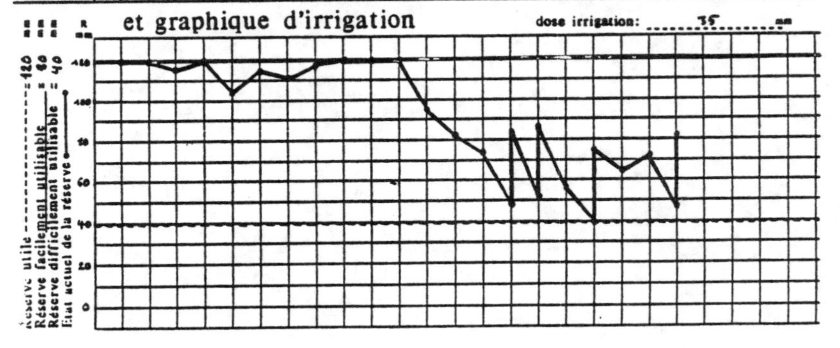

Bilan hydrique hebdomadaire à: Lochacker culture: Maïs Déc.

1984	avril					mai				juin				juillet					août				septembre			
lundi	2	9	16	23	30	7	14	21	28	4	11	18	25	2	9	16	23	30	6	13	20	27	3	10	17	24
ETP	15	10	17	20	32	17	19	21	19	19	26	36	31	30	29	39	29	27	34	15	26					
coef.cult.	0,1	0,1	0,1	0,1	0,5	0,6	0,6	0,7	0,7	0,7	0,7	0,7	0,7	0,8	0,9	0,9	1	1	1	1	1					
ETM	2	1	2	2	16	10	11	15	13	13	18	25	22	24	26	35	29	27	34	15	26					
pluies	12	2	0	6	0	21	7	22	25	21	29	0	9	16	1	2	0	10	24	22	1					
pluies-ETM	10	1	-2	4	-16	11	-4	7	12	8	11	-25	-13	-8	-25	-33	-29	-17	-10	7	-25					
Rés.avant	120	120	118	120	104	115	111	118	120	120	120	95	82	74	49	51	57	40	65	72	47					
irrigation	0	0	0	0	0	0	0	0	0	0	0	0	0	0	35	35	0	35	0	0	35					
Rés.aprés	120	120	118	120	104	115	111	118	120	120	120	95	82	74	84	86	57	75	65	72	82					

Figure 2

Mot clé IRRIG 00:41:21
BULLETIN DE L'IRRIGANT
MEYER Jean
SELESTAT
PARCELLES SUIVIES:
-1- HOERTLE AU 31.08
55 mm soit env 14 jours avant RS
-2- BEIM KAPUTZIENERWALD AU 31.0
60 mm soit env 15 jours avant RS
-3- AU MOULIN AU 30.09
36 mm soit env 28 jours avant RS
-4- CHEMIN VERT AU 30.09
40 mm soit env 31 jours avant RS
MESSAGE DEMONSTRATION
Pour les parcelles suivantes : SUITE
Pour avoir plus d'informations: GUIDE
Entrez votre choix et ENVOI :

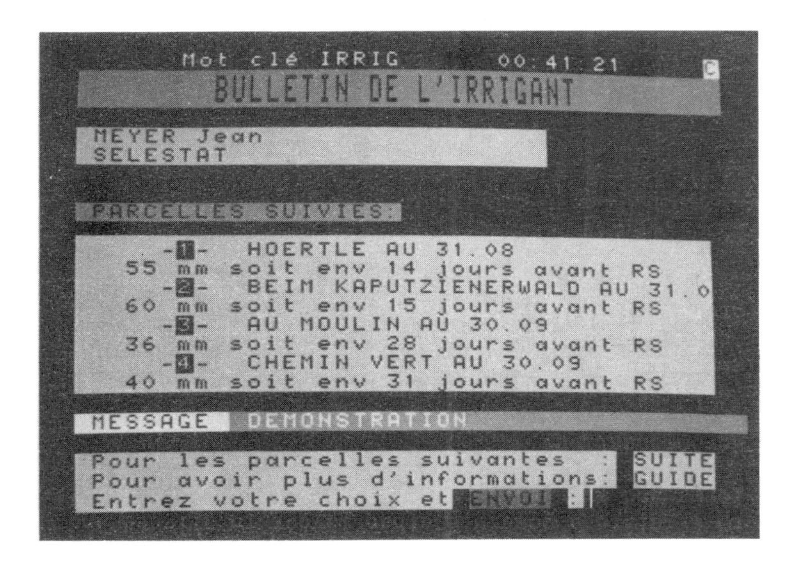

Figure 3

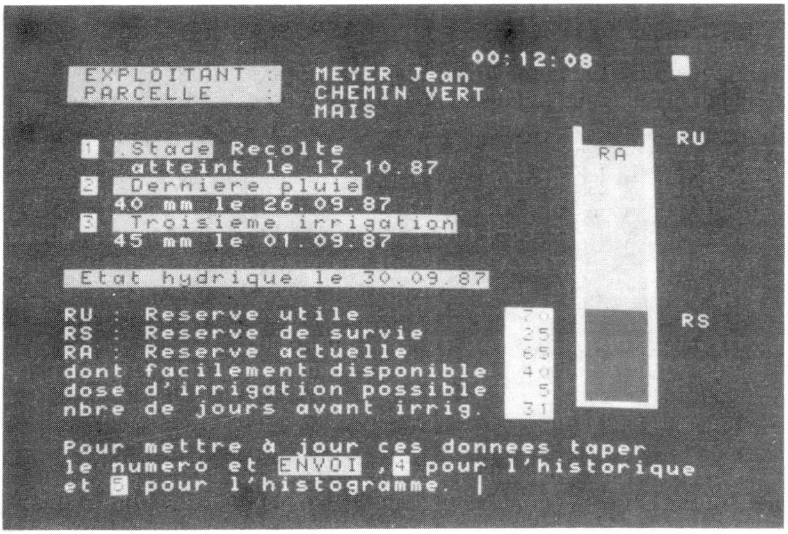

00:12:08
EXPLOITANT : MEYER Jean
PARCELLE : CHEMIN VERT
MAIS
1 Stade Recolte
atteint le 17.10.87
2 Derniere pluie
40 mm le 26.09.87
3 Troisieme irrigation
45 mm le 01.09.87
Etat hydrique le 30.09.87
RU : Reserve utile 70
RS : Reserve de survie 35
RA : Reserve actuelle 65
dont facilement disponible 40
dose d'irrigation possible 5
nbre de jours avant irrig. 31
RU
RA
RS
Pour mettre à jour ces donnees taper
le numero et ENVOI , 4 pour l'historique
et 5 pour l'histogramme.

Figure 4

00:17:04

BILAN HYDRIQUE JOURNALIER

NOM DE LA PARCELLE CHEMIN VERT

DATE	30.08	31.08	01.09	02.09	03.09
ETP	3,4	3	5,2	4,1	3
K	1	1	1	1	1
ETM	3,4	3,0	5,2	4,1	3,0
ETR	3,4	3,0	5,2	4,1	3,0
PLUIE	0	0	0	0	0
RES. AVANT	33,9	30,9	25,7	65,9	62,9
DOSE IRRIG	0	0	45	0	0
RES. APRES	34	31	70	66	63

ENVOI :reservoir RETOUR :5 jours prec.
JJ.MM.AA choix SUITE :5 j.suivants

Figure 5

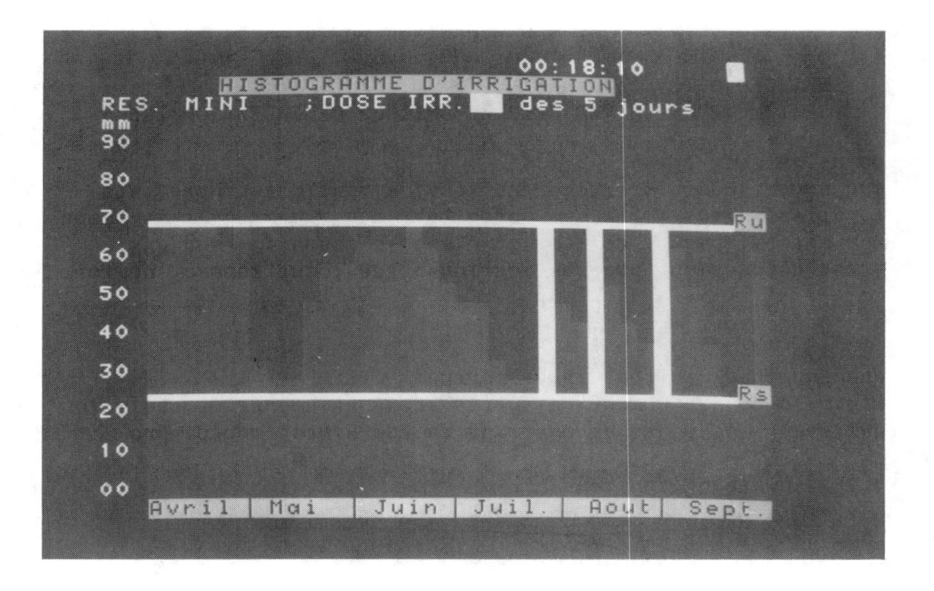

Figure 6

41

Bioclimatic analysis for the design of a town of 50,000 inhabitants at 4000 MASL – Pachachaca, Peru

Prof. Arch. T. Pesce, National Engineering University Faculty of Architecture, Town Planning and Arts Postgraduate and Secondary Specialization School, Lima – Peru

1. INTRODUCTION

Ever since the distant past, when mankind as an organized group needed to be on its guard to protect itself against the onslaughts of its environment, technological solutions were found which clearly identified dwellings and the subsequent outlines of towns; the similarity of these solutions was almost a constant throughout the world. There was almost a universal law that "the solution is the same where the climate is the same", despite the fact that there was no contact between the different groups in question.

Since the industrial revolution and the appearance of the automobile, cities have become ever more dense, both in population, and in construction; environmental pollution has also risen at a rapidly increasing rate. For this reason technology, making incredible and constant progress, provided a range of electromechanical solutions (fans, air conditioning, air filters and so on); this palliative response merely gave symptomatic relief and could not cure the disease: the problem was that cities had disrupted their natural ecological equilibrium, and designs for open, airy cities built of heat insulating materials right for the humid tropics had now become compact, hot and badly ventilated. Designs for dense cities, with dwellings jostling each other (the solution for conserving heat and humidity in a cold, dry climate), are now full of soot and the smell of burning oil, and there is a marked heat and humidity imbalance. Even though technology attempted to alleviate the pain, the disease was not cured; rather, it was advancing like a cancer as spatial and economic relationships had the upper hand over the need for a climatic technology for the city. With few exceptions, it has not been possible to do much in this direction. Indeed, the indiscriminate use of artificial means of providing comfort based on oil and other polluting sources of energy, which were therefore cheap, contributed to the progress of the diseases of the city.

Something unforeseen then occurred which unexpectedly benefited the city: the world oil crisis. The cities saw themselves threatened with having to bring their oil and petrol-driven automobiles to a standstill, and city-dwellers were soon profoundly affected by the threat of a lack of fuel against the cold. This, then, was the situation which gave rise to the need to seek other, more economic and less polluting, sources of energy. The sun was considered -sunlight was first used to heat water for domestic use and then to heat homes: "solar architecture" had come on the scene. A few short years passed, and two things became apparent in this technological development.

The first was that the use of solar energy had ceased to be an isolated fact; rather than using solar energy in isolation, the radiation phenomenon is taken side-by-side with suitable exploitation of wind power and an encompassing and thorough appreciation of all the meteorological phenomena which are the factors of climate. "Bioclimatic architecture" then made its appearance: the term is new, and the point will be to make the city a more human place. To take a complete overview of the problem, we are talking not only about a bioclimatic dwelling, but a bioclimatic city, in which biogas, wind and geothermal power will be in use, among other sources of energy.

The second was that we have started to look back through history - this time with our eyes open - at the solutions adopted by various past civilizations, in which, strange to tell (for there were no architects or town planners in the sense we normally use today), the inhabitants so adapted the city that they produced a coherent architectural and planning solution for the climatic environment, and this by means of a natural, ancestral knowledge inherent in each individual in the society. In only a few cases, where the civilization had begun to decline in accordance with its own historical law, did corrective solutions begin to be decreed: for example, the measures taken by Augustus in Rome, when he brought down the permissible height of buildings from 35 to 21 metres; or by Nero, who was concerned with the width of the streets. Both Augustus and Nero were concerned to provide better ventilation or more sunlight to people's homes. In any case, those ever-growing cities were never prepared for the car and even less so to house so many millions of inhabitants. In just the past hundred years, when urban development has reached its maximum development and complexity, plans for new cities have recently begun to take serious account of the climatic environment in which they are set and which they themselves generate.

2. BACKGROUND

It should be noted that Peru (population 18 million, 1.2 million km^2) has almost a third of its population concentrated in its capital (Lima, population 5.5 million) and 60% of its population in rural areas, and there is a marked degree of centralization in matters administrative, cultural, technological and so on. Because of this centralization, among other things, architectural production is concentrated in and around the capital and there are very few works in which the architect can play a part outside Lima: this has given rise to an architectural culture in Lima where there are major north American influences and perhaps a few European ones, and almost no coherence with the climate.

This, together with the climatic characteristics of Lima (its temperature of minimum 15°C, maximum 25°C, its humidity in winter of 98% during the nine overcast winter months, with a fine drizzle which never wets, with excessively sporadic and slight precipitation in summer - these may seem strange in a tropical location (between latitude 0° and 18°S), but are explained by the presence of the cold Humboldt current) means that the view taken of the climatic universe and the way it is interpreted are reduced to the area in and around Lima itself, which gives rise to a national confusion as to what constitutes summer and winter throughout Peru. This serious distortion, which is common at every level, means that in climatic architectural terms there are severe problems: when environments different from that of Lima are to be designed for, what happens is that the architectural output, which is designed more or less for the climatic realities of the capital, is grafted on completely regardless of the enormous variety of climates found in Peru.

There is no doubt that "international architecture", which also fails to remember the "Ten Books of Architecture" of Vitruvius (Rome, c. 15-12 BC), has an enormous influence on architectural production, with the result that in the capital itself architectural solutions are produced which bear no relation to the capital's climate. All these considerations are of enormous importance as it is possible that the same kind of thing is going on in many countries: this drives one to the conclusion that there is a need for the concept of the climate-city binomial to have wider currency, and for the nexus between climatologists and architects and town planners to be a matter of constant concern.

3. PACHACHACA

CENTROMIN PERU, the State body for mineral extraction and marketing in Peru, has seen the need to concentrate the scattered population around the rich mixed metal mines in the central range of the Peruvian Andes so as to provide it with a town for 50,000 inhabitants (approximately 10,000 families), and, under the auspices of the Peruvian College of Architects (the professional association for the trade), called a national town planning and architecture competition for multidisciplinary groups of professionals (town planners, architects, sociologists, economists, engineers and so on) to put forward ideas and proposals to this end.

3.1 The area and it population

The area proposed for the town is at 3890 masl, and is reached by road travelling west to east over the first range of the Andes through a gorge at 4980 masl. It is about 20 km from the Oroya, the main mining centre of the central Peruvian sierra. There is only a small settlement of under 1000 inhabitants, who have developed greater than normal thoracic capacities and cardiac volumes which enable them to play football after the day's work without getting any more tired than someone at sea level. Pachachaca has excellent conditions for a solar energy supply because of its altitude, as there is 4000 m less of the densest part of the atmosphere above it, which means that the irradiation arrives in nearly pure form; it also has the advantage of being at a tropical latitude, with the sun's rays striking the each almost vertically (see Figure 1 in the Figures Section on page 472 to 481).

3.2 The climate

The body running the competition provided only generic climatic data, which shows what a lack of climatic awareness there is, as we discussed in section 2. In particular, a minimum temperature of -9°C and a maximum of +19°C was mentioned, with no further data as to temperature. Nevertheless, we discovered that a meteorological station belonging to the firm and administered by the SENAMHI had been operating for the past year, and this provided us with extremely valuable information, particularly about insolation (annual daily mean: approximately 5 kWh/km^2); this is very unusual for meteorological stations in Peru. The information from this station is given below (Figure 2).

3.2.1 Precipitation

Monthly precipitation figures are low (15 mm) from June to September (winter) and relatively high (100 mm) in January, February and March (summer).

3.2.2 Relative humidity

Varies between 75% and 80%

3.2.3 Winds

In both January (summer) and June (winter) there is a marked daytime tendency for the wind to come up the valley from the ENE, and to blow down it from the WSW at night, with the aggravating factor that these night winds come down from the Andean snows.

3.2.4 Temperature

The absolute annual maximum is 16°C and the absolute annual minimum is -19°C, with a mean around 0°C. Diurnal temperature ranges were not available.

3.2.5 Hours of sunlight

Maxima of 200 hours of sunlight in the winter months and minima of 60 in summer, with the greatest amount of radiation around midday in the winter months.

3.2.6 Comfort

After corrections and applications, the local zone of comfort was established at between 22°C and 18°C and 19% and 82% relative humidity; it was noted that the annual daytime means and annual means of these two local climatic variables are below optimum (Figure 3).

4. DESIGN ANALYSIS AND RECOMMENDATIONS DEPENDENT ON THE METEOROLOGICAL VARIABLES

4.1 Solar radiation, temperature and humidity

4.1.1 Analysis

Solar radiation is high and can be used to counter the prevailing low temperature and humidity. Four possible passive solar heating systems are proposed for this (Figure 4).

- Water roof. A costly system, not very effective in buildings of more than one storey.

- Direct roof gain. The most economic system, but only a single storey can be heated.

- Direct gain through vertical windows. As the sun is almost overhead, solar penetration is lowest when the heat of the sun is greatest.

- Direct roof gain in offset buildings. This is the most effective, as all storeys receive sunlight.

- A study was carried out, using theoretical models, of the most favourable orientations towards the sun, and it was found the most advantageous were N, NE and E (Figure 7).

Combinations of these three were then studied comparatively using a suitable system of grading in order to see what were the effects of the shadows they cast (Figure 5).

4.1.2 Recommendations

- There should be a preference to build on the sides of the valley in order to let in the sunlight.

- Buildings should be designed low (one or two storeys) to avoid the negative effects of shadows which would necessitate the additional used of another form of energy (coal from the Goyarisquizga area, with optimized non-polluting burning systems) to heat the lower storeys (Figure 5).

- The distance between the models must be at a minimum to avoid heat losses, but still be sufficient to let in sunlight.

4.2 Winds

4.2.1 Analysis

A sequential study was carried out on a model which rotated through 360°, passing through 18 different orientations to determine the effects of the local winds on a wall taking into account the angle of incidence. Layouts were formed using the models to simulate a possible urban layout in aligned and offset forms and at various orientations (Figures 6 and 7).

4.2.2 Recommendations

- The walls facing the most frequent winds, especially the night-time winds from the E or SE which bring down masses of cold air from the Andean snows, must have special protection and insulation.

- The urban layout should be offset, with the orthogonality on the N-S axes.

5. ANALYSIS AND RECOMMENDATIONS ON THE SELECTION OF BUILDING MATERIALS

Of the materials which are readily available in the area, three were considered: brick, adobe and hollow concrete block, all of them in various forms. Despite the fact that earth walls (adobe) would apparently be the one to recommend most highly, given its thermal characteristics and low cost, we were obliged to select the hollow concrete block with a rendering based on cement and "diatomea" (a local material, with very good insulating properties). This choice was dictated by the fact that adobe cannot be used for buildings of more than one storey as the area is very earthquake-prone (Figure 8).

6. GENERAL RECOMMENDATIONS

The town and dwellings are to be a district where units are packed tightly together and are low in height to make solar heating possible, to obstruct the winds and to prevent the accumulated warmth being radiated out into the environment.

7. PROPOSAL

The town planning proposal is not perhaps the ideal response to the analysis and recommendations made; this is due to the fact that little experience has yet been gained in this respect. We should, however, state positively that the analysis and recommendations we have made are the first milestone along the way, and already indicate a change of attitude on the part of planners (Figure 9). We can also see from Figure 10 that the cheapest alternative domestic energy supply option is cooking with coal (improved low-pollution system) and hot water and heating using solar power.

8. CONCLUSIONS

It is important that countries should harmonize their activities towards town planning and architectural solutions which use non-polluting, passive energy concepts, as this not only improves the quality of life but is much more economic in the long term. There is no doubt that planners and specialists in handling meteorological information should embark now on a close relationship in order to discern and predict the climatic impacts which occur as cities grow. In order to achieve this, the first necessity is for there to be a major publicity and training campaign, particularly for architects, town planners and builders.

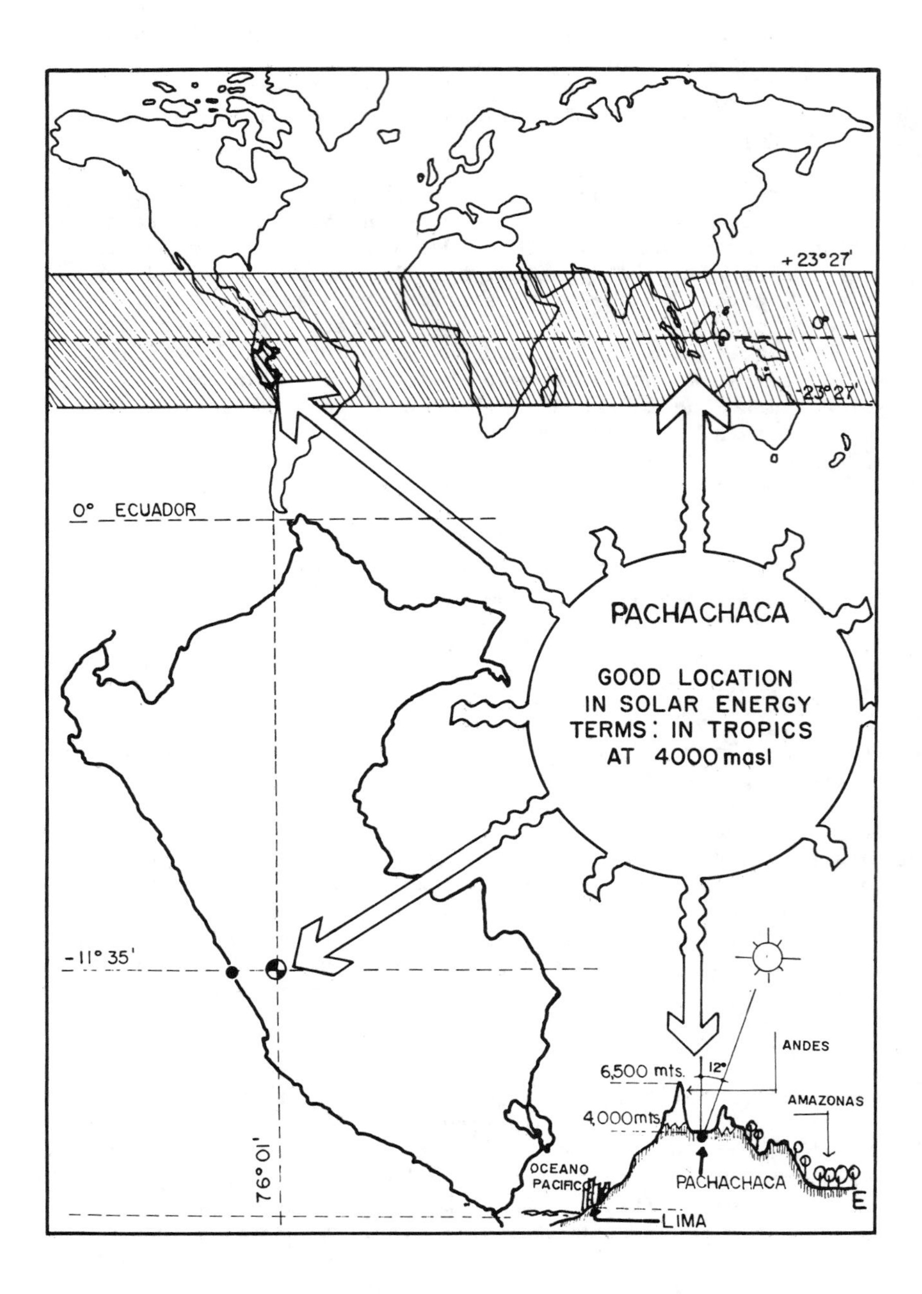

Figure 1

Geographical location of Pachachaca

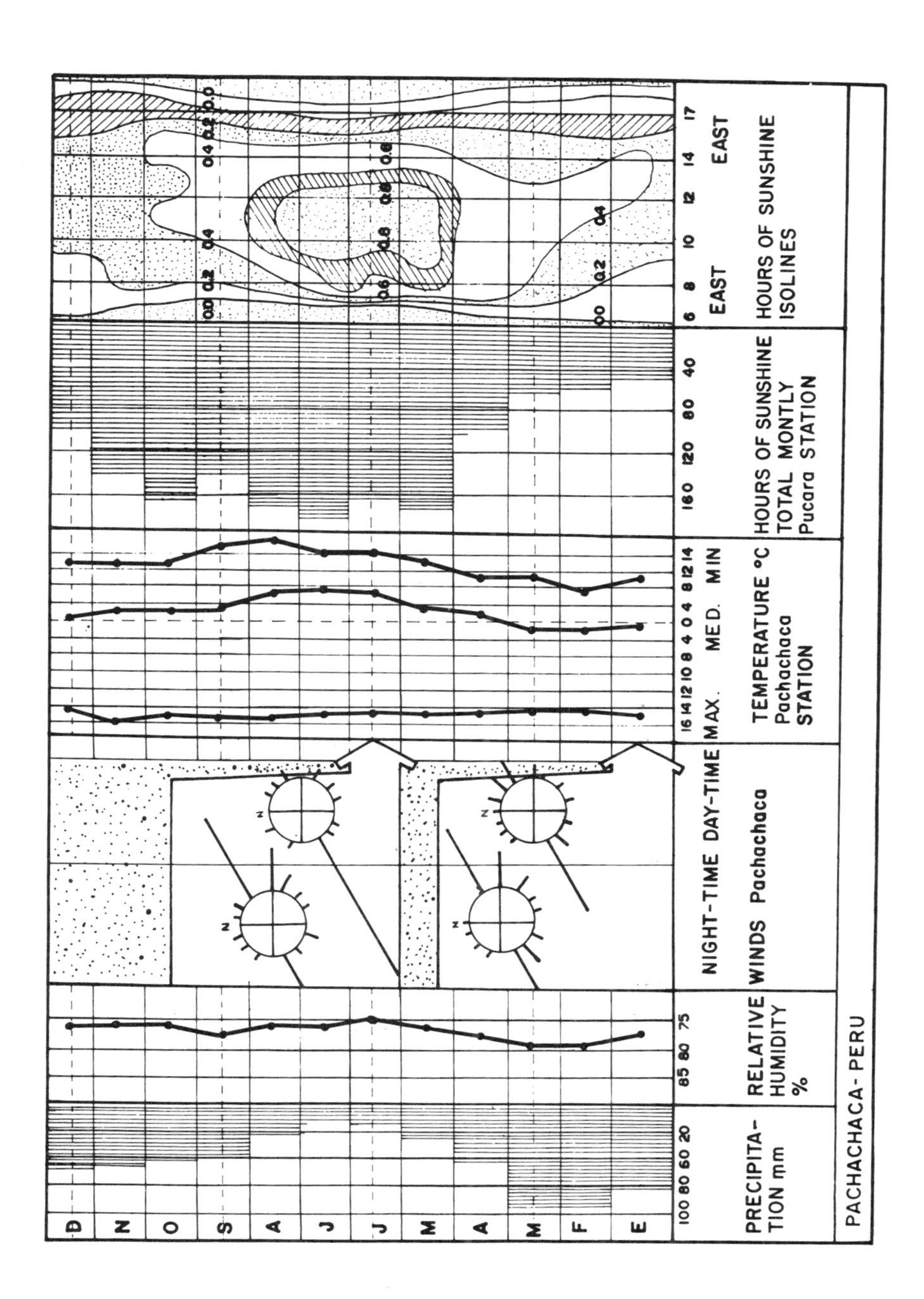

Figure 2

Meteorological information

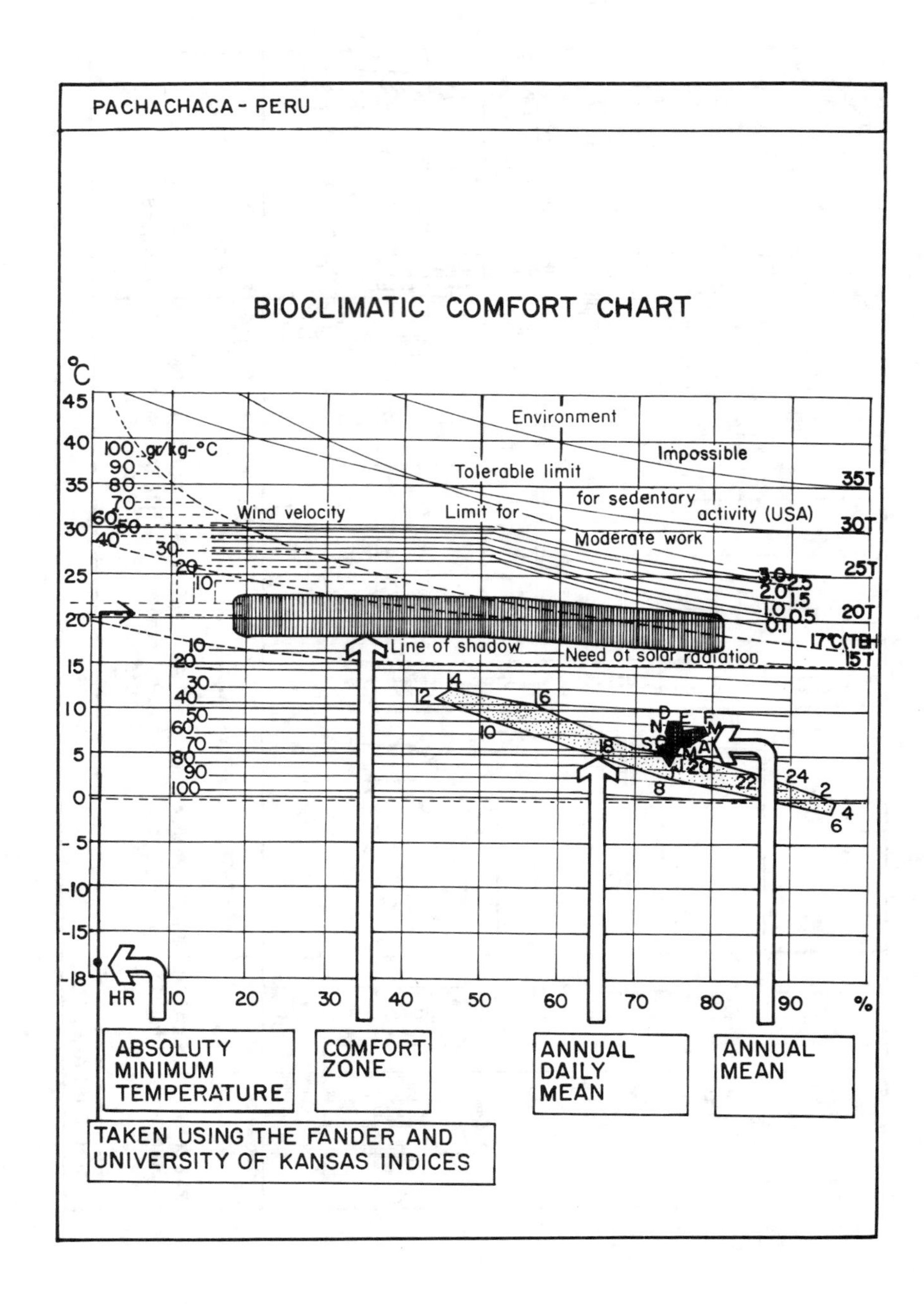

Figure 3

Bioclimatic comfort chart

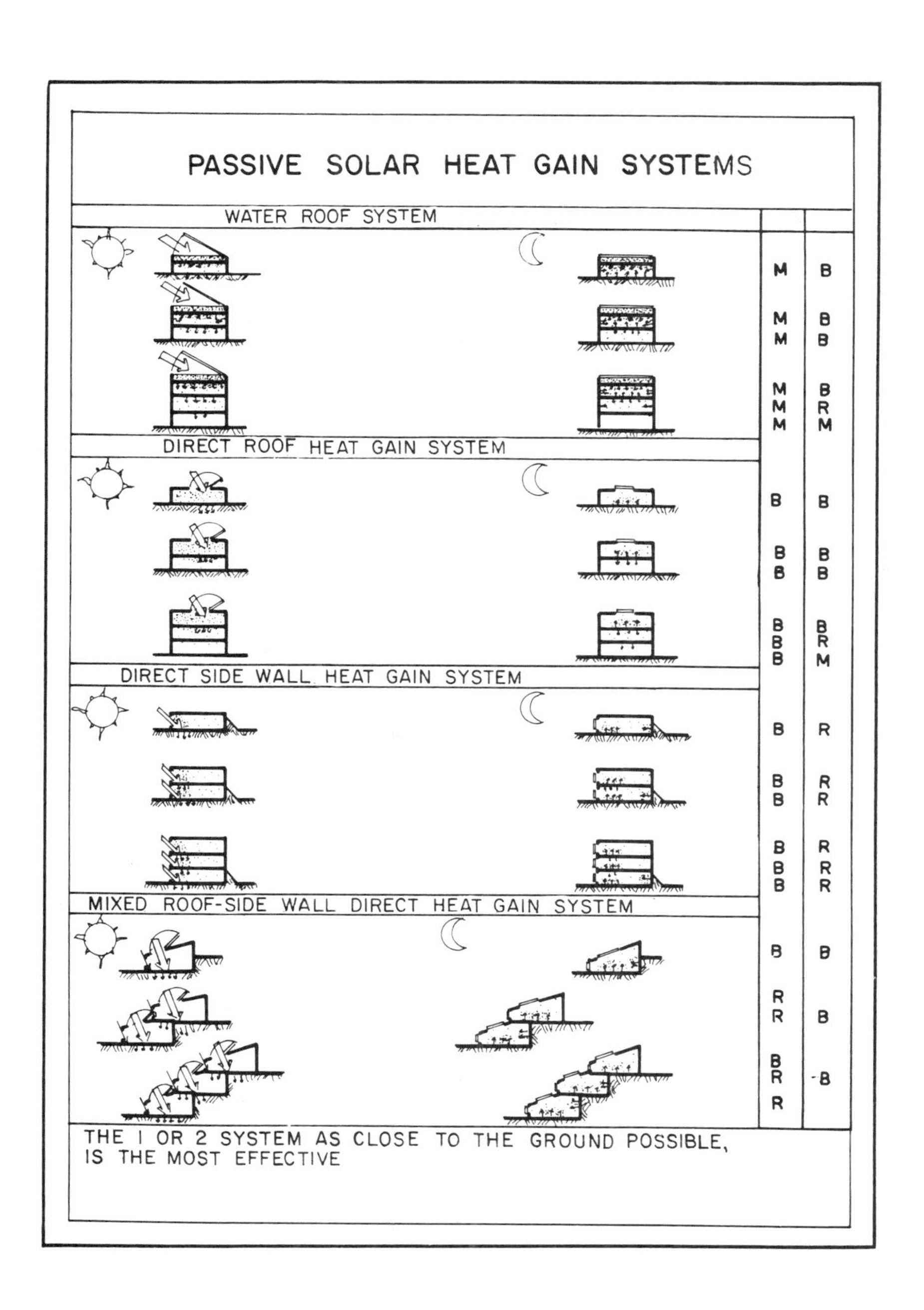

Figure 4

Passive solar heat gain systems

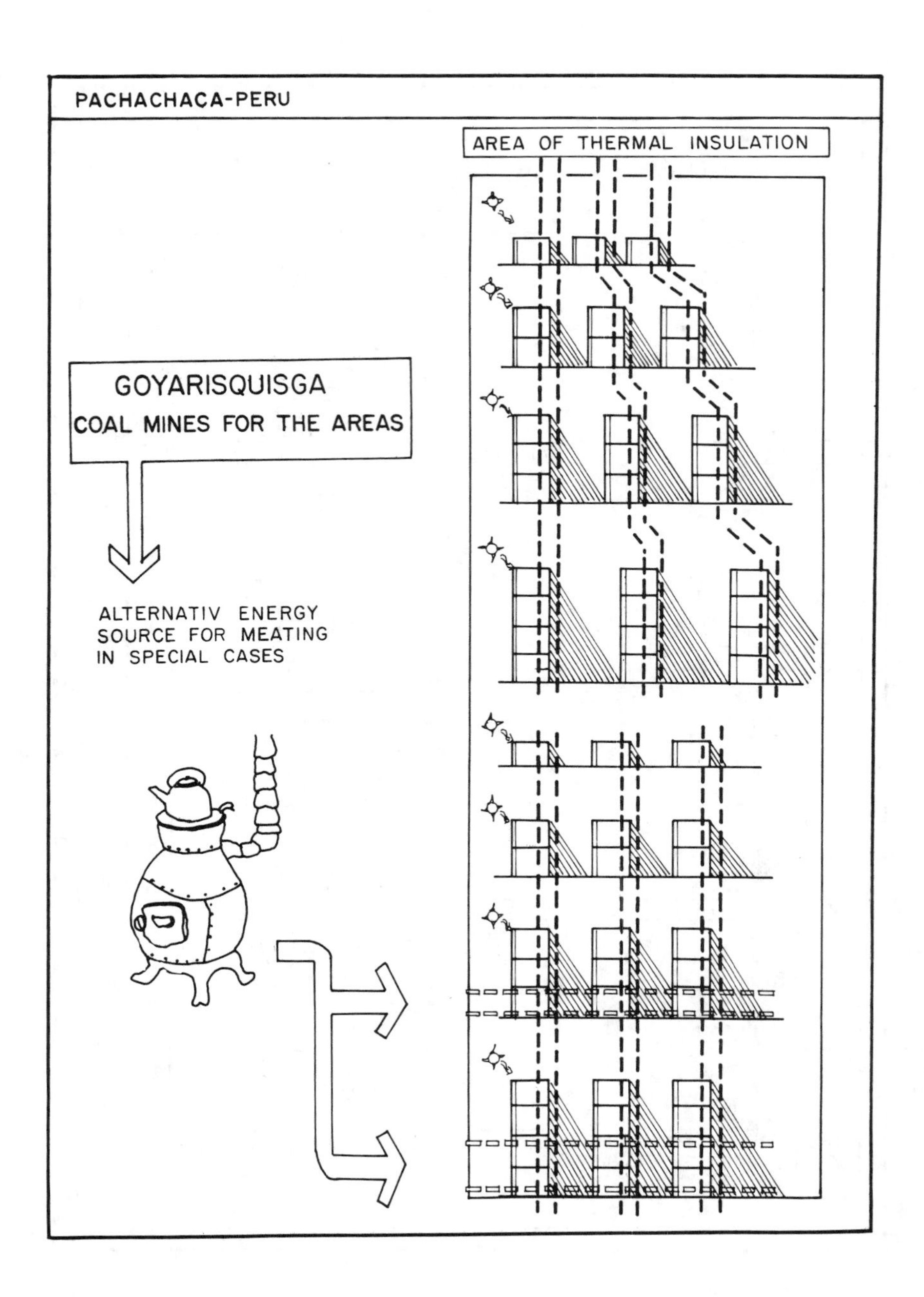

Figure 5

Height and distance between models

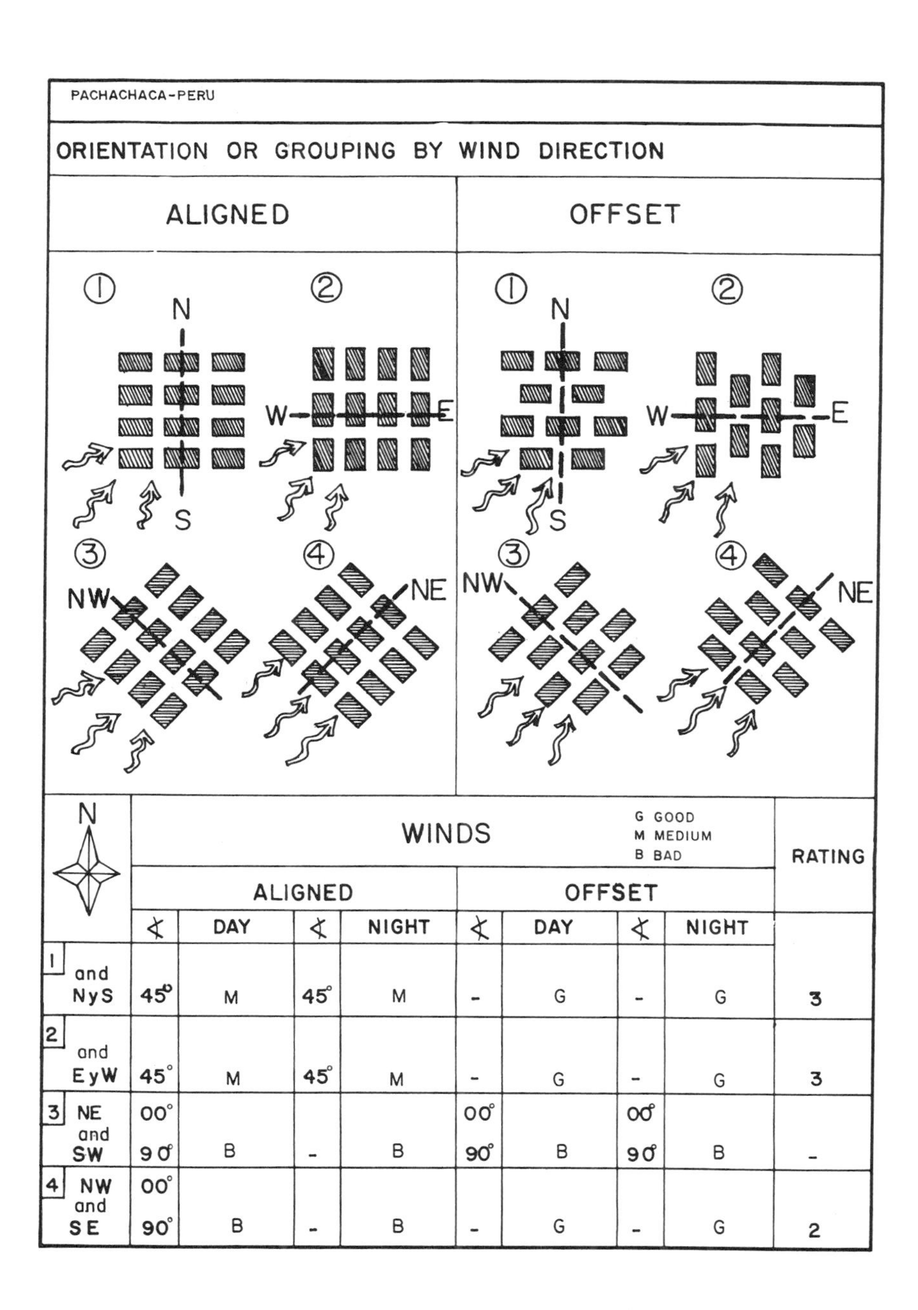

N	WINDS								RATING
	ALIGNED				OFFSET				
	∢	DAY	∢	NIGHT	∢	DAY	∢	NIGHT	
1 and NyS	45°	M	45°	M	-	G	-	G	3
2 and EyW	45°	M	45°	M	-	G	-	G	3
3 NE and SW	00° 90°	B	-	B	00° 90°	B	00° 90°	B	-
4 NW and SE	00° 90°	B	-	B	-	G	-	G	2

Figure 6

Orientation of grouping by wind direction

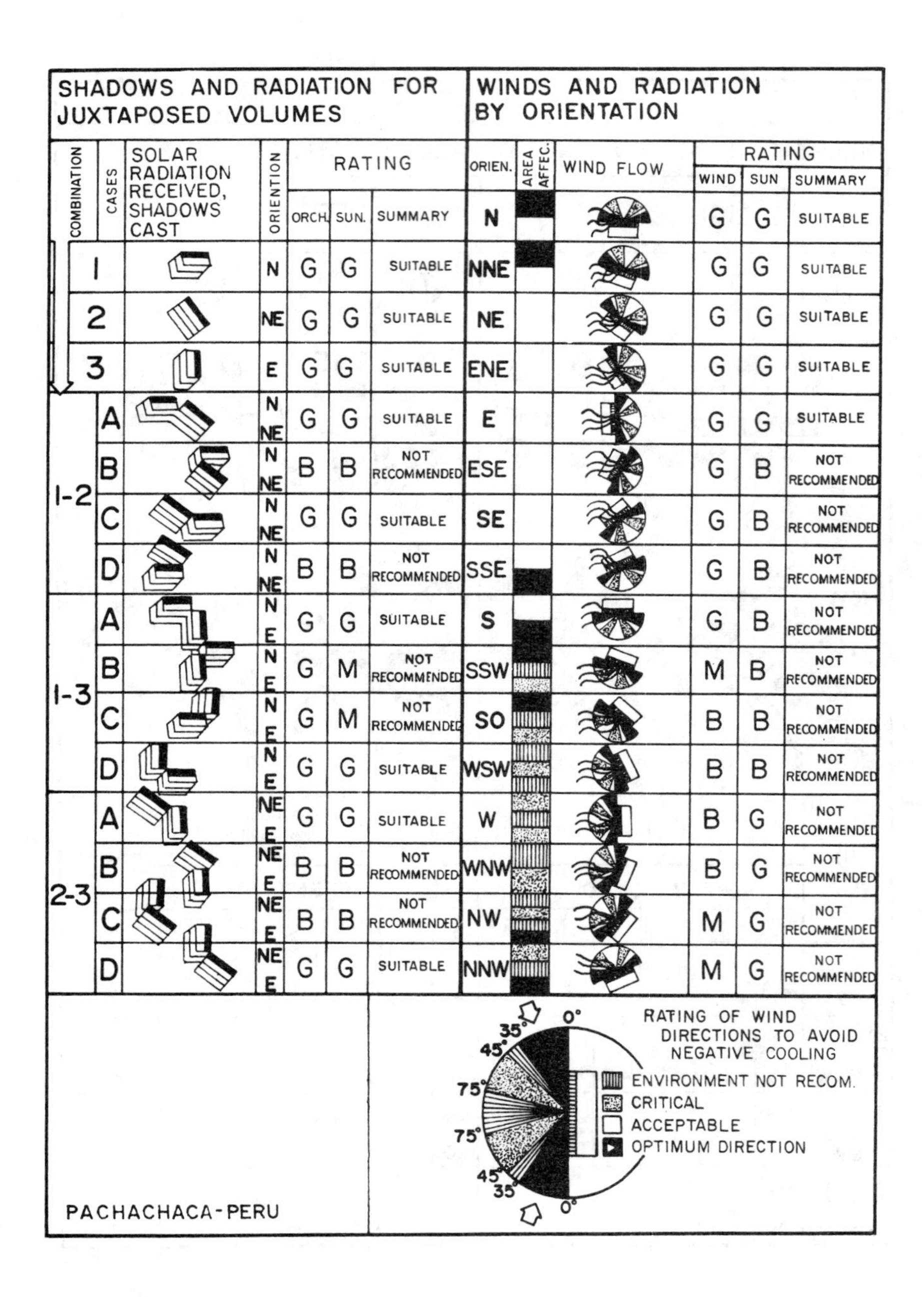

SHADOWS AND RADIATION FOR JUXTAPOSED VOLUMES

COMBINATION	CASES	SOLAR RADIATION RECEIVED, SHADOWS CAST	ORIENTION	RATING: ORCH.	RATING: SUN.	RATING: SUMMARY
	1		N	G	G	SUITABLE
	2		NE	G	G	SUITABLE
	3		E	G	G	SUITABLE
1-2	A		N NE	G	G	SUITABLE
	B		N NE	B	B	NOT RECOMMENDED
	C		N NE	G	G	SUITABLE
	D		N NE	B	B	NOT RECOMMENDED
1-3	A		N E	G	G	SUITABLE
	B		N E	G	M	NOT RECOMMENDED
	C		N E	G	M	NOT RECOMMENDED
	D		N E	G	G	SUITABLE
2-3	A		NE E	G	G	SUITABLE
	B		NE E	B	B	NOT RECOMMENDED
	C		NE E	B	B	NOT RECOMMENDED
	D		NE E	G	G	SUITABLE

WINDS AND RADIATION BY ORIENTATION

ORIEN.	AREA AFFEC.	WIND FLOW	RATING: WIND	RATING: SUN	RATING: SUMMARY
N			G	G	SUITABLE
NNE			G	G	SUITABLE
NE			G	G	SUITABLE
ENE			G	G	SUITABLE
E			G	G	SUITABLE
ESE			G	B	NOT RECOMMENDED
SE			G	B	NOT RECOMMENDED
SSE			G	B	NOT RECOMMENDED
S			G	B	NOT RECOMMENDED
SSW			M	B	NOT RECOMMENDED
SO			B	B	NOT RECOMMENDED
WSW			B	B	NOT RECOMMENDED
W			B	G	NOT RECOMMENDED
WNW			B	G	NOT RECOMMENDED
NW			M	G	NOT RECOMMENDED
NNW			M	G	NOT RECOMMENDED

Figure 7

Shadows and radiation for juxtaposed volumes, winds and radiation by orientation

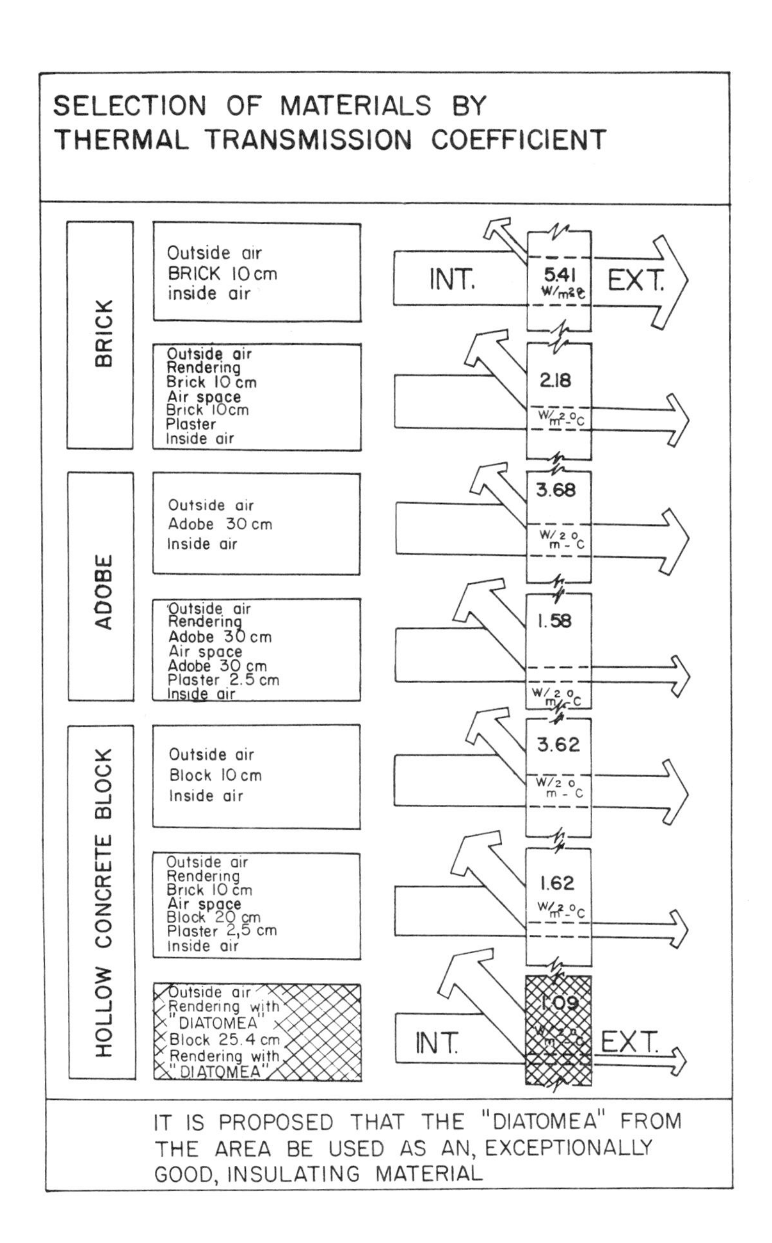

Figure 8

Selection of materials by thermal transmission coefficient

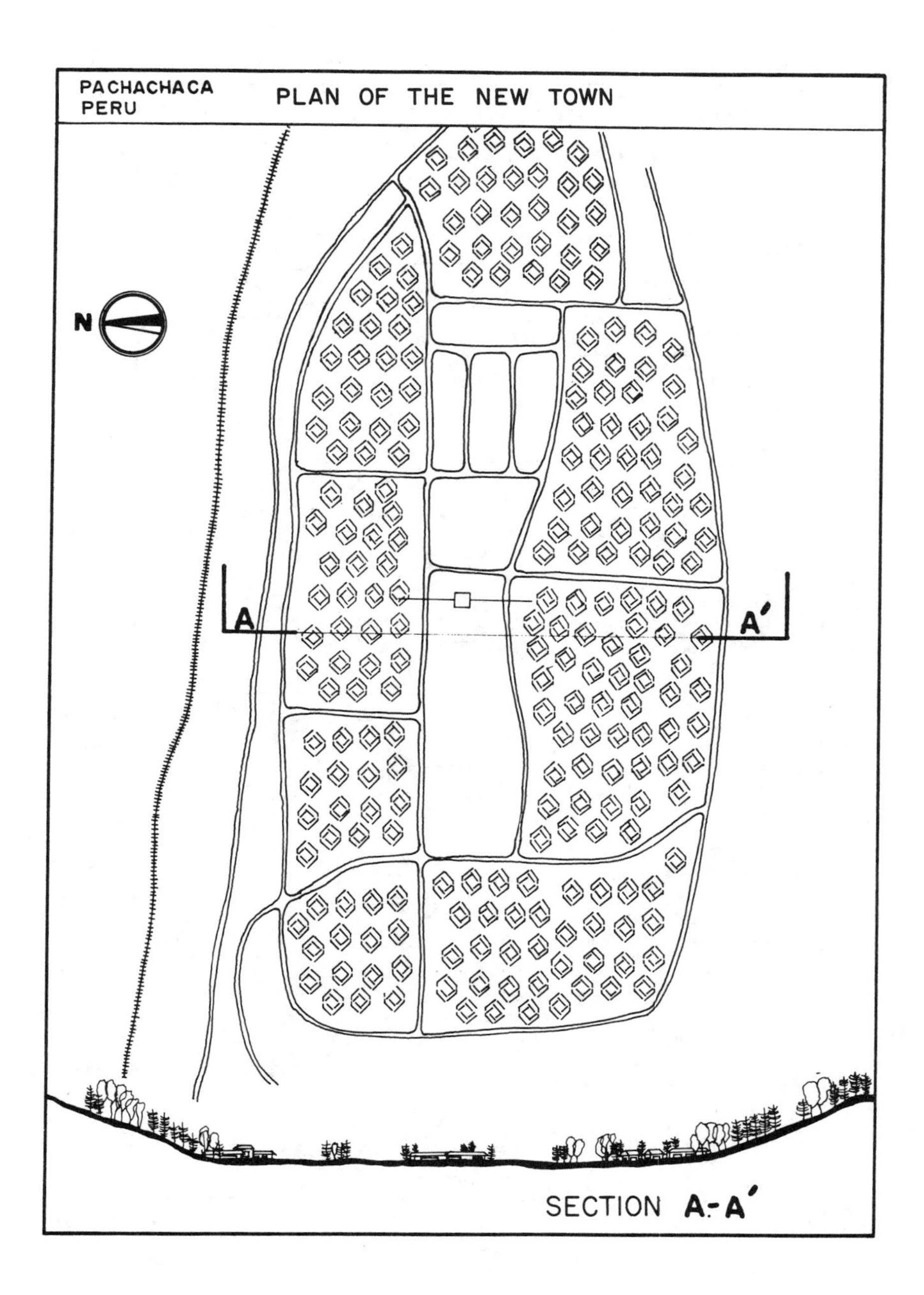

Figure 9

Plan of new town

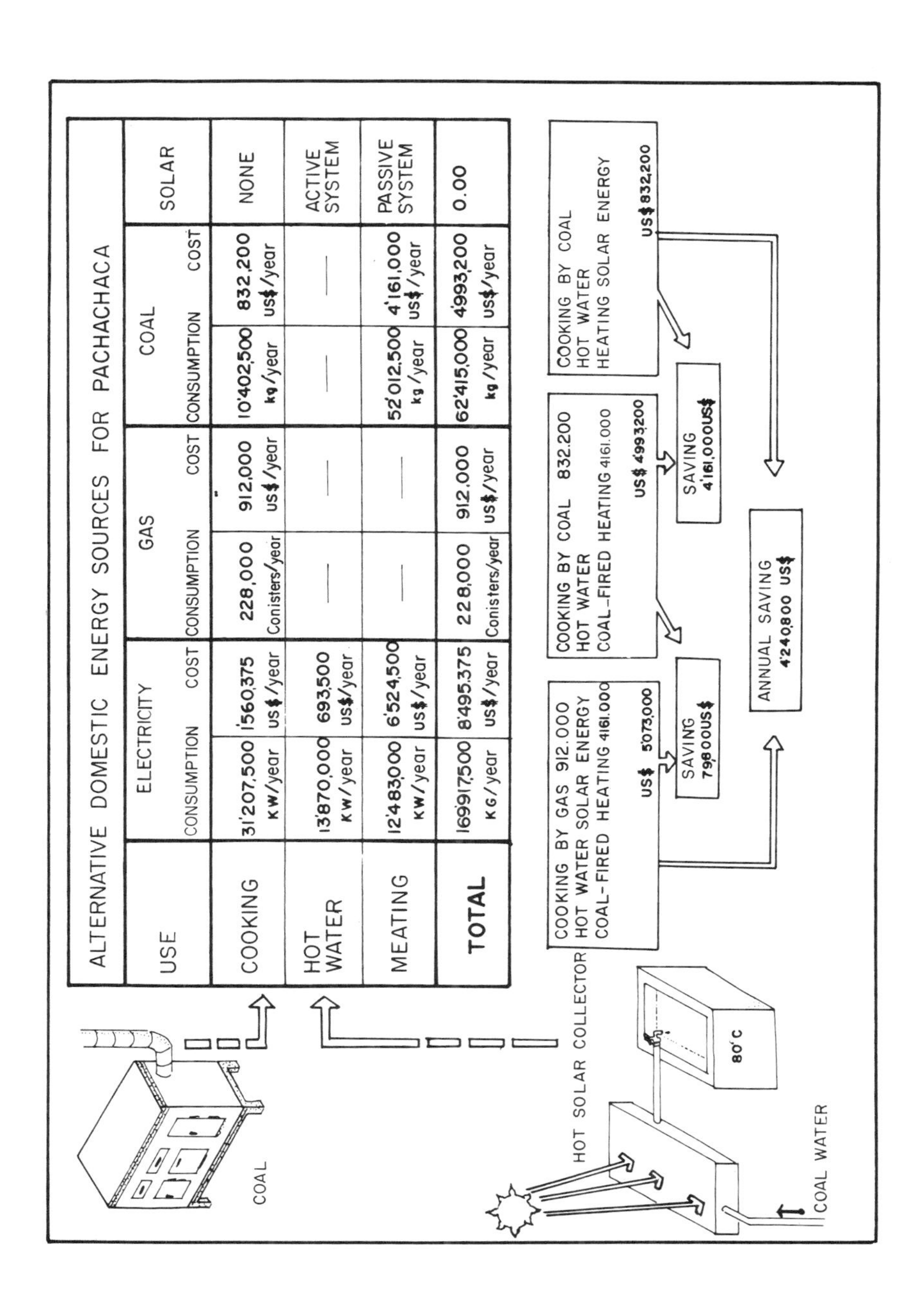

ALTERNATIVE DOMESTIC ENERGY SOURCES FOR PACHACHACA

USE	ELECTRICITY CONSUMPTION	ELECTRICITY COST	GAS CONSUMPTION	GAS COST	COAL CONSUMPTION	COAL COST	SOLAR
COOKING	31'207,500 KW/year	1'560,375 us$/year	228,000 Conisters/year	912,000 us$/year	10'402,500 kg/year	832,200 us$/year	NONE
HOT WATER	13'870,000 KW/year	693,500 us$/year	—	—	—	—	ACTIVE SYSTEM
MEATING	12'483,000 KW/year	6'524,500 us$/year	—	—	52'012,500 kg/year	4'161,000 us$/year	PASSIVE SYSTEM
TOTAL	169'917,500 KG/year	8'495,375 us$/year	228,000 Conisters/year	912,000 us$/year	62'415,000 kg/year	4'993,200 us$/year	0.00

Figure 10

Alternative domestic energy sources

42

Applications of climatology in land use planning

J. B. Williams, Land Resources Development Centre
Overseas Development Administration, UK

INTRODUCTION

This subject is concerned with detailed spatial variation of climate, whereas most applications of meteorology and climatology are, by and large, more interested in temporal variations at specified locations.

Climatological effects need be considered in both the optimal allocation of (limited) land for different uses, and the best realisation of the designated use once allocated; that is both *before* and *after* the planner's decision. These may require very different approaches.

In most countries the largest areas of land are used for agricultural purposes and forestry; but how much of the current distribution is the result of any formal scientific planning? Use of land for urban development, transport, industry, recreation and conservation is also affected by climatic considerations and more likely to be the subject of specific forward planning.

Users of climatic information working with other specialists in multidisciplinary teams investigating potential land use, need to be able to understand the detailed spatial variation of climate in areas where base data is often inadequate. The problem becomes one of intelligent interpolation, but this approach is limited in applicability, and one is prompted to ask if the data base and our understanding of climate processes is ever going to be adequate for this purpose?

Certain current developments in research will be of considerable assistance but the education of planners in more up-to-date concepts is needed.

VERY BRIEF HISTORY OF LAND USE PLANNING

It is only recently that the growing shortage of high quality land has forced development of methods for scientific evaluation of land quality, and economic appraisal of the different possibilities. Climate is usually a necessary but not sufficient conditions in the evaluation.

Early civilisations, for example those of China, the Indus Valley, Sri Lanka, Mesopotamia, Yemen, Egypt and the Mayas and Incas in South America were all based to an important extent on irrigation rather than rainfed agriculture, and consequently were insulated to a certain extent from climatic vagaries. However there are suggestions that subsequent longer term climate desiccation became a major problem in both Mesopotamia and Yemen for example, where water and land resources were limited relative to population.

The Romans were aware of the different potential value of lands for rainfed agriculture, and concentrated wheat production in North Africa where the winter rains were reliable. Julius Caesar is reported to have remarked on the great potential of eastern Britain for wheat production; only properly realised during the last ten years under the notoriously successful Common Agricultural Policy of the European Community.

However, in all such developments throughout history agricultural development in particular was very much a case of trial and error. At least as far as climate and land planning was concerned, there was little alternative as many important concepts have only been developed into useful numerical form over the last 50 years; evapotranspiration for example. And even as late as this there have been several large-scale disasters in agricultural planning. The almost complete failure of the 'planned' British ground-nut scheme in East Africa, and similar attempts at about the same time to organise maize cultivation on a large area of Russian steppe, are two famous examples, but there are many more. In the 1980's planned attempts were made to improve a certain nation's agriculture by preparing to grow wheat for export substitution, and rainfed rice in specific rice deficit areas of the country. Both government aid programmes were undertaken as if there was no climatic dimension. The wheat was grown in trial plots with no raingauges or other instrumentation adjacent. Interpretation of the results in terms of long-term potential productivity was far from obvious. The rice was an unmitigated disaster, since the rice deficit areas were such for good climatic reasons discovered much earlier by local farmers, and confirmed by available climatic data and examination of natural vegetation.

The purpose of mentioning these cases is to help explore reasons for the remarkable frequency with which land is developed without *apparent* adequate prior consideration of the climate. Why does it happen so often, when after all, potential agricultural production is climate constrained at the upper bound viz the FAO agroecological zones project? It is not denied that there have been many instances of climate analysis used in successful land use planning.

POSSIBLE REASONS FOR AN INADEQUATE APPRECIATION OF CLIMATOLOGICAL ASPECTS IN LAND USE PLANNING

1. <u>User requirements are often highly intricate and inadequately defined</u>

This is largely true in agriculture, but does it also apply to the location of nuclear power stations and local tornado incidence? ... or the siting of a new airport? When the route of the M25 around London was selected, were mist risk locations and frost hollows identified in advance or subsequent to the decision? How much is local weather and pollution hazard evaluated before industrial development is encouraged in an area.

"Optimum" climate varies with point of view, and never really exists for general agriculture (except perhaps the desert environment if abundant water supplies are available). But perceptions of greener grass abound between rain and radiation limiting environments. Climates suitable for low tech agriculture may not be so well suited for high tech agriculture.

In the design of an optimum climate for man's purposes we would have to include the needs of useful plants, and try to discriminate against insect pests, diseases and weeds. Single events can have a dominating effect and it is very probable that there are many subtle double or triple interactions like photo-thermo-periodism, varying between cultivars. The agroclimatic zones in virtually rainless Egypt clearly illustrate the sensitivities involved; how much more complicated are the multiple interactions with sporadic rainfall?

Crop models in CLICOM should go a long way towards helping climatologists and other interested individuals develop their awareness of the problems and requirements: this is expected to prove an invaluable tool for education and training, and for many different users besides those in meteorological services.

2. Spot measurements may not be representative of wider conditions and can distort perceptions of spatial variation

There can be strong local climatic effects especially in the tropics. As a consequence, the better sites tend to select themselves, so unless a policy of positive observational site selection is adopted significant bias may ensue.

What does a spot measurement tell us about the likely rainfall over the surrounding area? How many raingauges would be needed to be sure? Development of satellite estimation of tropical rainfall from satellite based cold cloud duration observations are starting to liberate us from limited concepts. They illustrate the rainfall processes occurring on their true spatial scale, and consequently developments will have unprecedented value for future understanding and training.

3. Statistics are emphasised at the expense of physics

Local climatic variation is an expression of the physical interaction between land and atmosphere. Statistical analysis of spatial variation is an expression of ignorance. Attempts to obtain a better understanding of the physical processes involved is essential. This will involve modelling of the surface interaction and rainfall distributions, and a combination of the two.

In large atmospheric circulations, it is often difficult to distinguish cause from effect, as changes in external inputs are usually small and slow acting; there is a large degree of interlinkage and self support in sustained dynamic processes. This is very important at all timescales, and is a good physical reason why a statistical approach to climatic variation may be inappropriate under some circumstances. If a particular state of regional climate tends to annul itself (negative feedback) it will disappear more rapidly than one that feeds itself positively. Thus the global atmosphere will tend to be comprised of mutually compatible self supporting regional climatic systems, and may lurch between these. It may be useful to view the southern oscillation in this manner. However, it is important that the student achieves an appreciation of the underlying systematic nature of climates, rather than an impression of a random series of events.

4. Climates are always changing and yet the concept of the static climatic norm still dominates perceptions

From the above analyses climate change becomes the norm, and means and normals highly unreliable. On an operational basis there is a 1/10th rule. Ten years rainfall data is usually adequate to provide an expected quantity (within a range) for the following year. Given 100 years data, one has a good idea of likely decadal mean rainfall and its variation ... and given 1000 years data ... etc. This assumes that climate changes are generally small and incremental, which is relevant to what is happening in the Sahel, and the river Nile, today. If there is a natural tendency for persistence in drought conditions in the Sahel, then man's activities may just be reinforcing this conditional stability.

CONCLUSION

Training and education have a vital role to play in the dissemination of modern concepts and techniques in climatology, in order that this powerful science can fulfil its potential for the benefit of mankind and his activities.

ISSUES

of the

Symposium on Education and Training in Meteorology with Emphasis on the Optimal Use of Meteorological Information and Products by all Potential Users

held at

Shinfield Park, United Kingdom
13 to 18 July 1987

1. There is an agreement that meteorological, hydrological and climatological information and associated products have improved considerably, and that they can contribute significantly to a country's economic and social development, productivity and human well-being. This improvement has increased the demand for, and use of, such products.

2. The application of meteorological services and products to minimize the impact of life-threatening situations must be considered a fundamental goal of national Meteorological and Hydrological Services. Much additional work and many improvements are still required, and it is essential that basic observation and telecommunication networks be adequately maintained if that goal is to be attained.

3. Professional meteorologists must give more attention to the preparation of educational material for the public. Public education in meteorology and hydrology is important at all levels of potential use for saving lives and reducing losses in property and productivity. Education should be encouraged, particularly in those sectors of greatest concern to local areas with specific local conditions.

4. Meteorologists must find out what each user's main requirements really are, and provide meteorological information in a format and language that the user can readily understand. There are large economic benefits to be gained

from applying these specialized services, and users should be encouraged to take advantage of them.

5. Experience has shown that problems preventing a full utilization of meteorological products arise from factors such as inadequate prediction systems, limited communication networks, linguistic differences and deficiencies in educational systems. Countries should identify such problems and strive to solve them. However, it is recognized that solutions to these problems may vary form country to country.

6. There are well-defined links between agricultural production, water management and urban development on the one hand, and the quality and safety of life on the other. These links should be strengthened, particularly in the developing countries. Such activities continue to be identified as being highly sensitive both to the weather and to information about the weather, they are vital and deserve our continuing attention.

7. The full utilization of meteorological products to boost economic productivity and reduce the impact of hazards on life and property will require new resources. These may be forthcoming through increased government expenditure or through private or local initiatives. Whatever mechanisms and policies are adopted to augment resources, care must be taken that policies do not discourage the use and application of meteorological information and data where they are needed for general economic improvement and development. It is recognized that each country will arrive at its own formula to deal with this issue.

8. Meteorology must be involved in promotion, marketing and commercial operations. Growing financial pressures on national Meteorological and Hydrological Services will accentuate the need for good customer relations, which in turn will lead to greater public awareness. In addition, the introduction or improvement by national Meteorological and Hydrological Services of marketing[1] activities applicable to local areas and circumstances will lead to a need for more resources by the Services.

9. Efficient promotion and marketing will necessitate (a) raising the level of meteorological literacy of the general public and the more specialized users, and (b) making meteorologists and hydrologists more sensitive to the needs of actual and potential users. A prerequisite for this is for meteorologists and hydrologists to have adequate skills in

---

[1]In this context, "marketing" refers to the act of determining, through close user interaction, the user's specific needs for meteorological products, including those not yet available.

communicating effectively with the public, the press, users and others.

10. Distance learning and multimedia methods are important for future education and training activities and increasing use should be made of them in training user-orientated professionals, including meteorologists.

67560

CKKD

Date Due

PRINTED IN U.S.A. CAT. NO. 24 161 BRODART